Current Topics in
Developmental Biology
Volume 36

Cellular and Molecular Procedures in Developmental Biology

Cellular and Molecular Procedures in Developmental Biology

Edited by

Flora de Pablo
Department of Cell and Developmental Biology
Centro de Investigaciones Biológicas
C.S.I.C.
Madrid, Spain

Alberto Ferrús
Instituto Cajal
C.S.I.C.
Madrid, Spain

Claudio D. Stern
Department of Genetics and Development
College of Physicians and Surgeons of Columbia University
New York, New York

Academic Press
San Diego London Boston New York Sydney Tokyo Toronto

Front cover photograph: In this dorsal view of an embryonic day 8 chick brain infected with CHAPOL, a retroviral library encoding alkaline phosphatase is shown. Each column of purple cells is an infected clone that was analyzed in a study of lineal relationships of the developing chick brain.

This book is printed on acid-free paper. 

Academic Press
a division of Harcourt Brace & Company
525 B Street, Suite 1900, San Diego, California 92101-4495, USA
http://www.apnet.com

Academic Press Limited
24-28 Oval Road, London NW1 7DX, UK
http://www.hbuk.co.uk/ap/

International Standard Book Number: 0-12-153136-8

PRINTED IN THE UNITED STATES OF AMERICA
97 98 99 00 01 02 BB 9 8 7 6 5 4 3 2 1

Contents

4

Use of Dominant Negative Constructs to Modulate Gene Expression

Giorgio Lagna and Ali Hemmati-Brivanlou

5

The Use of Embryonic Stem Cells for the Genetic Manipulation of the Mouse

Miguel Torres

6

Organoculture of Otic Vesicle and Ganglion

Juan J. Garrido, Thomas Schimmang, Juan Represa, and Fernando Giráldez

7

Organoculture of the Chick Embryonic Neuroretina

Enrique J. de la Rosa, Begoña Díaz, and Flora de Pablo

8

Embryonic Explant and Slice Preparations for Studies of Cell Migration and Axon Guidance

Catherine E. Krull and Paul M. Kulesa

9

Culture of Avian Sympathetic Neurons

Alexander v. Holst and Hermann Roher

10

Analysis of Gene Expression in Cultured Primary Neurons

Ming-Ji Fann and Paul H. Patterson

11

Selective Aggregation Assays for Embryonic Brain Cells and Cell Lines

Shinichi Nakagawa, Hiroaki Matsunami, and Masatoshi Takeichi

12

Flow Cytometric Analysis of Whole Organs and Embryos

José Serna, Belén Pimentel, and Enrique J. de la Rosa

13

Detection of Multiple Gene Products Simultaneously by *in Situ* Hybridization and Immunohistochemistry in Whole Mounts of Avian Embryos

Claudio D. Stern

14

Cloning of Genes from Single Neurons

Catherine Dulac

15

Methods for Detecting and Quantifying Apoptosis

Nicola J. McCarthy and Gerard I. Evan

16

Methods in *Drosophila* Cell Cycle Biology

Fabian Feiguin, Salud Llamazares, and Cayetano González

17

Single Central Nervous System Neurons in Culture

Juan Lerma, Miguel Morales, and María de los Angeles Vicente

18

Patch-Clamp Recordings from *Drosophila* Presynaptic Terminals

Manuel Martínez-Padrón and Alberto Ferrús

Contributors

Numbers in parentheses indicate the pages on which the authors' contributions begin.

Christopher Austin Merck Research Laboratories, West Point, Pennsylvania 19486 (51)

Constance L. Cepko Department of Genetics, Harvard Medical School, Boston, Massachusetts 02115 (51)

Enrique J. de la Rosa Department of Cell and Developmental Biology, Centro de Investigaciones Biológicas, Consejo Superior de Investigaciones Científicas, Velázquez 144, 28002 Madrid, Spain (133, 211)

Flora de Pablo Department of Cell and Developmental Biology, Centro de Investigaciones Biológicas, Consejo Superior de Investigaciones Científicas, Velázquez 144, 28006 Madrid, Spain (37, 133)

Begoña Díaz Department of Cell and Developmental Biology, Centro de Investigaciones Biológicas, Consejo Superior de Investigaciones Científicas, Velázquez 144, 28006 Madrid, Spain (133)

Catherine Dulac Department of Molecular and Cellular Biology, Harvard University, Cambridge, Massachusetts 02138 (245)

Elisabeth Dupin Institut d'Embryologie Cellulaire et Moléculaire du CNRS et du Collège de France, 94736 Nogent/Marne cedex, France (1)

Gerard I. Evan Biochemistry of the Cell Nucleus Laboratory, Imperial Cancer Research Fund Laboratories, London WC2A 3PX, United Kingdom (259)

Ming-Ji Fann Institute of Neuroscience, National Yang-Ming Medical College, Taipei, Taiwan 11221, Republic of China (183)

Fabian Feiguin EMBL, 69012 Heidelberg, Germany (279)

Alberto Ferrús Instituto Cajal, Consejo Superior de Investigaciones Científicas, Dr. Arce 37, 28002 Madrid, Spain (303)

Shawn Fields-Berry Harvard Medical School, Department of Genetics, Boston, Massachusetts 02115 (51)

Juan J. Garrido Instituto de Biología y Genética Molecular (IBGM), Facultad de Medicina, Universidad de Valladolid, 47005 Valladolid, Spain (115)

Fernando Giráldez Instituto de Biología y Genética Molecular (IBGM), Facultad de Medicina, Universidad de Valladolid, 47005 Valladolid, Spain (115)

Jeffrey Golden Department of Pathology, Children's Hospital of Philadelphia, Philadelphia, Pennsylvania 19104 (51)

Cayetano González EMBL, 69012 Heidelberg, Germany (279)

Ali Hemmati-Brivanlou Laboratory of Molecular Embryology, The Rockefeller University, New York, New York 10021-6399 (75)

Alexander v. Holst Max-Planck-Institute for Brain Research, D-60528 Frankfurt, Germany (161)

Catherine E. Krull Division of Biology, California Institute of Technology, Pasadena, California 91125 (145)

Paul M. Kulesa Division of Biology, California Institute of Technology, Pasadena, California 91125 (145)

Giorgio Lagna Laboratory of Molecular Embryology, The Rockefeller University, New York, New York 10021-6399 (75)

Nicole M. Le Douarin Institut d'Embryologie Cellulaire et Moléculaire du CNRS et du Collège de France, 94736 Nogent/Marne cedex, France (1)

Juan Lerma Instituto Cajal, Consejo Superior de Investigaciones Científicas, Dr. Arce 37, 28002 Madrid, Spain (293)

Salud Llamazares EMBL, 69012 Heidelberg, Germany (279)

Manuel Matínez-Padrón Instituto Cajal, Consejo Superior de Investigaciones Científicas, Dr. Arce 37, 28002 Madrid, Spain (303)

Hiroaki Matsunami Department of Neurobiology, Harvard Medical School, Boston, Massachusetts 02115 (197)

Nicola J. McCarthy Biochemistry of the Cell Nucleus Laboratory, Imperial Cancer Research Fund Laboratories, London WC2A 3PX, United Kingdom (259)

Aixa V. Morales Department of Cell and Developmental Biology, Centro de Investigaciones Biológicas, Consejo Superior de Investigaciones Científicas, Velázquez 144, 28006 Madrid, Spain (37)

Miguel Morales Instituto Cajal, Consejo Superior de Investigaciones Científicas, Dr. Arce 37, 28002 Madrid, Spain (293)

Shinichi Nakagawa Department of Biophysics, Faculty of Science, Kyoto University, Kyoto 606-01, Japan (197)

Paul H. Patterson Division of Biology 216-76, California Institute of Technology, Pasadena, California 91125 (183)

Belén Pimentel Department of Cell and Developmental Biology, Centro de Investigaciones Biológicas, Consejo Superior de Investigaciones Científicas, Velázquez 144, 28006 Madrid, Spain (211)

Juan Represa Instituto de Biología y Genética Molecular (IBGM), Facultad de Medicina, Universidad de Valladolid, 47005 Valladolid, Spain (115)

Hermann Rohrer Max-Planck-Institute for Brain Research, D-60528 Frankfurt, Germany (161)

Elizabeth Ryder Department of Biology and Biotechnology, W.P.I., Worcester, Massachusetts 01609 (51)

Thomas Schimmang Instituto de Biología y Genética Molecular (IBGM), Facultad de Medicina, Universidad de Valladolid, 47005 Valladolid, Spain (115)

José Serna Department of Cell and Developmental Biology, Centro de Investigaciones Biológicas, Consejo Superior de Investigaciones Científicas, Velázquez 144, 28006 Madrid, Spain (211)

Claudio D. Stern Department of Genetics and Development, College of Physicians and Surgeons, Columbia University, New York, New York 10032 (99, 223)

Masatoshi Takeichi Department of Biophysics, Faculty of Science, Kyoto University, Kyoto 606-01, Japan (197)

Miguel Torres Departamento de Immunología y Oncología, Centro Nacional de Biotecnología, Consejo Superior de Investigaciones Científicas, 28049 Madrid, Spain

María de los Angeles Vicente Instituto Cajal, Consejo Superior de Investigaciones Científicas, Dr. Arce 37, 28002 Madrid, Spain (293)

Catherine Ziller Institut d'Embryologie Cellulaire et Moléculaire du CNRS et du Collège de France, 94736 Nogent/Marne cedex, France (1)

Preface

In any future history of science, developmental biology will no doubt occupy a particularly prominent position in the last quarter of the twentieth century. A truly explosive growth has occurred in this field, nested at the interface between embryology and molecular biology. Although the central biological question of how a single cell, the egg, becomes a complex and highly hierarchical organism is an old one, methods to answer it have been slow to come. Recently, technical procedures have multiplied, finally allowing specific questions to be addressed in many different model systems at the molecular, cellular, and whole-organism levels. This book reflects this diversity by incorporating cellular, subcellular, genetic, and molecular levels of analysis of biological processes described in vertebrates and invertebrates. One of the great surprises of the last two decades is the degree to which developmental mechanisms appear to be conserved across animal phyla. At the same time, different model organisms can be more or less amenable to a particular approach. It has never been more important to learn from comparative studies, and the reader is encouraged to pay special attention to those chapters that deal with procedures different from his or her own specialty, to grasp new potential tools and approaches.

Because the field is currently so rich, no attempt has been made to cover all useful developmental models available to the researcher. Instead, updated procedures in a few classic models and new methods that can be applied to study developmental problems have been included. The first chapter revisits a powerful technique developed in the 1960s, embryonic chimeras with avian embryos, emphasizing new details to exploit its use. This is followed by four chapters on different approaches to interfering with the expression of genes using antisense oligonucleotides, retroviral vectors, dominant negative constructs, and mouse transgenesis. All of these approaches have advantages and limitations, and these have been noted by the authors to help readers decide whether the approaches are suitable complements to their research. Chapters 6 to 10 present several culture systems for modeling specific *in vivo* conditions using whole organs, slices, or primary cultures of neurons. Even more than with cell lines, these cultures usually require particularly well-controlled medium, substrate, and reservoir conditions to prolong life, allowing manipulations and a level of analysis not feasible *in vivo*. Chapters 11 to 16 present a set of technically advanced assays incorporating techniques with long traditions in other areas of biology, such as flow cytometry and differential cloning, together with some recent improvements allowing study of the expression of multiple gene products simultaneously, of apoptotic cell death, and of the cell cycle. The last two chapters deal with state-of-the-art technology in electrophysiology. Future experimental strategies in developmental biology will have to incorporate procedures for studying single molecular structures functioning *in vivo*. To date, the patch-clamp of cellular membranes is the only technique that offers a view of a molecular entity, an ion channel, operating in its native environment. The role of ion channels during development is beginning to be incorporated into the mainstream of developmental biology studies, and it is likely that future books of this type will contain more chapters of this kind.

Our aim has not been to be fully comprehensive; this would be an impossible task. Rather, we have endeavored to concentrate on areas of recent growth and those that have been less well represented in other texts, hoping that this collection of methods and insights by several experts in their respective fields will stimulate workers to diversify their approaches and to broaden their lines of inquiry. We are grateful to the authors that generously took time to write in detail reliable experimental protocols. While technology in the areas covered in the book is changing rapidly, we hope that its chapters remain valuable for many years, at least, as a useful point to begin.

Flora de Pablo
Alberto Ferrús
Claudio D. Stern

1

The Avian Embryo as a Model in Developmental Studies: Chimeras and *in Vitro* Clonal Analysis

Elisabeth Dupin, Catherine Ziller, and Nicole M. Le Douarin
Institut d'Embryologie Cellulaire et Moléculaire du CNRS et du Collège de France
94736 Nogent-sur-Marne, France

I. Introduction

The processes that lead from a single cell, the fertilized egg, to the pluricellular embryo and the functional adult organism, are essentially cell proliferation, cell migration, interactions between cells and their environment, and cell differentiation. Our understanding of these complex phenomena has greatly benefited from studies carried out on the avian embryo, which is accessible to experimentation throughout embryonic life.

The observation in the embryo of cells or groups of cells that move and differentiate requires precise and reliable cell labeling techniques that may be used *in vivo*. *In vitro* culture methods, however, are more appropriate for the study of the respective roles of environmental factors and intrinsic cellular properties in the differentiation of embryonic cells.

Here we describe two methods that have become classic in embryology and that, in combination with molecular techniques, have proven remarkably fruitful for the analysis and the elucidation of developmental problems. One is the

Current Topics in Developmental Biology, Vol. 36

0070-2153/98 $25.00

construction *in ovo* of chimeric embryos in which cells or rudiments from two species of birds, quail and chick, are combined and can be thereafter distinguished histologically at any time during ontogeny. The other method is the clonal culture *in vitro* of embryonic cells, which permits not only the establishment of differentiation potentials of individual precursor cells taken at various developmental stages, but the analysis of possible interactions of these cells or their progeny with environmental factors that can be manipulated at will *in vitro*.

II. Quail–Chick Embryonic Chimeras

A. Introduction

All quail cells, embryonic and adult, possess a natural and stable genetic marker, visible as a mass of heterochromatin associated with the nucleolus. As a consequence, this organelle is strongly stained by histochemical DNA-specific reactions. In chicken cells, as in most species, the heterochromatin is dispersed in the nucleoplasm (Le Douarin, 1969). During embryogenesis, cells or rudiments from one species can be transplanted to the other, resulting in interspecific chimeric embryos. Chick and quail are closely related in taxonomy; their developmental schedules are similar, at least during the first week of incubation, when most important events take place in organogenesis; and there is no rejection of the graft by the recipient embryo, because the immune system is not mature and functional during embryogenesis. In chimeric embryos, thanks to the difference between quail and chick cells, the behavior of definite embryonic rudiments can be followed dynamically, from the time of transplantation, through all intermediate stages, to their final fate in the mature animal.

This method has proven to be extremely successful in a great number of applications (Le Douarin *et al.*, 1996, for a review). The investigations carried out on the development of the nervous system, peripheral (Le Douarin, 1982, 1986) and central (Le Douarin, 1993), provide many excellent examples of the value of the technique. Here we describe two of its recent applications. The first is an example of isotopic transplantations that established the site of origin of definite cell types, their migration pathways, and their final fates in the organism: thus, the origin of oligodendrocytes in the spinal cord was examined in quail–chick chimeras resulting from isotopic transplantations of defined portions of neural tube. The second example illustrates the use of heterotopic transplantations, which are aimed at evaluating the degree of plasticity of an embryonic tissue or group of cells, and at analyzing the developmental potentials of the cells of interest and the influence of the environment on their final destiny; by displacing rhombomeres along the rostrocaudal axis, their plasticity with respect to *Hox* code and phenotypic expression was investigated.

B. Markers for Quail and Chick Cells

1. The Quail Nucleolus

During interphase, the nucleus of quail cells exhibits a very large and conspicuous nucleolus, which is strongly stained by DNA-specific dyes. This is due to the presence of a mass of heterochromatin associated with the nucleolus. In contrast, in most other species, including the chick, the nucleolus contains only minute amounts of chromatin. The most commonly used histochemical technique for the differential diagnosis of quail and chick cells in chimeric embryos is the Feulgen–Rossenbeck DNA staining method (Feulgen and Rossenbeck, 1924; Le Douarin, 1969, 1971, 1973a,b). The procedure involves fixation of tissues in Zenker's or Carnoy's fluids and subsequent embedment in paraffin. Staining is performed according to Gabe (1968) on 5- or 7-μm-thick sections.

The heterochromatic component of the quail nucleolus can also be evidenced by the fluorescent labels bisbenzimide (Hoechst 33258) (Franklin and Martin, 1980; Nataf *et al.*, 1993) and acridine orange (Fontaine-Pérus *et al.*, 1985) and by electron microscopy (Le Douarin, 1973a).

The Feulgen-Rossenbeck technique or fluorescent labeling with bisbenzimide can be combined with immunocytochemistry or with *in situ* hybridization: several antibodies or nucleic probes can be applied either on the same or on alternate paraffin sections, for the combined detection of specific gene expression and quail-versus-chick cells.

2. Antibodies

The chick anti-quail serum raised by Lance-Jones and Lagenaur (1987) recognizes a cell-surface antigen of all quail cells.

The monoclonal antibody QCPN, obtained by B. M. Carlson and J. A. Carlson (unpublished; available at Developmental Studies Hybridoma Bank, University of Iowa, Iowa City, IA, U.S.A.), reacts with a nuclear component of all quail cells at early embryonic stages. Both reagents are convenient for immunocytochemistry and can be combined either with other antibodies or with histochemical labelings and *in situ* hybridization.

3. Nucleic Probes

Some avian cDNA probes are specific for quail or chick. One such is the quail-specific SMP probe (Schwann cell myelin protein; Dulac *et al.*, 1988, 1992), which has been used in neural chimeras to distinguish oligodendrocytes from quail and chick (Cameron-Curry and Le Douarin, 1995). Avian probes for *Hox* homeobox genes hybridize both with chick and quail RNA. They have been used to investigate the induction of *Hox* gene expression in neural chimeras, by combining *in situ* hybridization with the quail–chick histochemical labeling tech-

Table I Species-Specific Antibodies and Nucleic Probes Used to Analyze Quail–Chick Chimeras

Cell type	Quail	Chick
All	Chick anti-quail serum (Lance-Jones and Lagenaur, 1987) Quail non-chick perinuclear (QCPN) (Carlson BM and Carlson JA, Developmental Studies Hybridoma Bank, University of Iowa)	
Neurons	Quail neuron (QN) (neurites) (Tanaka *et al.*, 1990)	37F5 (neuronal cell bodies) 39B11 (neurites) (Takagi *et al.*, 1989) Chick neuron (CN) (neurites) (Tanaka *et al.*, 1990)
Glial cells	Schwann cell myelin protein (SMP): nucleic acid probe (Dulac *et al.*, 1992)	
Hemangioblastic lineage	Quail hemopoïetic1 (MB1/QH1) (Péault *et al.*, 1983) (Pardanaud *et al.*, 1987)	
Major histocompatibility complex (MHC)	Thymic antigen caille1 (TAC1) (Cl II) (Le Douarin *et al.*, 1983)	Thymic antigen poulet1 (TAP1) (Cl II) (Le Douarin *et al.*, 1983)
T-cell markers		αTCR1 (γδ) (Chen *et al.*, 1988) αTCR2 (αβ) (Cihak *et al.*, 1988) αCT3 (Chen *et al.*, 1986) αCT4 (Chen *et al.*, 1988) αCT8 (Chen *et al.*, 1988)
Cells of organizer and primitive streak		goosecoid: nucleic acid probe (Izpisúa-Belmonte *et al.*, 1993)

nique (Grapin-Botton *et al.*, 1995). *In situ* hybridizations are performed as described by Eichmann *et al.* (1993).

Quail- and chick-specific antibodies and nucleic probes are summarized in Table I.

C. Material

1. Eggs

Fertilized eggs from chick and quail, purchased from commercial sources, are stored for 1 to 7 days at 15°C. The storage period should not exceed 1 week, because the rate of abnormal or retarded development increases rapidly thereafter.

2. Incubator

Incubators should be equipped with temperature (38°C ± 1°C) and humidity (45–75%) regulators and with a mechanism for gentle automatic rocking. The eggs are incubated either horizontally or vertically in holders of appropriate size and shape for chick and quail.

3. Dissecting Microscope

Microsurgery is performed under a stereomicroscope equipped with magnifications ranging from 6× to 40×. The opened egg (see discussion later) is placed on a wooden holder of appropriate shape and size. Illumination is obtained either from optic fibers or from a conventional lightbulb equipped with a heat filter to avoid harmful heating and drying of the embryo during the operation.

4. Instruments for Microsurgery

Small scissors (10 cm), sharp, straight, or curved
Forceps, Swiss jeweller type No. 5, straight or curved
Pascheff-Wolff scissors, 6 cm, with very delicate blades
Landolt transplantation spatula
Drilled spatula
Needle holders for microscalpels

Microscalpels are obtained by sharpening steel sewing needles on an Arkansas oil stone. Microscalpels are manufactured under the stereomicroscope. Microscalpels can also be made from tungsten needles sharpened by electrolysis (Conrad *et al.*, 1993).

Micropipettes are hand-drawn on a gas flame from the small end of Pasteur pipettes; they are held with a flexible plastic tube for mouth use.

5. Miscellaneous Supplies

Tyrode's solution or PBS (phosphate buffered saline) are used at room temperature to wash and store explanted embryos and tissues before operating. Other materials include the following:

Glass and plastic dishes of various sizes and shapes
India ink to visualize the embryo *in ovo*
Transparent adhesive tape to close the eggs after the operations
Glass dishes with a paraffin or plastic bottom made black by addition of charcoal for operating on explanted embryos
Entomology pins for holding the embryos on the paraffin or plastic bottom
Syringes (1 or 2 ml) and injection needles (0.8-mm diameter)
Enzyme pancreatin (Gibco) diluted 1:3 or 1:4 in Tyrode's solution

Tyrode's solution supplemented with 10% fetal calf serum (FCS), to wash the tissues after enzymatic treatment

D. Transplantation Technique

1. Preparation of the Recipient Embryo *in Ovo*

Eggs are incubated vertically with the large end (air chamber) up. The air chamber is perforated and the egg turned upside down. A window 1.5 to 2 cm in diameter, through which the operation will be made, is then opened in the shell on the long side of the egg. The air chamber empties, lowering the level of liquid within the egg. The blastoderm comes back to the top of the yolk, lying away from the shell. To visualize the embryos, India ink diluted in Tyrode's (1:1, v/v) is injected underneath the blastoderm with a micropipette connected to a plastic tube controlled with the mouth. This operation and the following ones are made under the dissecting microscope.

With a sharp microscalpel, the vitelline membrane covering the embryo is torn open, as precisely as possible over the region where the transplantation will be made. At this point, great care must be taken not to let the embryo get dry. To this end, a drop of Tyrode's may be deposited onto the blastoderm.

2. Explantation of the Donor Embryo

The egg is opened as described previously, or its whole contents gently poured into a dish containing Tyrode's solution. The blastoderm is dissected out with Pascheff-Wolff scissors, pulled away from the top of the yolk with forceps, washed in Tyrode's solution, and transferred with a drilled spatula to a dish with a soft black bottom of paraffin or Rhodorsyl, to which it is fixed with entomology pins.

The stages of the recipient and donor embryos must be established precisely. This can be done accurately once ink has been injected underneath the recipient and the donor has been pinned down to the black background of the operation dish. The main organs and rudiments can then be easily recognized. Stages are defined according to the tables of Hamburger and Hamilton (HH, 1951) for the chick and Zacchei (1961) for the quail. Before embryonic day 3 (E3), the segments (somites) can be counted, and their number is an excellent indication of the exact embryonic stage.

3. Neural Tissue Transplantation

Fragments of the neural primordium are exchanged either isotopically or heterotopically between chick and quail embryos, usually at E1–E2.

To excise the tissue from the host *in ovo*, slits are made with the microscalpel through the ectoderm, first bilaterally along the chosen region of the neural primordium, then transversally, rostrally, and caudally; finally, the piece to be excised is severed ventrally from the underlying tissues (essentially the notochord) using the microscalpel, sucked out with a micropipette, and discarded.

To prepare the transplant, the fragment of neural primordium is excised microsurgically from the explanted embryo as described for the removal of the host tissue and transferred with a micropipette to the host embryo, where it is inserted into the space left free by the ablation, using blunt microscalpels. If the transplantation to be performed is an isotopic replacement of a piece of host tissue by its equivalent from the donor embryo, care must be taken that the normal rostrocaudal and dorsoventral orientations of the graft are conserved. To this end, tiny slits can be created at one end of the graft with the microscalpel. Of course, depending on the experiment, the polarity of the graft versus the host can be reversed.

The transplant can also be prepared by enzymatic digestion. The region of the donor embryo comprising the fragment of neural tissue of interest plus the surrounding structures (i.e., ectoderm, endoderm, paraxial mesoderm [somites, if the segmented zone is concerned], and notochord) is dissected out with Pascheff-Wolff scissors, transferred to a pancreatin solution (1:4 v/v in Tyrode's). After 10–15 min at room temperature, the different tissues and organs can easily be separated with blunt microscalpels; the neural tissue graft is then transferred to Tyrode's solution supplemented with serum to stop proteolysis, and immediately put in place in the host embryo.

In isotopic transplantations, the graft and the tissue fragment excised from the host are exactly equivalent, taken from embryos at the same stage.

In heterotopic transplantations, pieces from any region of embryos at any stage are grafted. In the example described here (see later), fragments of neural primordium are taken from a rostrocaudal level that is different from the level where the neural tissue in the recipient is ablated.

As much as possible, the size of the graft should match the size of the groove resulting from the excision in the host embryo.

After grafting, the window in the shell is sealed with transparent adhesive tape and the egg is put back into the incubator.

E. Recent Applications of the Quail–Chick Transplantation Method

1. The Origin of Oligodendrocyte Precursors in the Spinal Cord

The origin of oligodendrocytes has been a controversial question. Studies performed *in vivo* and *in vitro*, mainly in rodents but also in avian embryos (LeVine and Goldman, 1988; Warf *et al.*, 1991; Noll and Miller, 1993; Yu *et al.*, 1994; Timsit *et al.*, 1995; Trousse *et al.*, 1995), suggested that oligodendrocyte precursors are generated exclusively in the ventral part of the ventricular zone of the neural tube region, from which they spread laterally and dorsally; however, there was no direct evidence that all the oligodendrocytes of the white matter in the spinal cord derive from this ventral pool of cells, nor was the ventrodorsal migration of the precursors demonstrated.

This question of the origin and migration of oligodendrocyte precursors was reinvestigated in quail–chick chimeras resulting from isotopic exchanges of neural tube sectors at E2 (Cameron-Curry and Le Douarin, 1995). The chimeric embryos were analyzed by labeling histologic sections with the classic techniques that permit distinction between quail and chick cells (Feulgen-Rossenbeck histochemical staining or QCPN immunolabeling) and by *in situ* hybridization with a quail-specific nucleic probe for the oligodendrocyte marker SMP (Dulac *et al.*, 1988; Cameron-Curry *et al.* 1989).

Experiments of two types were performed. In series A, unilateral dorsal sectors of approximately 45°, 90°, or 120° of the neural tube, extending over a length corresponding to 4 to 5 somites, were excised microsurgically from host quail embryos at the 19- to 26-somite stage (Zacchei 11–13) and replaced by equivalent sectors of neural tube from chick donors of the same stage (HH 13–15). Thus, the potential of precursors from the ventral part of the neural tube to give rise to oligodendrocytes in the dorsal part was evaluated (Fig. 1A,B).

In series B, the reciprocal procedure was carried out: the chick host embryos received dorsal portions of quail neural tube. In these chimeras, the potential of dorsal territories of the neural tube to yield oligodendrocytes was tested (Fig. 1A,C,D).

In both cases, quail cells of any kind in the nervous system were recognized by the nucleolar marker (Fig. 1D) or by QCPN immunoreactivity (Fig. 1B), and quail oligodendrocytes were identified by the quail-specific SMP nucleic probe (Fig. 1C). Most chimeric embryos of both series were examined at E12–E14, that is, when the white matter of the spinal cord is well developed. In both types of experiments, Feulgen-Rossenbeck staining and QCPN immunolabeling revealed that considerable mixing of quail and chick cells had occurred, except in the ventricular epithelium, where the graft–host limits remained clearly defined. This demonstrates that the cells migrate extensively during the development of the spinal cord (Fig. 1B,C,D).

Analyses with the quail SMP probe by *in situ* hybridization showed, in series A, the presence of quail oligodendrocytes not only in the ventral host territory (quail) but in the lateral and dorsal regions of the spinal cord originating from the chick graft. Thus, oligodendrocyte precursors had migrated from the ventral neural tube and colonized its dorsal part.

In series B, where the grafted dorsal part of the neural tube was from a quail donor embryo, quail oligodendrocytes were found both in the dorsal quail territory and in the ventral chick region of the spinal cord (Fig. 1C), demonstrating the existence of oligodendrocyte precursors in the dorsal part of the neural tube.

These data clearly (1) confirm the presence of a ventrally located oligodendrocyte precursor population, as expected; (2) demonstrate in addition that an oligodendrogenic population also exists in the dorsal part of the avian neural

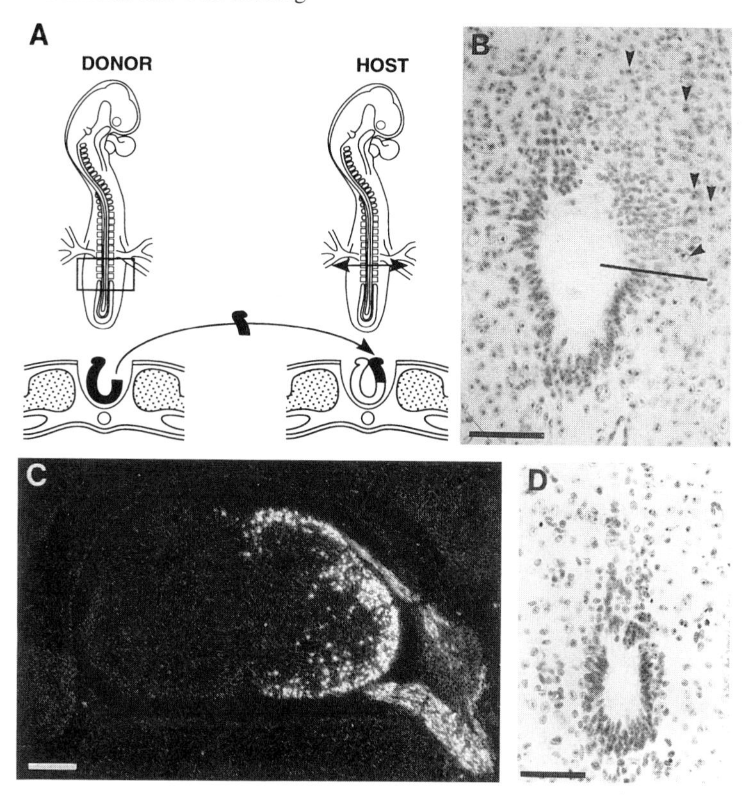

Fig. 1. (A) Schematic representation of the grafting procedure. Unilateral fragments of neural tube extending over a length corresponding to 4–5 somites, were exchanged isotopically between quail and chick embryos at the 19–26 somite stage, at the level of the last-formed somites and unsegmented plate. (B–D) Sections of chimeric embryos at E9–E13. (B) Type A graft: transverse section, in the region of the ventricular epithelium, of a chimeric spinal cord. QCPN staining. The limits of the grafted territory from the chick donor are clearly visible in the epithelium. Quail cells of the recipient have migrated dorsally into the chick territory (arrowheads; bar = 50 μm). (C) Type B graft: *in situ* hybridization with quail oligodendrocyte-specific SMP probe of a transverse section of a chimeric spinal cord. Quail oligodendrocytes are present throughout the section (bar = 200 μm). (D) Type B graft: Feulgen-Rossenbeck staining of a transverse section, in the region of the ventricular epithelium, of a chimeric spinal cord (bar = 50 μm).

tube; and (3) show that extensive migration of oligodendroblasts takes place both in the ventrodorsal and in the dorsoventral directions.

The experimental approach used in this work also revealed the existence of significant (although limited in space) longitudinal anteroposterior and postero-anterior migrations, whereas contralateral cell movements were barely detectable.

The significant and novel finding of this study was the discovery, in avians, of a laterodorsal source of oligodendrocytes, in addition to the well known ventral one. It may be assumed that this fact is also valid in mammals, where it is possible that the ventrally located oligodendroblasts mature earlier than the dorsal precursors, explaining why in mammals only the ventral origin of oligodendrocytes has been demonstrated so far.

2. Plasticity of Rhombomeres

The hindbrain (rhombencephalon) of the vertebrate embryo is transiently subdivided, at early developmental stages, into a series of segments, the rhombomeres (reviewed by Wilkinson, 1993). It has been shown that the identity of these segments along the rostrocaudal axis is defined by a specific expression of *Hox* homeobox genes, whose different combinations generate a *Hox* code (Krumlauf, 1994). Alterations of the *Hox* code either by inactivation or by experimental ectopic expression of certain *Hox* genes in the rhombomeres cause phenotypic modifications of the structures they give rise to during development, demonstrating the role played by these regulatory genes in establishing rhombomere identity (Krumlauf, 1994, for a review; Condie and Capecchi, 1994; Zhang *et al.*, 1994; Manley and Capecchi, 1995; Alexandre *et al.*, 1996). This influence of *Hox* genes is not restricted to the central nervous system: it extends to the rhombencephalic neural crest (NC) cells, which express the same *Hox* code as the corresponding rhombomeres and whose derivatives are severely affected when *Hox* genes are disrupted by targeted mutations in mouse (for reviews, McGinnis and Krumlauf, 1992; Krumlauf, 1994).

In view of the considerable developmental importance of the *Hox* code, it was of interest to know how it is established and maintained in the hindbrain during rhombomere formation and differentiation, and to see whether it is irreversibly determined in the neuroepithelium of a given rhombomere. Autonomy was supported by heterotopic transplantations of the anterior rhombomeres 2 and 4 that did not modify the normal expression pattern of the *Hoxb-1* gene in this area (Guthrie *et al.*, 1992; Kuratani and Eichele, 1993; Prince and Lumsden, 1994; Simon *et al.*, 1995).

Taking advantage of the quail–chick chimera technique and of the availability of avian probes for several *Hox* genes, Grapin-Botton and coworkers (1995) reinvestigated the question of cell autonomy in the hindbrain by examining *Hox* gene expression in defined rhombomeres that had been heterotopically transposed along the rostrocaudal axis. The genes under scrutiny were mainly *Hoxb-4* and its paralogues *Hoxa-4* and *Hoxd-4*, and also *Hoxb-1*, *Hoxa-3*, and *Hoxb-3*. Transplantations were performed at the five-somite stage, before boundaries have formed between rhombomeres and before *Hox* gene expression has actually started in the hindbrain. A precise fate map of the rhombencephalon had been previously established, using the vital dye and carbon particles to mark the precise anteroposterior limits of each presumptive rhombomere.

Most transplantations were unilateral, with the contralateral portion of the neuroepithelium serving as a control.

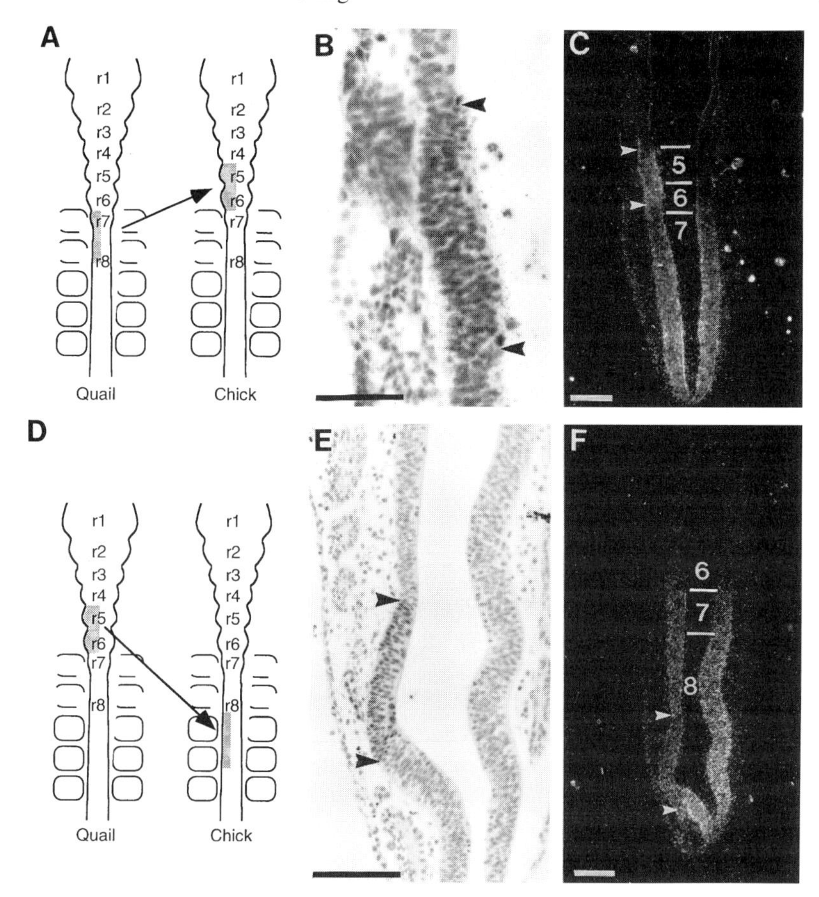

Fig. 2. (A–C) Caudorostral transplantation of rhombomeres. *In situ* analysis of *Hoxb-4* expression after transplantation at the five-somite stage of rhombomeres 7–8 to the level of rhombomeres 5–6 on the left side. (A) Schematic representation of the operation. (B,C) The embryo has been sectioned frontally, 1 day after the graft. The grafted tissue covers rhombomere 5 and part of rhombomere 6. (B) Hematoxylin staining (bar = 50 μm). (C) Same section as (B). *Hoxb-4* expression is clearly visible in the transplant (arrowheads; bar = 100 μm). (D–F) Rostrocaudal transplantation of rhombomeres. (D) Schematic representation of the operation. The rhombomeres 5–6 are transposed to the level of rhombomere 8, facing somites 3–4. (E,F) Adjacent frontal sections. (E) QCPN staining. (F) *Hoxb-4* expression is induced in the transplanted tissue, on the left side (arrowheads; bars = 100 μm).

When the territory of the presumptive hemirhombomeres 7–8 taken from a quail embryo was transplanted to the region corresponding to the future rhombomeres 5–6 of a chick host (caudorostral transplantation), it expressed *Hoxb-4* at stage 15 HH (25 somites) in the resulting chimera, as it would have done if left at its level of origin (Fig. 2A–C). Therefore, it appears that rhombomeres 7–8 are programmed to express *Hoxb-4* as early as the five-somite stage and they do so autonomously even in an environment from which this gene is always absent, as is the case of the level of rhombomeres 5–6. Similar observations were made with *Hoxa-4* and *Hoxd-4*, which were maintained in rhombomeres 7 and 8 transposed to the level of rhombomeres 5–6.

The result of the reciprocal experiment (rostrocaudal transplantation) was the opposite: rhombomeres 1–2, 3–4, or 5–6, which normally do not express *Hoxb-4*, became *Hoxb-4*-positive when they were transplanted to the level of rhombomeres 7–8. In this case, *Hoxb-4* expression was induced in the caudally transposed rhombomeres (Fig. 2D–F).

Whereas the first experiment (caudal to rostral) reflects autonomy of the rhombomeres regarding *Hox* gene expression, the second case (rostral to caudal) reveals inducibility of *Hox* gene expression and plasticity of the rhombomeres. In other words, posteriorization of the neuroepithelium is possible but anteriorization is not. One interpretation of this result is that at the five-somite stage, the inducing signal has already reached the hindbrain, long before *Hox* gene expression actually starts in the rhombomeres.

Further transplantation experiments showed that the signal diffuses horizontally through the neuroepithelium, by planar transmission, from posterior to anterior regions of the neural tube. The paraxial mesoderm is also able to provide the signal, but only in regions posterior to somite 5 (Itasaki *et al.*, 1996; Grapin-Botton *et al.*, unpublished data).

Thanks to quail–chick labeling in the chimeras, it was possible to investigate whether the genetic conversion of caudally transposed rhombomeres was correlated with phenotypic modifications of the resulting structures in the hindbrain. Chick embryos in which hemirhombomeres 5–6 of quail had been grafted at the level of rhombomere 8 were analyzed at E8, when the nuclei of the hindbrain were already well differentiated. It appeared that the nuclei that had developed in the quail graft were identical to the contralateral chick nuclei, showing that anteroposterior transposition of rhombomeres induces homeotic transformation consecutive to and in agreement with changes in *Hox* gene expression. Thus, it was demonstrated that the phenotype of the transplanted rhombomeres is modified and converted from type 5–6 to type 8. This result supports the contention that the *Hox* code is responsible for segment identity in the vertebrate hindbrain.

III. *In Vitro* Cloning of Neural Crest Cells

A. Introduction: Analysis of Cell Potentialities

One of the most important issues in developmental biology is how cell type diversity emerges and is maintained during embryogenesis of multicellular organisms. To address this problem, we must be able to identify changes from undifferentiated precursor cells to specialized cell types. Such changes produce cell lineages, that is, the branching successions of immature cell types through which mature cells develop. One of the problems in detecting successive steps in generating a given lineage is to be able to identify different immature cell types. For this purpose, only a few criteria, such as morphologic characteristics or the capacity to bind cell type-specific antibodies, are usually available at early developmental stages. It is thus easier and more informative to classify precursor cells by the range of differentiated cell types they generate among their progeny. So, two distinct approaches allow the development of the progeny of individual cells to be studied. One is lineage mapping *in vivo* that defines the developmental fate of an embryonic cell by revealing the topographic distribution of mature cells arising from a vitally labeled precursor cell after undisturbed development. The alternative approach analyzes the developmental potentiality of one precursor cell, that is, its ability to differentiate along one or several particular pathways when exposed to a modified environment, after either cell transplantation *in vivo* or in *in vitro* cultures of isolated cells.

Of particular interest in such studies are pluripotent cellular systems in which immature cells are accessible for isolation and well characterized markers of differentiation are available. The vertebrate neural crest (NC), as a pluripotent embryonic cell population, provides an interesting and appropriate model system to investigate how cell lineages emerge and stabilize during embryogenesis. At the end of neurulation, the NC bilaterally forms at the dorsal aspect of the neural folds. Its component cells then detach from the neural tube and start migrating to colonize distant and elected sites in the embryo. The important contribution of NC cells to organogenesis of diverse vertebrate structures was unraveled through the analysis of quail–chick embryonic chimeras (for a review, see Le Douarin, 1982; Le Douarin and Ziller, 1993; and this chapter, Section II). Mapping precisely all NC derivatives at every axial level has proven the multiple fate of the NC cell population. The melanocytes, the Schwann cells lining the peripheral nerves, and the ganglionic satellite glial cells, as well as the peripheral neurons of the sensory, autonomic, and enteric types and certain paraendocrine cells, were shown to originate from the NC. In addition, the NC in the head and neck contributes, besides pigment cells and peripheral nervous system (PNS) cells, to ectomesenchymal derivatives forming muscle cells of the wall of large arteries, connective tissues, and most bones and cartilages of the skull and face (Couly and Le Douarin, 1987; Le Douarin *et al.*, 1993). The NC derivatives are regionalized along the anteroposterior axis; however, heterotopic transplantations of fragments of the neural primordium showed that such diversification is not predetermined, at least for the PNS (see Le Douarin, 1982, for a review). Thus, local influences encountered by NC cells during migration govern the final differentiation to diverse types of neurons. The large developmental potencies and the phenotypic plasticity of the NC cell population were also evident *in vitro* (for references, see Le Douarin and Smith, 1988). Therefore, the question arises of how and when the appropriate topographic distribution of specific cell types becomes established in NC derivatives. In a given derivative, if the NC is heterogeneous, what mechanism selects for the cells that are committed to the phenotypes expressed in that derivative? Conversely, if NC cells are multipotent, how do local factors influence the choice of their final differentiation? Answering these questions required the study of the developmental potential of individual cells. *In vivo* tracing of cell lineages and *in vitro* clonal cultures applied to the avian NC have provided similar conclusions and demonstrated multipotency of crest cells. On the one hand, tracing the fate of the daughter cells derived from a single precursor cell labeled with a fluorescent dye has permitted the identification of several distinct precursors in the premigratory and migratory trunk NC (Bronner-Fraser and Fraser, 1988, 1989; Fraser and Bronner-Fraser, 1991). Most single cells thus generated both ganglionic neurons and non-neuronal cells, whereas others contributed cells of only one NC derivative. These experiments uniquely revealed the multiple fate of crest cells *in vivo*; however, they are not suitable to investigate the role of environmental factors in cell diversification, nor to follow the emergence of late-differentiating cell types because of dye dilution in cycling labeled cells. On the other hand, the complementary approach of cloning cells *in vitro*, although not probing the actual cell fates, offered the possibility of revealing and challenging the developmental potentials of one cell under controlled culture conditions.

B. Cloning Avian Neural Crest Cells *in Vitro*

Cohen and Konigsberg (1975) pioneered *in vitro* clonal techniques to study the prospective fate of NC cells. By analysis of avian trunk NC cells isolated from cultures of the whole neural primordium, they unraveled heterogeneity of the NC cell population and demonstrated that melanocytes and unpigmented cells could arise either from a common progenitor or from distinct committed precursors. Later, improved by probing more diversified phenotypes (e.g., adrenergic and sensory neurons), colony assay of trunk NC cells further confirmed the pluripotency of crest cell precursors (Sieber-Blum and Cohen, 1980; Sieber-Blum, 1989, 1991). Sieber-Blum and collaborators extended the study to crest cells removed from different axial levels and at different developmental stages, including cells that have colonized NC derivatives (Ito and Sieber-Blum, 1991, 1993; Duff *et al.*, 1991; Richardson and Sieber-Blum, 1993).

All of these experiments involved limiting dilution to plate crest cells at low density on plastic or plastic coated with a synthetic substrate (e.g., collagen). Accordingly, repeated microscopic inspections of the cultures (just after plating and then 2 and 18 hr after attachment) were needed to establish whether colonies develop from single cells. In addition, because of the statistical distribution of cells during plating, the efficiency of crest cells in forming clones was often difficult to quantify under these culture conditions.

For these reasons, we chose to use another method of cell seeding that ensures clonality of every culture and allows accurate determination of cloning efficiency. A procedure was thus devised wherein single cells are plated individually in separate wells under microscopic control, and new culture conditions were designed to improve the survival and differentiation of single cells and their progeny (Baroffio *et al.*, 1988; see following for details). Thus, because these experiments were aimed at revealing the whole cell developmental repertoire, it was crucial to provide crest cells with culture conditions that avoid selecting for particular progenitors and biasing crest-derived phenotypes. Such permissive conditions were obtained by supplementing the culture medium with mitogens, hormones, and growth factors, and by growing single crest cells on a feeder layer of 3T3 fibroblasts, according to a method modified from the original procedure of Barrandon and Green (1985) to clone human keratinocytes *in vitro*. Under those conditions, each culture actually develops from a single-plated cell, so that colony efficiency is given by the percentage of clones from the total number of plated NC cells. The proliferation potential of individual crest cells is compared after counting the total number of cells in the colonies. To reveal the differentiation potentials of clonogenic crest cells, cell phenotypes in the clones are analyzed by testing a set of cell-specific markers. So, the resulting number of cell types detected in the progeny of a given cell determines *a posteriori* the degree of potency of that founder cell: by definition, colonies containing two or more different cell types arise from bipotent and multipotent cells, respectively, whereas colonies with a unique cell type derive from unipotent progenitors, which can be considered as committed cells, at least in the environment dictated by those culture conditions.

The next sections describe the cloning procedure in detail, including preparation of feeder layers, seeding of individual cells, and how to analyze the clones. Various methods for isolating avian NC cells are also presented. We then indicate some modifications of the general procedure that were designed to study the fate of particular subsets of crest cells. Finally, results and perspectives provided by the studies of avian NC cells in clonal cultures are discussed.

C. Materials and Methods for Neural Crest Clonal Cultures on 3T3 Cells

1. Materials

Materials for explanting NC and NC-derived tissues of quail embryos (i.e., eggs, incubator, dissecting microscope, instruments for microsurgery) are essentially as described in the previous section concerning transplantations *in vivo* (§ II. C.). Miscellaneous requirements include Ca^{2+},Mg^{2+}-free PBS (CMF-PBS), pancreatin solution for isolating embryonic tissues, a solution of 0.05% trypsin and 0.2% ethylenediamine tetraacetic acid (EDTA) (1×; Gibco) for cell dissociation, and a hemacytometer to count cells in suspension. Micropipettes or fire-pulled pipettes of about 100- to 150-μm diameter and plastic tubes (e.g., Tygon) are needed to handle single cells. Cell cloning is performed under the control of an inverted phase-contrast microscope (e.g., Olympus CK2).

Disposable materials for culture include 10-ml tubes for centrifuging cells, culture flasks for 3T3 cell cultures, 35-mm diameter plastic culture dishes, and eight-well Lab-Tek glass chamber slides (Nunc) for single-cell cultures.

Dulbecco's modified Eagle's medium (DMEM) supplemented with 10% heat-inactivated FCS is used as basic culture medium in experimental steps preceding cloning (e.g., preparation of cell suspensions and feeder layers). A complex culture medium (cloning medium) is required for growing single-cell cultures (Baroffio *et al.*, 1988). It consists of Ham's F-12 nutrient mixture/DMEM/BGjb medium (6:3:1 [vol/vol/vol]) supplemented with 10% FCS, 2% 11-day chick embryo extract (Ziller *et al.*, 1987), and gentamicin sulfate (10 μg/ml). The following mitogens, hormones, and growth factors (all from Sigma) were added: adenine (24.3 ng/ml), isoproterenol (0.25 μg/ml), choleragen (8.4 ng/ml), hydrocortisone (0.4 μg/ml), insulin (5 μg/ml), triiodothyronine (13 ng/ml), transferrin (10 μg/ml), and epidermal growth factor (10 ng/ml). Fresh media are made at least every 2 weeks.

2. Preparation of 3T3 Cell Feeder Layers

Mouse Swiss 3T3 fibroblasts were initially provided by H. Green and coworkers (Barrandon and Green, 1985); however, Swiss 3T3 cells from commercial sources may also be used. 3T3 cells are grown in DMEM supplemented with 10–15% FCS and passaged every third day by replating at the initial concentration of 10^6 cells per 75 mm^2 flask. The day before NC cell cloning, monolayers of 3T3 cells are prepared as follows: Confluent cultures of 3T3 cells are incubated for 3 hr at 37°C with culture medium containing 4 μg/ml mitomycin-C (Sigma) to stop cell division. Cells are then rinsed with DMEM, and detached after briefly incubating at 37°C with 1× trypsin–EDTA. Cells are then pelleted by centrifugation for 10 min at 1000g and resuspended in cloning medium at the dilution of 2 × 10^6 cells/ml. Finally, growth-arrested 3T3 cells are plated on eight-well Lab-Tek glass chambers slides (250 μl per well) that were previously coated with rat tail collagen (Biomedical Technologies, Inc.). Cells attach after a few hours, forming a monolayer to be used as a feeder layer for crest cell cultures.

3. Isolation of Neural Crest Cells for *in Vitro* Cloning

Various methods to isolate avian NC cells at different times of their migration from the neural primordium, either at the trunk or cephalic axial levels, have been reviewed previously (for details of the procedures, see Dupin and Le Douarin, 1993). We describe here only briefly those techniques that were used in previously reported clonal analysis of avian crest cells.

The most commonly used technique to isolate crest cells consists of culturing the whole trunk neural primordium, which is separated enzymatically from the surrounding structures at a donor stage before emigration of NC cells starts (Cohen and Konigsberg, 1975; Sieber-Blum and Cohen, 1980). Crest cells then form an outgrowth onto the substratum *in vitro* by migrating out from the dorsal region of the explanted neural tube. After 24 or 48 hr of primary culture, the neural tubes are scraped away and the crest cells left behind can be detached after brief incubation with 0.5× trypsin–EDTA solution. Cells are then prepared for cloning experiments as described in the next section.

Removing the neural primordium together with premigratory NC, as described previously, may be performed in theory at every level of the anteroposterior axis, with care taken that, at the level considered, the stage of neural tube removal precedes that of NC cell migration (e.g., see Ito and Sieber-Blum, 1991, for isolation of quail rhombencephalic NC).

Few methods are available to purify NC cell populations at later developmental stages, when crest cells have already left the neural tube, because migratory cells disperse rapidly in the embryo to become intermingled with noncrest cells, such as in the presumptive skin and the somitic mesenchyme. Therefore, these subpopulations of crest cells must be identified and sorted out to make further *in vitro* study of individual crest cell fate possible. By contrast, NC cells at the cephalic level anterior to the first somite have been shown to migrate dorsolaterally under the ectoderm; such a superficial location until the 12- to 13-somite stage has permitted the development of a surgical procedure for isolating migrating mesencephalic–metencephalic crest cells from E2 quail embryos at the 9- to 12-somite stage, without contamination by central nervous system and mesodermal cells (Smith *et al.*, 1979; Ziller *et al.*, 1983). Mesencephalic crests forming a multilayered sheet under the ectoderm are removed bilaterally by microsurgery and incubated with 0.25× trypsin–EDTA at 37°C for 7 min. After replacing the enzyme solution by serum-containing medium, the ectodermal layer can be removed and crest cells are dissociated to single cells by trituration with a micropipette (Fig. 3) (Ziller *et al.*, 1983). This procedure leads to isolation of about 1000 to 2000 crest cells per embryo.

4. Neural Crest Cell Cloning Procedure

The method to perform single-cell cultures is identical whatever the source of NC or NC-derived cells under study, and suitable for the analysis of small cell populations, because as few as 1000 isolated cells are needed for these experiments (Fig. 3). All steps are carried out in the presence of cloning culture medium.

Cells to be cloned are prepared at a dilution of 1000 to 2000 cells per milliliter and plated in a 35-mm Petri dish. After a few minutes, cells fall to the bottom of the dish. The resulting cell suspension, when examined at 100× magnification

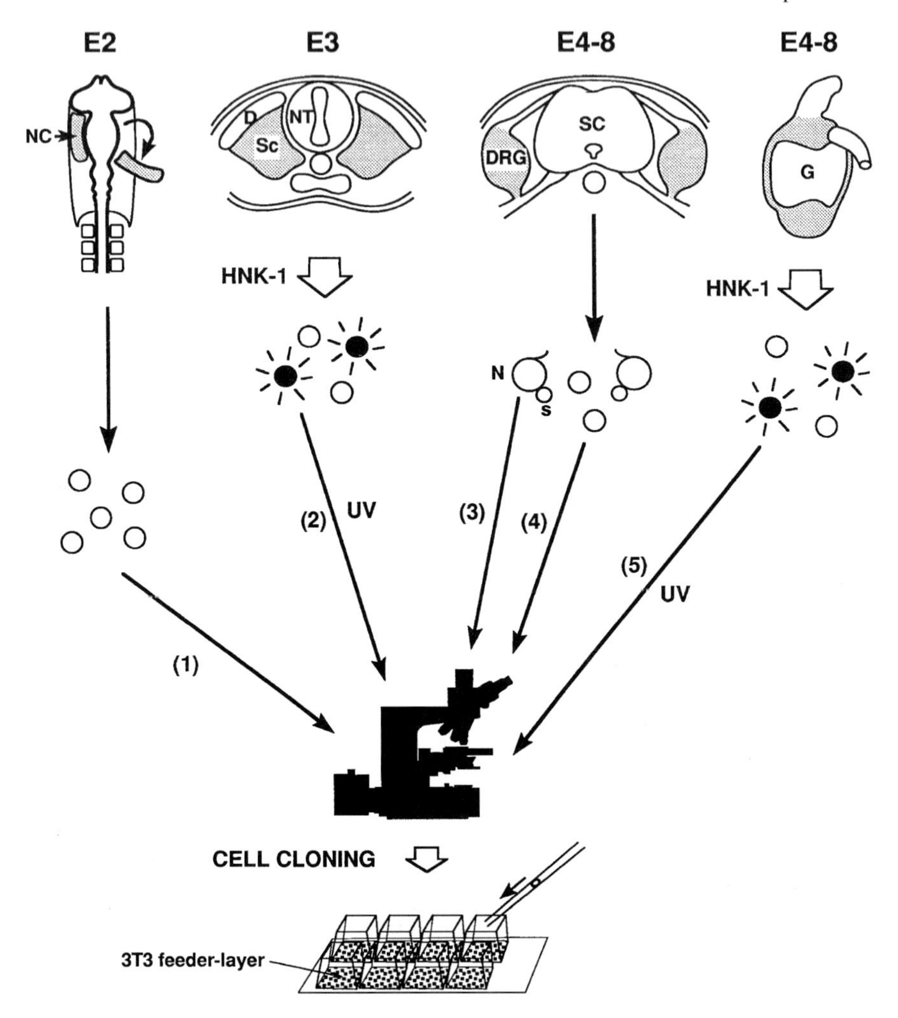

Fig. 3. Schematic representation of the isolation and cloning of neural crest (NC) and NC-derived cells. Migratory cephalic NC cells were taken from E2 quails. Trunk NC cells located in the somites of E3 quails were prepared after isolating the sclerotomes (Sc) from surrounding structures. Trunk NC-derived cells of the dorsal root ganglia (DRG) and gizzard (G) containing vagal NC-derived cells were removed at E4–E8 by microsurgery. Cephalic NC explants were dissociated to single cells for subsequent cloning (1). DRG cells after enzymatic dissociation included doublets of neurons (N) and satellite glial cells (s) (3) as well as single cells (4). Before cloning, crest-derived cells in dissociated sclerotomes (2) and gizzards (5) were labeled with human natural killer 1 (HNK-1) Mab to be selected under fluorescence microscopy from HNK-1-negative mesenchymal cells of the somites and gut. Single cells prepared in this way were removed from the suspension with a micropipette under microscopic control and then settled individually in culture wells containing a previously established feeder layer of 3T3 fibroblasts. NT, neural tube; SC, spinal cord; D, dermomyotome; UV, ultraviolet light.

(with 10× optics) under phase-contrast microscopy, reveals a small number of cells per microscopic field. Hence, under such conditions, a randomly chosen individual cell can be easily picked up from the suspension by aspiration with a micropipette; control of this procedure under the microscope ensures that only a single cell is taken. Then, the single cell-containing drop of medium is settled into a Lab-Tek well containing a previously established culture of growth-inhibited 3T3 cells (see Preparation of 3T3 Cell Feeder Layers).

Preparation of dissociated cells and then plating individual cells in separate wells until two entire racks of Lab-Tek chambers (i.e., 128 cultures) are complete, takes less than 2 hr. To optimize cell survival during the procedure, two identical dishes of NC cell suspension are alternatively used, one to pick out cells under a sterile atmosphere, while the other is kept at 37°C in the culture incubator. Finally, when cell cloning is complete, NC cells remaining in the suspension are pelleted, resuspended in cloning medium, and then distributed in two or four Lab-Tek wells with 3T3 cell cultures, as described previously for single cells. These cultures can be used as controls for subsequent analysis of NC cell proliferation and differentiation. Cultures are maintained at 37°C in an atmosphere of 5% CO_2/95% air. The day after cloning seeding, 250 μl of fresh cloning medium per well is added to the cultures. Half the medium volume is then replaced every third day.

5. Modified Procedure for the Study of Subsets of Neural Crest Cells

We have described the protocol for cloning cells from a homogeneous suspension of pure crest cells. However, isolating a pure NC cell population may be precluded by the fact that migratory NC cells often develop within a tissue together with different, noncrest cells. Such is the case for migrating trunk NC cells that disperse among mesenchymal cells of the rostral somite and the subdermal region, or of enteric nerve cells located in the gut wall. A strategy to circumvent this problem has been to culture NC cells separated from their noncrest cell counterparts on the basis of expression of early surface markers. One such marker currently available is the sulfated carbohydrate moiety recognized by the human natural killer 1 (HNK-1) monoclonal antibody (Mab) (Abo and Balch, 1981). This epitope, carried by several proteins and lipids, is expressed by most avian NC cells at early stages of migration, and later by crest-derived cells of the PNS, whereas it is lost by pigment cells and ectomesenchymal derivatives (Vincent and Thiery, 1984). Thus, after immunofluorescence labeling of living cells with HNK-1 Mab, HNK-1-reactive NC cells may be segregated from negative cells by separation methods such as fluorescence-activated cell sorting (Maxwell *et al.*, 1988; Maxwell and Forbes, 1991) or immunopanning (Hennig and Maxwell, 1995). Both methods lead to enrichment of the positive population; however, the NC cell populations thus sorted out are incompletely depleted of nonreactive cells, so they are not suitable for direct plating as single cells. Therefore, one way to study the development *in vitro* of NC cells purified from a mixed cell population is to choose fluorescent HNK-1-positive cells in the cell suspension observed under epifluorescence microscopy, when single cells are taken out from the suspension and then cloned. Sextier-Sainte-Claire Deville and collaborators (1992, 1994) applied this method to isolate NC cells from the somites and the gut that cannot be removed from the surrounding noncrest cells by dissection (Fig. 3). Thus, trunk NC cells migrating in the sclerotomes and crest-derived cells populating the enteric plexuses of the gut wall, both of which are HNK-1-positive cells, can be isolated during the cloning procedure from HNK-1-negative, mesenchymal cells that develop *in situ* in their vicinity.

6. Clone Analysis

Cultures derived from single NC cells are fixed after 7 to 16 days with 4% paraformaldehyde in CMF-PBS for 1 hr. First, one has to identify NC cells growing onto their feeder layers of 3T3 fibroblasts: detecting clones and then counting crest cells of a given colony may be somewhat difficult using simple microscopic observation; nevertheless, an obvious distinction between quail and mouse cell nuclei is drawn from fluorescence labeling of DNA by staining cultures with bisbenzimide (Hoechst 33342, Serva) for 10 min (1 μg/ml) (Fig. 4). Once identified, the clones are phenotypically analyzed for the presence of different cellular types. All colonies are thus tested with the same set of cell-specific markers, the choice of which depends on the experiment. In experiments wherein a large diversity of crest-derived cell types is expected, one has to combine several markers suitable for the identification of various phenotypes in one colony. For example, for those studies aimed at revealing the whole repertoire of cephalic NC cells, we tested a battery of markers representative of the major NC-derived lineages (Table II): expression of a glial marker, SMP (Dulac *et al.*, 1988), and that of one or two neuronal markers was analyzed by immunocytochemistry, whereas cartilage and pigment cells were detected simply by microscopy (Baroffio *et al.*, 1991). Among other markers available to characterize avian cell types, melanoblast–melanocyte early marker (MelEM) (Nataf *et al.*, 1993) or tyrosine hydroxylase (Fauquet and Ziller, 1989) may be used to assess the presence of unpigmented melanoblasts and adrenergic cells, respectively (Dupin and Le Douarin, 1995).

D. Results of Avian Neural Crest Clonal Studies

1. The Developmental Repertoire of Cephalic Quail Neural Crest Cells

Using clonal cultures on 3T3 cells, we first investigated the developmental potentials of quail NC cells from the mesencephalic level, a cell population endowed with a large range of cell fates [i.e., giving rise to mesenchymal derivatives in

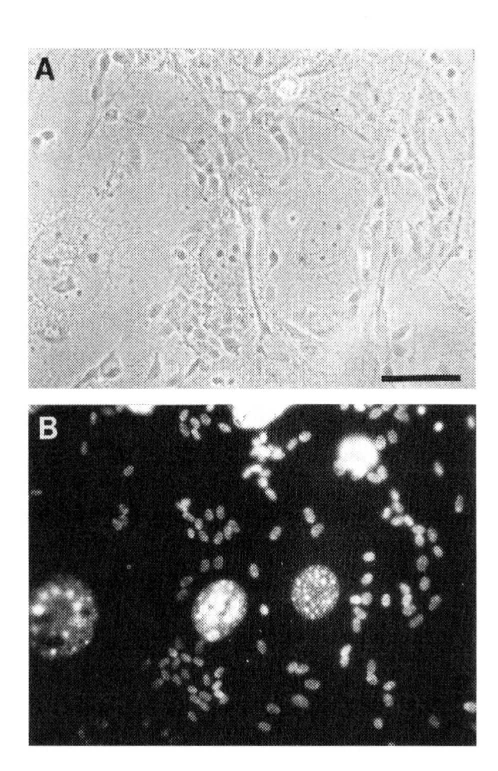

Fig. 4. Part of a 10-day clone of trunk NC cells grown on 3T3 fibroblasts. (A) Phase-contrast micrograph. (B [same field as A]) Fluorescence microscopy. Hoechst nuclear staining shows that the small nuclei of quail NC cells can be distinguished from the large spotted nuclei of 3T3 fibroblasts (bar = 50 μm).

Table II Cell Type Markers for Phenotypic Analysis of Clones Derived from Quail NC Cells

Cell type	Marker	Reference/source
Glial cells	Anti-SMP mouse IgG1	Dulac *et al.*, 1988; DSHB
Neurons	Anti-NF 200 kD rabbit serum	Sigma Chemical Company
Adrenergic	Anti-quail TH mouse IgG2a	Fauquet and Ziller, 1989; DSHB
Peptidergic	Anti-VIP rabbit serum	Garcià-Arraràs *et al.*, 1987
	Anti-SP rat hybridoma	Sera-Lab
Melanoblasts	MelEM mouse IgG1	Nataf *et al.*, 1993; DSHB
Melanocytes	Melanin pigment	
Cartilage	Morphology	
Most crest cells	HNK-1/NC1 mouse IgM	Vincent and Thiery, 1984; Sigma Chemical Company

NF, neurofilament protein; SMP, Schwann cell myelin protein; TH, tyrosine hydroxylase; VIP, vasoactive intestinal peptide; SP, substance P; MelEM, melanocyte–melanoblast early marker; HNK-1/NC1, human natural killer 1/neural crest 1; DSHB, Developmental Studies Hybridoma Bank, University of Iowa.

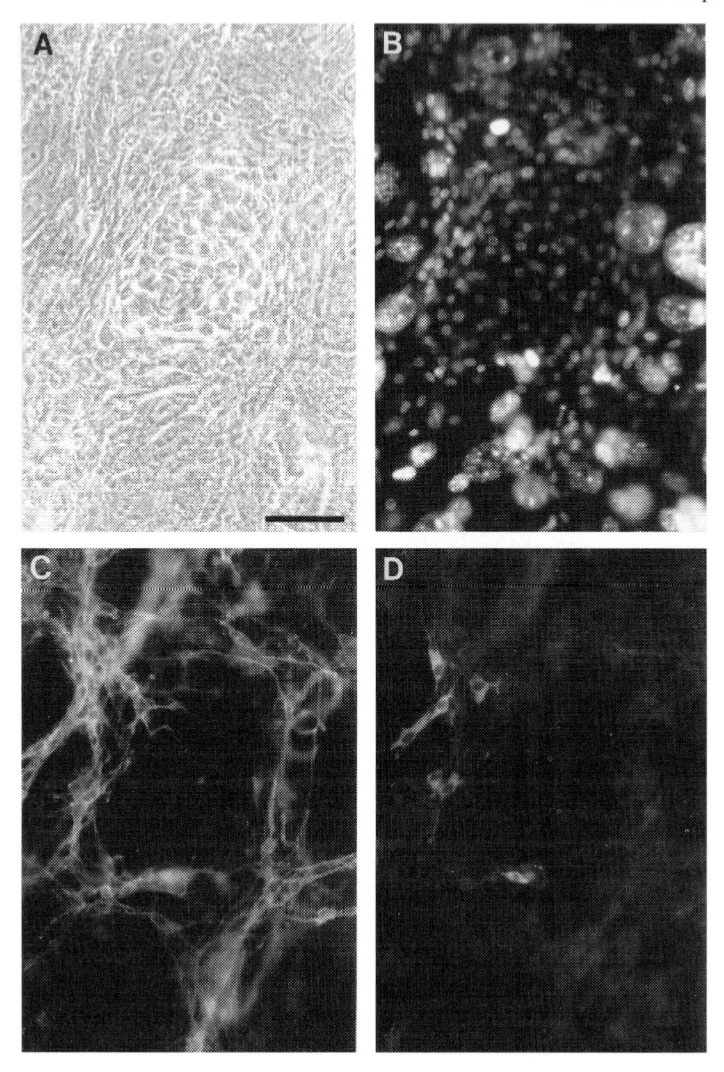

Fig. 5. Part of a 10-day clone of cephalic NC cells. (A–D [same microscopic field]). (A) Phase-contrast micrograph. (B) Fluorescence microscopy after Hoechst DNA staining. (C,D) Immunofluorescence labeling with anti-SMP (C) and anti-tyrosine hydroxylase (D) Mabs. A nodule of cartilage cells is surrounded by SMP-positive glial cells and adrenergic cells (bar = 50 μm).

addition to pigment cells and PNS cell types (Baroffio *et al.*, 1988, 1991; Dupin *et al.*, 1990)]. This population was chosen to test pluripotency of migratory cells and investigate whether mesectodermal precursors are segregated in the head NC from the cells generating PNS neural cells and melanocytes. These experiments were designed with the goal of culturing single cells in optimal conditions so that the largest possible repertoire of sublineages inherent to each progenitor could be expressed. Moreover, to identify as many phenotypes as possible in the progeny of individual cells, it was necessary to apply several cell type-specific markers to the same culture. We thus analyzed more than 500 colonies for the presence of neural (neurons and glial) cells, melanocytes, and cartilage (Baroffio *et al.*, 1988, 1991; Dupin *et al.*, 1990).

The results provide evidence for a heterogeneity among migrating cephalic NC cells concerning both their proliferation and differentiation potentials. Clones of variable size and containing between one and four distinct cell types were found. Thus, the cephalic NC population comprises multipotent cells with various degrees of restriction such as tripotent (neuronal, glial, melanogenic) and bipotent (neuronal, glial or melanogenic, glial) progenitors, as well as unipotent precursors that are committed to giving rise only to neurons or glial cells or cartilage. Interestingly, cartilage also differentiated in the progeny of pluripotent cells (Fig. 5), implying that cells with the capacity to form mesectodermal derivatives are not completely segregated from neural and pigment cell precursors in the migratory NC. Furthermore, a rare progenitor was characterized that produced all the cell types under scrutiny, that is, glial and neuronal (adrenergic and non-adrenergic) cells, cartilage, and melanocytes (Baroffio *et al.*, 1991). It was therefore suggested that this highly multipotent founder cell is a stem cell able to give rise to all the major crest-derived lineages, by analogy to hematopoietic stem cells that can generate all blood cell types (see Metcalf, 1989, for a review). In addition, as for the hematopoietic system, precursors of various intermediate potentials between highly multipotent precursors and committed cells were identified. Statistical analysis of phenotype frequency in the cephalic NC clones further supports the idea that such intermediate precursors are generated stochastically rather than in a sequential order (Baroffio and Blot, 1992). Altogether, data are consistent with a model according to which multiple lineages arise through progressive restrictions of the developmental repertoire of multipotent stem cells. The next question concerned how and when restrictions to final sublineages of the NC are established *in vivo*.

Previous studies suggested that developmental restrictions are likely to occur as NC cells proliferate during the migratory phase (Ziller *et al.*, 1983; Le Douarin, 1986 and Le Douarin *et al.*, 1994 for reviews). However, in back-transplantation experiments, the non-neuronal cell population of PNS ganglia may be induced to express phenotypes that are never found in normal development (e.g., Ayer-Le Lièvre and Le Douarin, 1982; Schweizer *et al.*, 1983; Dupin, 1984; Rothman *et al.*, 1990 and references therein). These findings suggest that, at final sites of differentiation, some NC cells may retain pluripotency and that precursors with partial developmental restrictions are generated asynchronously in NC derivatives.

2. Emergence of Precursors with Restricted Developmental Potentials; Clonal Analysis of Neural Crest Cells during Gangliogenesis

To investigate how and when various differentiated cell types are produced from pluripotent precursors during NC ontogeny, one must study the dynamics of the emergence of restricted precursors in developing NC derivatives. *In vitro* clonal cultures of NC cells taken at different times of their migration from a given level of the neural primordium offer the possibility of examining the time course of the changes in the state of cell commitment leading to the expression of diversified phenotypes. Therefore, by using the same clonal culture method as for the cephalic NC, we studied the progeny of trunk NC cells isolated at different developmental stages, that is, when crest cells are in the course of migration, and later on, when they have formed the dorsal root ganglia (DRG) (Sextier-Sainte-Claire Deville *et al.*, 1992; Cameron-Curry *et al.*, 1993).

Crest-derived cells of E3 quails that have migrated ventrally from the trunk neural tube into the sclerotome mesoderm were distinguished from the surrounding somitic cells by means of HNK-1 immunoreactivity and thereafter cloned under fluorescence microscopy (Fig. 3). The colonies so obtained were compared with the progeny derived from DRG cells isolated at increasing developmental stages (E4–E6). In an attempt to investigate the pluripotency of DRG cells, differentiated neurons, recognizable in the DRG cell suspension by the remnants of their processes, were avoided when cloning (Sextier-Sainte-Claire Deville *et al.*, 1992). Analysis of the colonies derived from sclerotomal crest cells and DRG cells indicates that both the proliferation capacity and the differentiation abilities of trunk NC-derived cells decrease with increasing stage of donor embryos. In particular, the neuronoglial precursors that were evident in the sclerotomes disappeared in DRG, where postmitotic sensory neurons and non-neuronal precursors instead developed.

In a different study, Cameron-Curry *et al.* (1993) further analyzed *in vitro* the differentiation of DRG non-neuronal precursors and examined the regulation of the expression of the glial cell marker SMP (Dulac *et al.*, 1988, 1992). In these experiments, doublets of one neuron and its adhering satellite cell were isolated from E8 quail DRG cells obtained after partial enzymatic digestion and then cultured on 3T3 cells as previously described (Fig. 3). Glial satellite ganglion cells, which are SMP-negative *in vivo*, were thus shown to give rise to a progeny of SMP-expressing cells, and therefore can be converted *in vitro* to a Schwann cell-like phenotype (Cameron-Curry *et al.*, 1993). The induction of SMP synthesis also implies that multiple glial cell types arise from a common precursor and that expression of a particular glial fate is regulated by local environmental factors. The gut thus provides an inhibitory signal that prevents SMP expression by enteric glia; such an inhibition can be relieved when enteric cells are withdrawn from their normal environment by transplantation or in culture (Dulac and Le Douarin, 1991; Sextier-Sainte-Claire Deville *et al.*, 1994).

The developmental potentials of enteric NC cells colonizing the developing gizzard were analyzed in detail after selecting the crest cells by prelabeling with HNK-1 (Fig. 3); these cells in clonal cultures gave rise to various types of colonies, with a high proportion of clones containing SMP-expressing cells (Sextier-Sainte-Claire Deville *et al.*, 1994). Besides induction of SMP, the latter *in vitro* experiment triggered the expression of another phenotype that is never expressed *in situ* by enteric cells, the adrenergic neuronal phenotype. Such a phenotype could be revealed in the descendants of some founder cells from quail gizzards up to E6. In addition, analysis of the clones has demonstrated that the enteric precursors able to yield adrenergic cells are bipotent cells. Therefore, impairment of adrenergic cell development in enteric ganglia *in vivo* is not the result of a negative selection exerted on committed adrenergic cells that might have been inappropriately located in the gut. Rather, local environmental factors are likely to act on bipotent enteric NC cells by preventing their differentiation. Experimental conditions such as those of clonal cultures led to suppression of the gut inhibitory signal, and revealed the adrenergic potentiality of some enteric cells that remained in a latent state.

3. Role of Environmental Factors in Neural Crest Cell Fate

From the studies described earlier, it became increasingly apparent that the cells reaching the final sites where NC derivatives develop are multipotent as well as more restricted. Hence, local factors may be necessary there to promote differentiation to specific options and to select for those progenitors that are committed to phenotypes appropriate to a given derivative. A number of studies have been devoted to the search for such factors capable of controlling the spatiotemporal production of the diverse differentiated cells from the NC. Several defined growth factors were found to regulate the development of particular sublineages *in vitro* (for references, see Le Douarin and Dupin, 1993; Sieber-Blum *et al.*, 1993; Stemple and Anderson, 1993). Primary cultures have been instrumental in testing specific factors for their effects on early NC cells. Certain factors previously known to affect the function of particular cell types, such as neurotrophin-3 for neurons or endothelin-3 for endothelial cells, were shown to act as mitogens for early NC cells (Kalcheim *et al.*, 1992; Lahav *et al.*, 1996). These and other factors such as basic fibroblast growth factor (Kalcheim, 1989; Brill *et al.*, 1992), epidermal growth factor (Erickson and Turley, 1987), and stem cell factor (Lahav *et al.*, 1994) may in addition promote the differentiation of cultured quail NC cells to particular phenotypes. In these experiments performed on heterogeneous cell populations, it was often difficult to characterize NC-responsive precursors and to assess the relative contribution of mitogenic, trophic, and differentiative activities of a given factor.

Clonal culture systems clearly provide a means to identify subsets of NC precursor cells that are the target of environmental factors and to understand how these signals actually influence the development of responsive cells. So far, only a few investigations have used the *in vitro* clonal approach to study the effects of growth factors. Satoh and Ide (1987) reported the promoting action of pituitary α-melanocyte-stimulating hormone (α-MSH) on quail NC cells *in vitro*. Clonal analysis revealed that α-MSH stimulates the development of melanogenic unipotent precursors, while not affecting cells yielding mixed (unpigmented and pigmented) colonies.

Evidence for a trophic activity of brain-derived neurotrophic factor (BDNF) during the initial formation of DRG *in vivo* (Kalcheim *et al.*, 1987) led to further *in vitro* investigations as to how several members of the neurotrophin family may influence the development of sensory neurons from NC cells in clonal culture. BDNF, but not nerve growth factor, increased sensory neuronal differentiation in the progeny of pluripotent crest cells (Sieber-Blum, 1991). Because every cell in a colony was scored and categorized, it could be determined that an increase of sensory neurons was at the expense of adrenergic and undifferentiated cells in the colonies. These data therefore argue for a role for BDNF in the commitment of trunk NC cells to the sensory neuronal phenotype.

As for the differentiation of crest-derived autonomic neurons, the active factors are poorly understood, although it has long been established that the development of adrenergic cells from cultured avian NC cells is promoted by addition of high concentrations of chick embryo extract (CEE) (Howard and Bronner-Fraser, 1985; Ziller *et al.*, 1987). In one study, we showed that CEE could be replaced by the vitamin A derivative all-*trans* retinoic acid (RA) (Dupin and Le Douarin, 1995). Like CEE, RA *in vitro* promoted the differentiation of adrenergic cells and melanocytes in quail NC cultures (Fig. 6). Clonal cultures revealed that RA increased the number of adrenergic cells and pigment cells in the progeny of a common pluripotent precursor also able to generate glial cells (Dupin and Le Douarin, 1995). In fact, RA promoted melanin synthesis by melanoblasts but did not affect the production of the latter cells expressing MelEM. In contrast, RA increased the number of adrenergic cells in the colonies, without showing a proliferative effect, suggesting that it may favor the commitment of pluripotent cells to the adrenergic lineage (Fig. 7). The mechanisms underlying RA actions on NC cells remain to be understood. Developing NC cells *in vivo* and *in vitro* express nuclear receptors for RA and other retinoids (Rowe *et al.*, 1991, 1994), which are known to regulate the transcription of a number of different genes (see Linney, 1992, for references). Therefore, it is conceivable that RA modulates NC cell fates by influencing the activity of multiple target genes such as transcription factors, extracellular matrix molecules, or receptors to trophic factors.

E. Conclusion

In the absence of available techniques for *in vivo* single-cell transplants in avian embryos, *in vitro* clonal cultures have been instrumental in testing the commit-

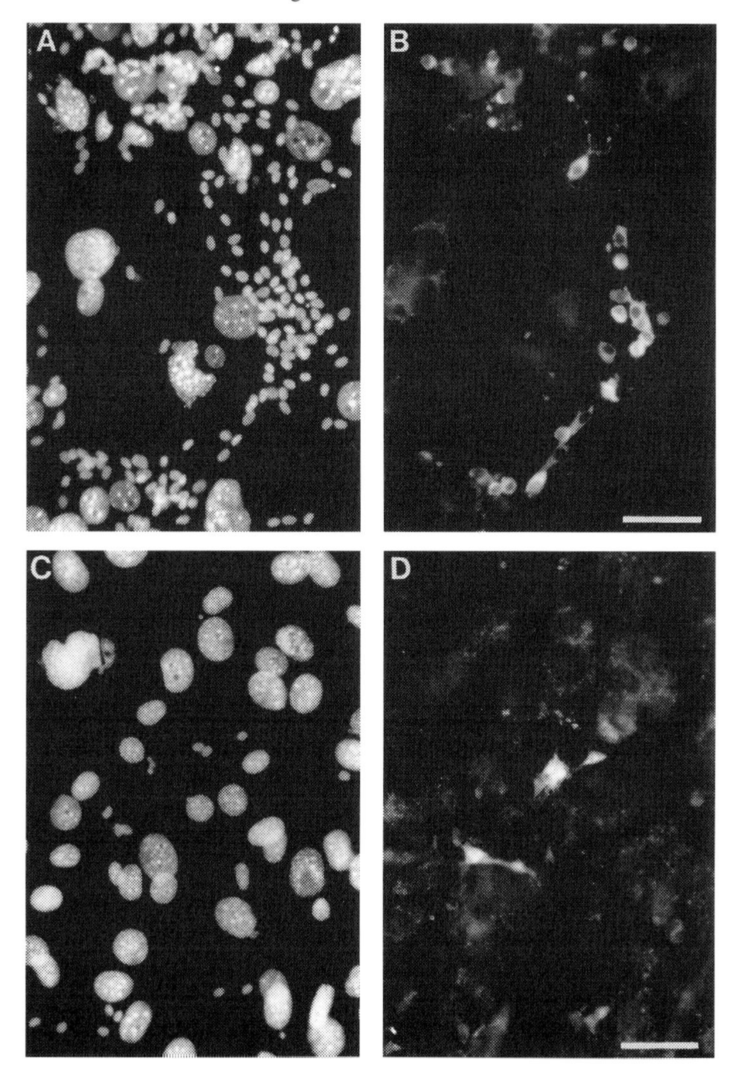

Fig. 6. Clonal cultures of trunk NC cells grown for 8 days in the presence of 100 n*M* RA. (A,C) Hoechst nuclear staining. (B,D) fluorescence microscopy after labeling with anti-tyrosine hydroxylase (B [same field as A]) and MelEM Mab (D [same field as C]) to reveal adrenergic neurons and melanoblasts, respectively (bars = 50 μm).

ment of individual NC cells. Growing single quail NC cells under culture conditions that allow the expression of the whole spectrum of crest-derived phenotypes has thus further clarified how diversity arises from the pluripotent crest cell population. Clonal cultures first revealed the developmental repertoire of mesencephalic NC cells, demonstrating the pluripotency of individual precursors as

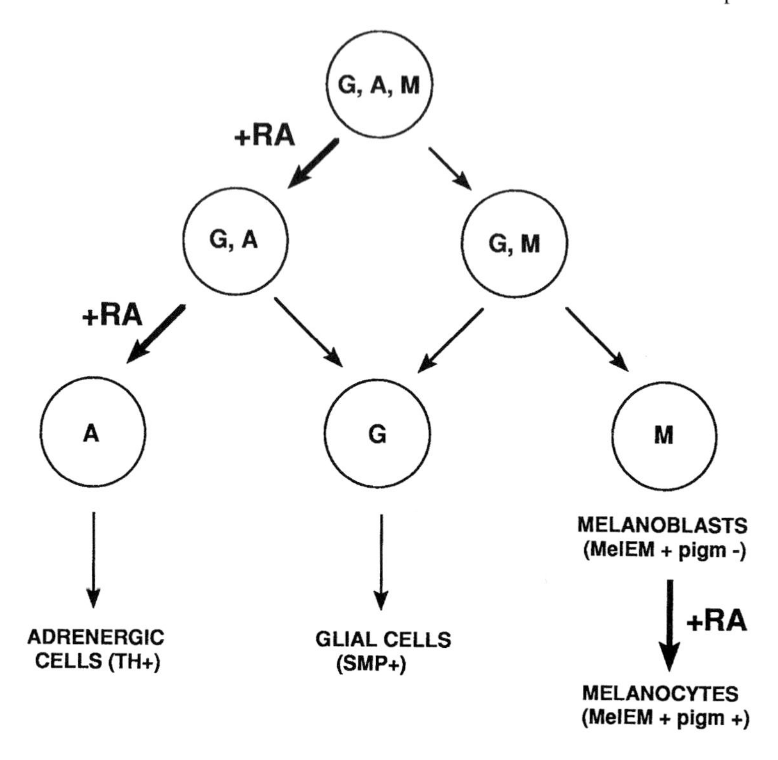

Fig. 7. Model for the action of retinoic acid (RA) on the differentiation of NC precursor cells. The precursors of glial cells (G), adrenergic neuronal cells (A), and melanoblasts (M) were studied in NC clonal cultures. Analysis of colonies grown in the absence or in the presence of 100 n*M* RA showed two different effects of RA (Dupin and Le Douarin, 1995): first, RA enhances the development of pluripotent NC cells, giving rise to adrenergic cells; second, it stimulates pigment synthesis by melanocytic precursors without affecting the emergence of unpigmented melanoblasts in the colonies. MelEM, melanoblast–melanocyte early marker; SMP, Schwann cell myelin protein; TH, tyrosine hydroxylase.

well as a possible common origin of mesectodermal and neural cell types in the embryonic head. The emergence of progressively more restricted precursors during NC-derivative ontogeny was deduced from analyzing the colonies derived from cells taken during and at the end point of migration. In addition, single NC cell cultures have been exploited to assess the role of defined factors in the development and segregation of NC-derived lineages.

Although such experiments have their limitations, we believe that they are suitable for further study of avian NC cell development *in vitro*, after improving and extending the clonal culture system. One of the major problems in interpreting data often is the reliance on the heterogeneity of the NC cell population. It would be useful to work on more defined NC cell populations, particularly when studying the relative contribution of cell interactions and activities of external cues on precursor development. Therefore, new cell surface markers are required to purify identified subsets of precursors among heterogeneous NC cells for further analysis. Besides the activities of already known growth factors, the action of environmental signals such as membrane-bound factors or receptors needs to be investigated. Our coculture system of NC cells and 3T3 fibroblasts could be useful in this respect by permitting investigation of whether the NC cell repertoire is altered in the presence of modified feeder-layer cells after transfection of 3T3 cells with genes of interest. An alternative possibility for examining the function of developmental regulatory genes is to analyze *in vitro* the differentiation potentials of single NC cells that have been previously genetically modified by transfection or retroviral infection in primary culture.

IV. Summary

The avian embryo is a model in which techniques of experimental embryology and cellular and molecular biology can converge to address fundamental questions of development biology. The first part of the chapter describes two examples of transplantation and cell labeling experiments performed *in ovo*. Thanks to the distinctive histologic and immunocytochemical characteristics of quail and chick cells, the migration and development of definite cells are followed in suitably constructed chimeric quail–chick embryos. Isotopic transplantations of neural tube portions between quail and chick, combined with *in situ* hybridization with a nucleic probe specific for a quail oligodendrocyte marker, allowed study of the origin and migration of oligodendroblasts in the spinal cord. Heterotopic transplantations of rhombomeres were performed to establish the degree of plasticity of these segments of the hindbrain regarding *Hox* gene expression, which was revealed by labeling with chick-specific nucleic probes. The second part describes *in vitro* cell cloning experiments devised to investigate cell lineage segregation and diversification during development of the NC. An original cloning procedure and optimal culture conditions permitted analysis of the developmental potentials of individual NC cells taken at definite migration stages. The results revealed a striking heterogeneity of the crest cell population, which appeared to be composed of precursors at different states of determination. Clonal cultures also provide a means to identify subsets of cells that are the target of environmental factors and to understand how extrinsic signals influence the development of responsive cells.

Acknowledgments

We thank M.-F. Simon for preparing the manuscript, and F. Viala and S. Gournet for the illustrations. This work was supported by the Centre National de la Recherche Scientifique, The Institut National de la Santé et de la Recherche Médicale, and by grants from the Association pour la Recherche contre le Cancer and AMGEN, Inc.

References

Abo, T., and Balch, C. M. (1981). A differentiation antigen of human NK and K cells identified by a monoclonal antibody (HNK-1). *J. Immunol.* **127,** 1024–1029.

Alexandre, D., Clarke, J. D. W., Oxtoby, E., Yan, Y. L., Jowett, T., and Holder, N. (1996). Ectopic expression of Hoxa-1 in the zebrafish alters the fate of the mandibular arch neural crest and phenocopies a retinoic acid-induced phenotype. *Development* **122,** 735–746.

Ayer-Le Lièvre, C. S., and Le Douarin, N. M. (1982). The early development of cranial sensory ganglia and the potentialities of their component cells studied in quail–chick chimeras. *Dev. Biol.* **94,** 291–310.

Baroffio, A., and Blot, M. (1992). Statistical evidence for a random commitment of pluripotent cephalic neural crest cells. *J. Cell Sci.* **103,** 581–587.

Baroffio, A., Dupin, E., and Le Douarin, N. M. (1988). Clone-forming ability and differentiation potential of migratory neural crest cells. *Proc. Natl. Acad. Sci. U.S.A.* **85,** 5325–5329.

Baroffio, A., Dupin, E., and Le Douarin, N. M. (1991). Common precursors for neural and mesectodermal derivatives in the cephalic neural crest. *Development* **112,** 301–305.

Barrandon, Y., and Green, H. (1985). Cell size as a determinant of the clone-forming ability of human keratinocytes. *Proc. Natl. Acad. Sci. U.S.A.* **82,** 5390–5394.

Brill, G., Vaisman, N., Neufeld, G., and Kalcheim, C. (1992). BHK-21-derived cell lines that produce basic fibroblast growth factor, but not parental BHK-21 cells, initiate neuronal differentiation of neural crest progenitors. *Development* **115,** 1059–1069.

Bronner-Fraser, M., and Fraser, S. E. (1988). Cell lineage analysis reveals multipotency of some avian neural crest cells. *Nature* **335,** 161–164.

Bronner-Fraser, M., and Fraser, S. E. (1989). Developmental potential of avian trunk neural crest cells *in situ*. *Neuron* **3,** 755–766.

Cameron-Curry, P., and Le Douarin, N. M. (1995). Oligodendrocyte precursors originate from both the dorsal and the ventral parts of the spinal cord. *Neuron* **15,** 1299–1310.

Cameron-Curry, P., Dulac, C., and Le Douarin, N. (1989). Expression of SMP antigen by oligodendrocytes in the developing avian central nervous system. *Development* **107,** 825–833.

Cameron-Curry, P., Dulac, C., and Le Douarin, N. M. (1993). Negative regulation of Schwann cell myelin protein gene expression by the dorsal root ganglionic microenvironment. *Eur. J. Neurosci.* **5,** 594–604.

Chen, C. L., Ager, L., Gartland, L., and Cooper, M. (1986). Identification of a T3/T cell receptor complex in chickens. *J. Exp. Med.* **164,** 375–380.

Chen, C. L., Cihak, J., Losch, U., and Cooper, M. D. (1988). Differential expression of two T cell receptors, TcR1 and TcR2, on chicken lymphocytes. *Eur. J. Immunol.* **18,** 539–543.

Cihak, J., Ziegler-Heitbrock, H. W., Trainer, H., Schranner, I., Merkenschlager, M., and Losch, U. (1988). Characterization and functional properties of a novel monoclonal antibody which identifies a T cell receptor in chickens. *Eur. J. Immunol.* **18,** 533–537.

Cohen, A. M., and Konigsberg, I. R. (1975). A clonal approach to the problem of neural crest determination. *Dev. Biol.* **46,** 262–280.

Condie, B. G., and Capecchi, M. R. (1994). Mice with targeted disruptions in the paralogous genes *hoxa-3* and *hoxd-3* reveal synergistic interactions. *Nature* **370,** 304–307.

Conrad, G. W., Bee, J. A., Roche, S. M., and Teillet, M. A. (1993). Fabrication of microscalpels by electrolysis of tungsten wire in a meniscus. *J. Neurosc. Methods* **50,** 123–127.

Couly, G. F., and Le Douarin, N. M. (1987). Mapping of the early neural primordium in quail–chick chimeras: II. The prosencephalic neural plate and neural folds: Implications for the genesis of cephalic human congenital abnormalities. *Dev. Biol.* **120,** 198–214.

Duff, R. S., Langtimm, C. J., Richardson, M. K., and Sieber-Blum, M. (1991). *In vitro* clonal analysis of progenitor cell patterns in dorsal root and sympathetic ganglia of the quail embryo. *Dev. Biol.* **147,** 451–459.

Dulac, C., and Le Douarin, N. M. (1991). Phenotypic plasticity of Schwann cells and enteric glial cells in response to the microenvironment. *Proc. Natl. Acad. Sci. U.S.A.* **88,** 6358–6362.

Dulac, C., Cameron-Curry, P., Ziller, C., and Le Douarin, N. M. (1988). A surface protein expressed by avian myelinating and nonmyelinating Schwann cells but not by satellite or enteric glial cells. *Neuron* **1,** 211–220.

Dulac, C., Tropak, M. B., Cameron-Curry, P., Rossier, J., Marshak, D. R., Roder, J., and Le Douarin, N. M. (1992). Molecular characterization of the Schwann cell myelin protein, SMP: Structural similarities within the immunoglobulin superfamily. *Neuron* **8,** 323–334.

Dupin, E. (1984). Cell division in the ciliary ganglion of quail embryos *in situ* and after back-transplantation into the neural crest migration pathways of chick embryos. *Dev. Biol.* **105,** 288–299.

Dupin, E., and Le Douarin, N. M. (1993). Culture of avian neural crest cells. *In* "Essential Developmental Biology: A Practical Approach" (C. D. Stern and P. W. H. Holland, eds.), pp. 153–166. Oxford University Press, New York.

Dupin, E., and Le Douarin, N. M. (1995). Retinoic acid promotes the differentiation of adrenergic cells and melanocytes in quail neural crest cultures. *Dev. Biol.* **168,** 529–548.

Dupin, E., Baroffio, A., Dulac, C., Cameron-Curry, P., and Le Douarin, N. M. (1990). Schwann-cell differentiation in clonal cultures of the neural crest, as evidenced by the anti-Schwann cell myelin protein monoclonal antibody. *Proc. Natl. Acad. Sci. U.S.A.* **87,** 1119–1123.

Eichmann, A., Marcelle, C., Bréant, C., and Le Douarin, N. M. (1993). Two molecules related to the VEGF receptor are expressed in early endothelial cells during avian embryonic development. *Mech. Dev.* **42,** 33–48.

Erickson, C. A., and Turley, E. A. (1987). The effects of epidermal growth factor on neural crest cells in tissue culture. *Exp. Cell Res.* **169,** 267–279.

Fauquet, M., and Ziller, C. (1989). A monoclonal antibody directed against quail tyrosine hydroxylase: Description and use in immunocytochemical studies on differentiating neural crest cells. *J. Histochem. Cytochem.* **37,** 1197–1205.

Feulgen, R., and Rossenbeck, H. (1924). Mikroskopisch-chemischer Nachweis einer Nukleinsäure von Typus der Thymonukleinsäure und die darauf beruhende elektive Färbung von Zellkernen in mikroskopischen Präparaten. *Hoppe-Seyler's Zeitschrift für Physiologische Chemie* **135,** 203–248.

Fontaine-Pérus, J., Chanconie, M., and Le Douarin, N. M. (1985). Embryonic origin of substance P containing neurons in cranial and spinal sensory ganglia of the avian embryo. *Dev. Biol.* **107,** 227–238.

Franklin, R. M., and Martin, M. T. (1980). Staining and histochemistry of undecalcified bone embedded in a water-miscible plastic. *Stain Technol.* **55,** 313–321.

Fraser, S. E., and Bronner-Fraser, M. (1991). Migrating neural crest cells in the trunk of the avian embryo are multipotent. *Development* **112,** 913–920.

Gabe, M. (1968). "Techniques Histologiques," Masson, Paris.

Garciá-Arrarás, J. E., Chanconie, M., Ziller, C., and Fauquet, M. (1987). *In vivo* and *in vitro* expression of vasoactive intestinal polypeptide-like immunoreactivity by neural crest derivatives. *Dev. Brain Res.* **430,** 255–265.

Grapin-Botton, A., Bonnin, M. A., McNaughton, L. A., Krumlauf, R., and Le Douarin, N. M. (1995). Plasticity of transposed rhombomeres: *Hox* gene induction is correlated with phenotypic modifications. *Development* **121,** 2707–2721.

Guthrie, S., Muchamore, I., Kuroiwa, A. Marshall, H., Krumlauf, R., and Lumsden, A. (1992). Neuroectodermal autonomy of *hox*-2.9 expression revealed by rhombomere transpositions. *Nature* **356,** 157–159.

Hamburger, V., and Hamilton, H. L. (1951). A series of normal stages in the development of chick embryo. *J. Morphol.* **88,** 49–92.

Hennig, A. K., and Maxwell, G. D. (1995). Persistent correlation between expression of a sulfated carbohydrate antigen and adrenergic differentiation in cultures of quail trunk neural crest cells. *Differentiation* **59,** 299–306.

Howard, M. J., and Bronner-Fraser, M. (1985). The influence of neural tube-derived factors on differentiation of neural crest cells *in vitro*: I. Histochemical study on the appearance of adrenergic cells. *J. Neurosci.* **5,** 3302–3309.

Itasaki, N., Sharpe, J., Morrison, A., and Krumlauf, R. (1996). Reprogramming *Hox* expression in the vertebrate hindbrain: Influence of paraxial mesoderm and rhombomere transposition. *Neuron* **16,** 487–500.

Ito, K., and Sieber-Blum, M. (1991). *In vitro* clonal analysis of quail cardiac neural crest development. *Dev. Biol.* **148,** 95–106.

Ito, K., and Sieber-Blum, M. (1993). Pluripotent and developmentally restricted neural crest-derived cells in posterior visceral arches. *Dev. Biol.* **156,** 191–200.

Izpisúa-Belmonte, J. C., de Robertis, E. M., Storey, K. G., and Stern, C. D. (1993). The homeobox gene *goosecoid* and the origin of organizer cells in the early chick blastoderm. *Cell* **74,** 645–659.

Kalcheim, C. (1989). Basic fibroblast growth factor stimulates survival of nonneuronal cells developing from trunk neural crest. *Dev. Biol.* **134,** 1–10.

Kalcheim, C., Barde, Y. A., Thoenen, H., and Le Douarin, N. M. (1987). *In vivo* effect of brain-derived neurotrophic factor on the survival of developing dorsal root ganglion cells. *EMBO J.* **6,** 2871–2873.

Kalcheim, C., Carmeli, C., and Rosenthal, A. (1992). Neurotrophin 3 is a mitogen for cultured neural crest cells. *Proc. Natl. Acad. Sci. U.S.A.* **89,** 1661–1665.

Krumlauf, R. (1994). *Hox* genes in vertebrate development. *Cell* **78,** 191–201.

Kuratani, S. C., and Eichele, G. (1993). Rhombomere transplantation repatterns the segmental organization of cranial nerves and reveals cell-autonomous expression of a homeodomain protein. *Development* **117,** 105–117.

Lahav, R., Lecoin, L., Ziller, C., Nataf, V., Carnahan, J. F., Martin, F. H., and Le Douarin, N. M. (1994). Effect of the *Steel* gene product on melanogenesis in avian neural crest cell cultures. *Differentiation* **58,** 133–139.

Lahav, R., Ziller, C., Dupin, E., and Le Douarin, N. M. (1996). Endothelin 3 promotes neural crest cell proliferation and mediates a vast increase in melanocyte number in culture. *Proc. Natl. Acad. Sci. U.S.A.* **93,** 3892–3897.

Lance-Jones, C. C., and Lagenaur, C. F. (1987). A new marker for identifying quail cells in embryonic avian chimeras: A quail-specific antiserum. *J. Histochem. Cytochem.* **35,** 771–780.

Le Douarin, N. (1969). Particularités du noyau interphasique chez la Caille japonaise (*Coturnix coturnix japonica*): Utilisation de ces particularités comme “marquage biologique” dans les recherches sur les interactions tissulaires et les migrations cellulaires au cours de l’ontogenèse. *Bulletin Biologique de France et de Belgique* **103,** 435–452.

Le Douarin, N. (1971). Comparative ultrastructural study of the interphasic nucleus in the quail (*Coturnix coturnix japonica*) and the chicken (*Gallus gallus*) by the regressive EDTA staining method. *C. R. Acad. Sci. III* **272,** 2334–2337.

Le Douarin, N. (1973a). A biological cell labeling technique and its use in experimental embryology. *Dev. Biol.* **30,** 217–222.

Le Douarin, N. M. (1973b). A Feulgen-positive nucleolus. *Exp. Cell Res.* **77,** 459–468.

Le Douarin, N. (1982). "The Neural Crest," Cambridge University Press, Cambridge.

Le Douarin, N. M. (1986). Cell line segregation during peripheral nervous system ontogeny. *Science* **231,** 1515–1522.

Le Douarin, N. M. (1993). Embryonic neural chimeras in the study of brain development. *Trends Neurosci.* **16,** 64–72.

Le Douarin, N. M., and Dupin, E. (1993). Cell lineage analysis in neural crest ontogeny. *J. Neurobiol.* **24,** 146–161.

Le Douarin, N. M., and Smith, J. (1988). Development of the peripheral nervous system: Cell line segregation and chemical differentiation of neural crest cells. *In* "Handbook of Chemical Neuroanatomy, Vol. 6: The Peripheral Nervous System" (A. Björklund, T. Hökfelt, and C. Owman, eds.), pp. 1–50. Elsevier, Amsterdam.

Le Douarin, N. M., and Ziller, C. (1993). Plasticity in neural crest cell differentiation. *Curr. Opin. Cell Biol.* **5,** 1036–1043.

Le Douarin, N. M., Guillemot, F., Oliver, P., and Péault, B. (1983). Distribution and origin of a-positive cells in the avian thymus analyzed by means of monoclonal antibodies in heterospecific chimeras. *In* "Progress in Immunology V" (Y. Yamamura and T. Tada, eds.), pp. 613–631. Academic Press, New York.

Le Douarin, N. M., Ziller, C., and Couly, G. F. (1993). Patterning of neural crest derivatives in the avian embryo: *In vivo* and *in vitro* studies. *Dev. Biol.* **159,** 24–49.

Le Douarin, N. M., Dupin, E., and Ziller, C. (1994). Genetic and epigenetic control in neural crest development. *Curr. Opin. Genet. Dev.* **4,** 685–695.

Le Douarin, N., Dieterlen-Lièvre, F., and Teillet, M. A. (1996). Quail–chick transplantations. *Methods Cell Biol.* **51,** 23–61.

LeVine, S. M., and Goldman, J. E. (1988). Embryonic divergence of oligodendrocyte and astrocyte lineages in the developing rat cerebrum. *J. Neurosci.* **8,** 3992–4006.

Linney, E. (1992). Retinoic acid receptors: Transcription factors modulating gene regulation, development, and differentiation. *Curr. Top. Dev. Biol.* **27,** 309–350.

Manley, N. R., and Capecchi, M. R. (1995). The role of *Hoxa-3* in mouse thymus and thyroid development. *Development* **121,** 1989–2003.

Maxwell, G. D., and Forbes, M. E. (1991). Spectrum of *in vitro* differentiation of quail trunk neural crest cells isolated by cell sorting using the HNK-1 antibody and analysis of the adrenergic development of HNK-1+ sorted subpopulations. *J. Neurobiol.* **22,** 276–286.

Maxwell, G. D., Forbes, M. E., and Christie, D. S. (1988). Analysis of the development of cellular subsets present in the neural crest using cell sorting and cell culture. *Neuron* **1,** 557–568.

McGinnis, W., and Krumlauf, R. (1992). Homeobox genes and axial patterning. *Cell* **68,** 283–302.

Metcalf, D. (1989). The molecular control of cell division, differentiation commitment and maturation in haemopoetic cells. *Nature* **339,** 27–30.

Nataf, V., Mercier, P., Ziller, C., and Le Douarin, N. M. (1993). Novel markers of melanocyte differentiation in the avian embryo. *Exp. Cell Res.* **207,** 171–182.

Noll, E., and Miller, R. H. (1993). Oligodendrocyte precursors originate at the ventral ventricular zone dorsal to the ventral midline region in the embryonic rat spinal cord. *Development* **118,** 563–573.

Pardanaud, L., Altmann, C., Kitos, P., Dieterlen-Lièvre, F., and Buck, C. A. (1987). Vasculogenesis in the early quail blastodisc as studied with a monoclonal antibody recognizing endothelial cells. *Development* **100,** 339–349.

Péault, B. M., Thiery, J. P., and Le Douarin, N. M. (1983). Surface marker for hemopoietic and endothelial cell lineages in quail that is defined by a monoclonal antibody. *Proc. Natl. Acad. Sci. U.S.A.* **80,** 2976–2980.

Prince, V., and Lumsden, A. (1994). *Hoxa-2* expression in normal and transposed rhombomeres: Independent regulation in the neural tube and neural crest. *Development* **120,** 911–923.

Richardson, M. K., and Sieber-Blum, M. (1993). Pluripotent neural crest cells in the developing skin of the quail embryo. *Dev. Biol.* **157,** 348–358.

Rothman, T. P., Le Douarin, N. M., Fontaine-Pérus, J. C., and Gershon, M. D. (1990). Developmental potential of neural crest-derived cells migrating from segments of developing quail bowel back-grafted into younger chick host embryos. *Development* **109,** 411–423.

Rowe, A. Eager, N. S., and Brickell, P. M. (1991). A member of the RXR nuclear receptor family is expressed in neural crest-derived cells of the developing chick peripheral nervous system. *Development* **111,** 771–778.

Rowe, A., Sarkar, S., Brickell, P. M., and Thorogood, P. (1994). Differential expression of RAR-beta and RXR-gamma transcripts in cultured cranial neural crest cells. *Roux's Archives for Developmental Biology* **203,** 445–449.

Satoh, M., and Ide, H. (1987). Melanocyte-stimulating hormone affects melanogenic differentiation of quail neural crest cells *in vitro*. *Dev. Biol.* **119,** 579–586.

Schweizer, G., Ayer-Le Lièvre, C., and Le Douarin, N. M. (1983). Restrictions of developmental capacities in the dorsal root ganglia during the course of development. *Cell Differentiation* **13,** 191–200.

Sextier-Sainte-Claire Deville, F., Ziller, C., and Le Douarin, N. M. (1992). Developmental potentialities of cells derived from the truncal neural crest in clonal cultures. *Dev. Brain Res.* **66,** 1–10.

Sextier-Sainte-Claire Deville, F., Ziller, C., and Le Douarin, N. M. (1994). Developmental potentials of enteric neural crest-derived cells in clonal and mass cultures. *Dev. Biol.* **163,** 141–151.

Sieber-Blum, M. (1989). Commitment of neural crest cells to the sensory neuron lineage. *Science* **243,** 1608–1611.

Sieber-Blum, M. (1991). Role of the neurotrophic factor BDNF and NGF in the commitment of pluripotent neural crest cells. *Neuron* **6,** 949–955.

Sieber-Blum, M., and Cohen, A. M. (1980). Clonal analysis of quail neural crest cells: They are pluripotent and differentiate *in vitro* in the absence of noncrest cells. *Dev. Biol.* **80,** 96–106.

Sieber-Blum, M., Ito, K., Richardson, M. K., Langtimm, C. J., and Duff, R. S. (1993). Distribution of pluripotent neural crest cells in the embryo and the role of brain-derived neurotrophic factor in the commitment to the primary sensory neuron lineage. *J. Neurobiol.* **24,** 173–184.

Simon, H., Hornbruch, A., and Lumsden, A. (1995). Independent assignment of antero-posterior and dorso-ventral positional values in the developing chick hindbrain. *Curr. Biol.* **5,** 205–214.

Smith, J., Fauquet, M., Ziller, C., and Le Douarin, N. M. (1979). Acetylcholine synthesis by mesencephalic neural crest cells in the process of migration *in vivo*. *Nature* **282,** 853–855.

Stemple, D. L., and Anderson, D. J. (1993). Lineage diversification of the neural crest: *In vitro* investigations. *Dev. Biol.* **159,** 12–23.

Takagi, S., Tsuji, T., Kinutani, M., and Fujisawa, H. (1989). Monoclonal antibodies against species-specific antigens in the chick central nervous system: Putative application as transplantation markers in the chick–quail chimeras. *J. Histochem. Cytochem.* **37,** 177–184.

Tanaka, H., Kinutani, M., Agata, A., Takashima, Y., and Obata, K. (1990). Pathfinding during spinal tract formation in quail–chick chimera analysed by species-specific monoclonal antibodies. *Development* **110,** 565–571.

Timsit, S., Martinez, S., Allinquant, B., Peyron, F., Puelles, L., and Zalc, B. (1995). Oligodendrocytes originate in a restricted zone of the embryonic ventral neural tube defined by DM-20 mRNA expression. *J. Neurosci.* **15,** 1012–1024.

Trousse, F., Giess, M. C., Soula, C., Ghandour, S., Duprat, A.-M., and Cochard, P. (1995). Notochord and floor plate stimulate oligodendrocyte differentiation in cultures of the chick dorsal neural tube. *J. Neurosci. Res.* **41,** 552–560.

Vincent, M., and Thiery, J. P. (1984). A cell surface marker for neural crest and placodal cells: Further evolution in peripheral and central nervous system. *Dev. Biol.* **103,** 468–481.

Warf, B. C., Fok-Seang, J., and Miller, R. H. (1991). Evidence for the ventral origin of oligodendrocyte precursors in the rat spinal cord. *J. Neurosci.* **11,** 2477–2788.

Wilkinson, D. G. (1993). Molecular mechanisms of segmental patterning in the vertebrate hindbrain and neural crest. *Bioessays* **15,** 499–505.

Yu, W. P., Collarini, E. J., Pringle, N. P., and Richardson, W. D. (1994). Embryonic expression of myelin genes: Evidence for a focal source of oligodendrocyte precursors in the ventricular zone of the neural tube. *Neuron* **12,** 1353–1362.

Zacchei, A. M. (1961). Lo sviluppo embrionale della quaglia giapponese (*Coturnix coturnix japonica*). *Arch. Ital. Anat. Embriol.* **66,** 36–62.

Zhang, M. B., Kim, H. J., Marshall, H., Gendron-Maguire, M., Lucas, D. A., Baron, A., Gudas, L. J., Gridley, T., Krumlauf, R., and Grippo, J. F. (1994). Ectopic *Hoxa-1* induces rhombomere transformation in mouse hindbrain. *Development* **120,** 2431–2442.

Ziller, C., Dupin, E., Brazeau, P., Paulin, D., and Le Douarin, N. M. (1983). Early segregation of a neuronal precursor cell line in the neural crest as revealed by culture in a chemically defined medium. *Cell* **32,** 627–638.

Ziller, C., Fauquet, M., Kalcheim, C., Smith, J., and Le Douarin, N. M. (1987). Cell lineages in peripheral nervous system ontogeny: Medium-induced modulation of neuronal phenotypic expression in neural crest cell cultures. *Dev. Biol.* **120,** 101–111.

2

Inhibition of Gene Expression by Antisense Oligonucleotides in Chick Embryos *in Vitro* and *in Vivo*

Aixa V. Morales and Flora de Pablo
Department of Cell and Developmental Biology
Centro de Investigaciones Biológicas
Consejo Superior de Investigaciones Científicas
E-28006 Madrid, Spain

I. Introduction

Strategies using synthetic oligonucleotides to inhibit expression of specific genes have now been used for over a decade (see Van der Krol *et al.*, 1988, for early studies). This versatile tool has helped us to understand the biologic role of many eukaryotic genes in diverse cellular and developmental processes. As opposed to the powerful antigene strategy based on homologous recombination, which is very demanding in time for preparing the DNA vectors and in space for animal facilities, oligonucleotide design, synthesis, and delivery is reasonably simple. The use of oligonucleotides to disturb gene function also allows modulation in a cell-type- and time-specific manner. However, this strategy varies greatly in efficiency and many controls are required to obtain reliable results. In this chapter, we briefly consider essential aspects for the effective use of antisense oli-

Current Topics in Developmental Biology, Vol. 36

0070-2153/98 $25.00

gonucleotides as gene expression inhibitors and present protocols tested in early chick embryos.

Antisense oligodeoxynucleotides (ODNs) are short sequences of single-stranded DNA, usually less than 30 nucleotides in length, produced by chemical synthesis. Their sequences are complementary to specific intracellular target mRNAs, to which they hybridize. Synthetic antisense RNAs have more recently become available, although the high cost has limited their widespread use.

Because they are synthesized chemically, ODNs must be applied either externally or microinjected into cells or embryos and, as a consequence, their effects normally are transient. Repeated administration is possible only when they are applied extracellularly. They offer, however, the important advantage that the full power of organic chemistry can be used to obtain ODNs with particular modified linkages or terminal groups, which improve their stability *in vivo* and, therefore, their effectiveness.

II. Designing Antisense Oligodeoxynucleotides

The goal is to obtain an ODN that hybridizes to a specific mRNA target sequence and inhibits its translation. The main criteria that must be taken into account are discussed in the following sections (see Toulmé, 1992, for a detailed review).

A. Specificity

The specificity of action depends on the specificity of the ODN hybridization, and this increases if the target sequence is unique in the cell. The uniqueness of the targeted mRNA is related to its length and to its sequence, as well as to the complexity of the genome. Considering the nonrandom distribution of nucleotides in, for example, human genomic DNA, and that only 0.5% is transcribed, the minimal length of an antisense ODN should be 11–15 nucleotides. Many investigators have chosen to target the translation initiation site (including the AUG region) of an mRNA, on the assumption that this region is important and accessible. Recent studies indicate that most regions of mRNAs are in fact accessible to ODNs, except for those with strong secondary structure (Wagner, 1994).

To avoid results biased by the choice of target sequence selection, it is important to show that the same effect is produced by more than one antisense ODN sequence from the same mRNA. Frequently, simultaneous addition of two ODNs results in synergistic effect (Nieto *et al.*, 1994).

B. Stability

Conventional, unmodified ODNs are susceptible to quick degradation by nucleases when they are used as regulatory agents in biologic environments (intact cells, tissue slices, whole embryo, etc.). This problem can be minimized by various modifications of the ODN molecule.

Oligonucleotide analogs can be synthesized with the following backbone modifications: phosphodiester internucleoside linkages have been substituted by phosphotriester, methyl phosphonate, phosphoramidate, phosphorodithioate, and phosphorothioate (where -S replaces -O). Because of their nuclease resistance and their ability to elicit RNase H activity, phosphorothioate analogs have been widely used in intact cells. To overcome the nonspecific association of the phosphorothioates with cellular proteins, including reverse transcriptase, the modifications are only introduced in two to three nucleotides on both ends of the ODN.

C. Uptake and Transport

Oligodeoxynucleotides cross the cellular membranes, despite their high negative charge density, by a still unclear mechanism, either endocytosis or passive diffusion. Nevertheless, several strategies have been reported greatly to enhance the potency of antisense ODNs. They include the use of cationic liposomes or conjugation of ODNs to fusogenic peptides or to a fragment of the Antennapedia protein, to improve delivery of the ODNs into cells (Gewirtz *et al.*, 1996; Prochiantz, 1996; Lewis *et al.*, 1996). In all of these delivery-supported ODN applications, it is important carefully to evaluate the increase in toxicity, a factor that limits their usefulness.

Additional modifications to improve oligonucleotide membrane penetration include the use of 3′-conjugated groups, such as polylysine and cholesterol. Reactivity with RNA can be increased by modifications including acridine rings, alkylating and cross-linking moieties, and metal complexes (Colman, 1990; Gewirtz *et al.*, 1996).

When a complex tissue or whole embryo is treated with ODNs, the accessibility of the target area needs to be monitored (see III.C). The larger the ODN, the lower the concentration that appears to be effective, but the worse its penetration in the tissues.

Most of these parameters and criteria are more easily monitored in cell culture systems. In fact, the vertebrate early embryo has not been considered unanimously a suitable target for antisense ODN technology. The examples included in this chapter illustrate that it is possible to modify cellular events by applying antisense ODNs to whole avian embryos in culture and *in ovo*.

III. Controls to Assess the Specificity of Antisense Oligodeoxynucleotide Effects

Control experiments should cover all important aspects of ODN strategy, namely uptake, specificity, and effectiveness (see also Wagner, 1994).

A. Direct Measurement of the Target mRNA or Protein Levels after Treatment

The specific decrease in target mRNA can be measured by RNase protection assay (Austin *et al.*, 1995), *in situ* hybridization (Nieto *et al.*, 1994), Northern blotting, or reverse transcription (RT)-polymerase chain reaction. In all cases the selected mRNA must be compared with an internal control mRNA that remains unaffected by the ODN.

The total protein level can be assessed by immunoblot (Souza *et al.*, 1994; Osen-Sand *et al.*, 1993), immunohystochemistry (Sariola *et al.*, 1991), or by immunoprecipitation followed by polyacrylamide gel electrophoresis. In some cases, the ODNs can decrease the protein content up to 80% (Souza *et al.*, 1994), if the protein is moderately abundant. Biosynthetic labeling allows the measurement of new protein synthesis, which can be decreased to undetectable levels (Morales *et al.*, 1997).

B. Control Sequences for Oligodeoxynucleotides

Many studies use a sense oligonucleotide complementary to the antisense sequence, whereas others use ODNs of the same length as the antisense ODN, but composed of a random mixture of all four nucleotides. Because the ODNs may be degraded in culture depending on length and composition, the best control sequences are those that contain the same base composition as the antisense sequence in a random order. The ideal control ODN differs from the antisense sequence by the minimum necessary to prevent specific hybridization to target mRNA. An elegant example of how a few base differences among ODNs can affect its effect is shown in Figure 1 (Austin *et al.*, 1995). These authors demonstrate that the efficiency of a 23-mer antisense ODN against c*Notch* in raising the percentage of retina ganglion cells decreases progressively with increasing sequence mismatches. Consequently, the oligonucleotide with five mismatches (M5) is already very ineffective, close to the level obtained with the sense oligonucleotide.

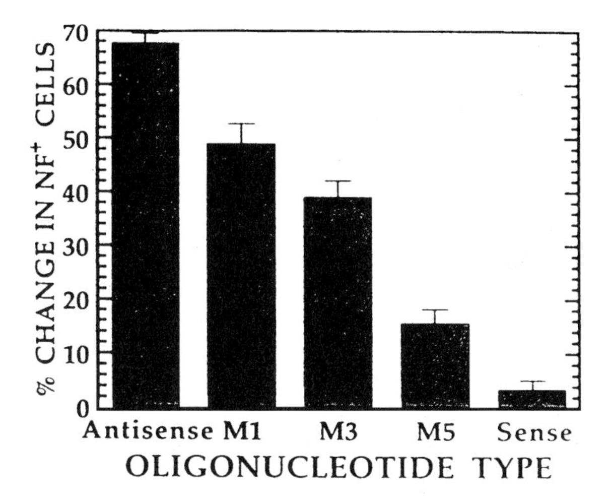

Fig. 1. Control of the specificity of antisense ODNs. Nucleotide mismatches decrease the ability of an ODN antisense against c*Notch* mRNA to increase numbers of NF+ cells in explant cultures of retina. M1, M3, and M5 indicate one, three, or five mismatches, respectively. (Reproduced from Austin *et al.*, 1995, with permission from the Company of Biologists, Ltd.).

C. Demonstration of Oligodeoxynucleotide Uptake

The efficiency of ODN uptake by a whole embryo or a whole organ can be evaluated using radiolabeled ODNs. For an E1.5 chick embryo in culture (see IV.A) we add 7×10^5 cpm (ca. 30 fmole) [^{33}P]-labeled ODN mixed with 12.5 nmole nonlabeled ODN. The ODN is labeled with [γ-^{33}P]ATP using T4 polynucleotide kinase and a DNA tailing kit using standard protocols (Sambrook *et al.*, 1989). The [^{33}P]-labeled ODN is then applied under the embryo. After 6–8 hr in culture, the embryo is fixed in 4% (w/v) paraformaldehyde and 0.2% (v/v) glutaraldehyde and embedded in Historesin (Jung; Heraeus Kulzer, GmbH). For testing penetration of a 15-mer, we used sections, 25 μm thick, placed on slides and apposed to Hyperfilm-Beta Max (Amersham) for 1 month (Fig. 2B). In a whole-embryo culture system, the [^{33}P]-labeled ODN penetrated all the embryo layers, but it was slightly more concentrated in the ventral region.

When an E2 chick embryo has been treated *in vivo* to reach a retinal mRNA, external application over the eye has also provided a higher concentration of ODN in the area (Fig. 2C). The distribution of the radiolabeled ODN on the whole embryo can be assessed by placing 10^4 cpm (ca. 1 fmole) [^{32}P]-labeled

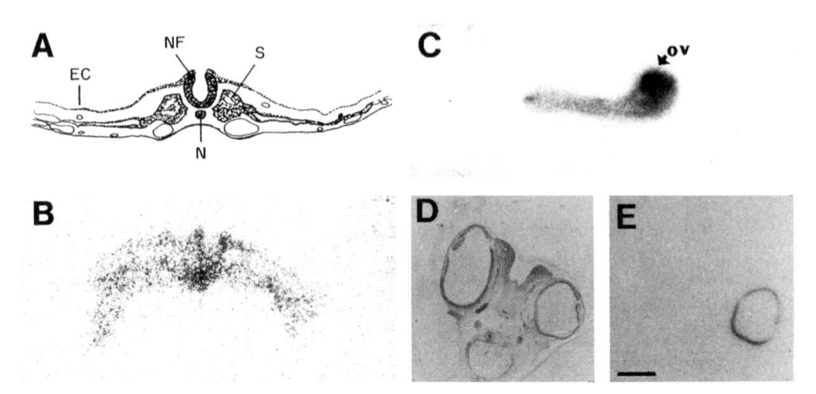

Fig. 2. Control of the uptake of antisense ODNs in neurulating embryos and in E5 retina. (A) Schematic representation of a transverse section through the midgut of an E1.5 embryo. The section is oriented dorsal to the top. EC, ectoderm; N, notochord; NF, neural fold; S, somite. (B) Film autoradiogram of a transverse section at the same level and with the same orientation as (A) of an embryo cultured *in vitro* for 6 hr after application of [^{33}P]-labeled antisense ODN. A wide distribution of the ODN through the three germ layers can be appreciated, with slightly higher concentration in the ventral side. (C) Autoradiogram of a E2.5 whole embryo, oriented rostral to the right, after treatment with labeled ODN for 6 hr. The ODN is mainly retained in the optic vesicle (OV), next to which it was deposited. (D) Phase-contrast image of a horizontal section of the head of an E5 embryo injected *in vivo* in the right eye inside the vitreous with a [^{32}P]-labeled ODN and further incubated for 6 hr. (E) Autoradiogram of the same section as (D); note the distribution of the ODN throughout the retina, with lower uptake by the lens, and good retention within the injected eye. Scale bars: B, 75 μm; C, 2 mm; D, E, 1.7 mm.

ODN mixed with 6 nmole nonlabeled ODN under the vitelline membrane just beside the eye. After 6 hr of incubation at 38.2°C, the embryo is dissected and fixed in paraformaldehyde 4% (w/v) in 0.1 *M* phosphate buffer, pH 7.1 overnight, placed as a whole mount on slides, dried at 60°C for 30 min, and apposed to Hyperfilm-Beta Max for 15 days.

A similar method can be used to test uptake by the chicken retina after injecting ODNs *in vivo* (described in IV.B.2). We mix 10^4 cpm (ca. 1 fmole) [^{32}P]-labeled ODN with 2 nmole nonlabeled ODN and inject it in the vitreous of an E5 chick embryo (Fig. 2D,E). After incubation for 6 hr at 38.2°C, whole heads are fixed overnight in paraformaldehyde 4% (w/v)/sucrose 11% (w/v) in 0.1 *M* phosphate buffer, pH 7.1. Then the heads are further infiltrated in sucrose 30% (w/v) in 0.01 *M* phosphate buffer, pH 7 for at least 12 hr and embedded in Tissue-Tek (OCT compound; Miles, Inc., Elkhart, IN). Sections 16 μm thick are placed on slides and apposed to Hyperfilm-Beta Max for 1 month.

IV. Application of Oligodeoxynucleotides to Early Chick Embryos

A. Application to the Chick Embryo in Culture

To study the effect of antisense ODNs blocking the expression of a soluble growth factor, (pro)insulin, we have used a defined medium culture system for whole embryos in neurulation (E1.5) (Pérez-Villamil *et al.*, 1994; Morales *et al.*, 1997), a modification of the New method (New, 1955).

1. Culture Procedure

The eggs are incubated for 36–40 hr at 38.2°C until the embryos reach stage 9–10 of Hamburger and Hamilton (1951). The shell is cleaned by wiping it with cotton soaked in ethanol 70% (v/v). Then, holding the egg horizontally, the blunt end is gently punctured and 3 ml of albumen is removed with a 5-ml syringe and an 18–21-gauge needle introduced ~1 cm. To collect the embryo, place the egg on its side and cut on the top part of the shell a window of approximately 2 × 2 cm to expose the embryo on the yolk upper side. The extraembryonic membranes are adhered to a nitrocellulose ring placed around the embryo, and the membranes are cut around the ring (Fig. 3A). The nitrocellulose ring carrying the embryo and the surrounding extraembryonic membranes is floated on warm phosphate-buffered saline for 1–2 min to clean it of yolk particles. It is then transferred to a well of a 24-well plate, previously filled with 0.3 ml of culture medium (see later) containing 0.5% (w/v) agarose (low melting point) to form a soft gel. The ring and the embryonic pack are immobilized on the gel with two metal weights placed on the ring (Fig. 3B). The maintenance of membrane tension is essential for adequate growth in culture. The wells are then filled with 0.3 ml of F12/Dulbecco's modified Eagle medium (DMEM-F12), supplemented with 100 μg/ml transferrin, 16 μg/ml putrescin, 6 ng/ml progesterone, 5.2 ng/ml sodium selenite, 50 μg/ml gentamicin (all from Sigma), and 0.7% (w/v) methylcellulose (Serva, Heidelberg, Germany). Insulin is an optional supplement, excluded in our studies. This type of culture can be incubated up to 24 hr at 37°C, with 5% CO_2 in a standard cell culture incubator. The embryo develops with normal morphology for at least 18 hr, adding a new somite every 3–5 hr and progressing in other aspects of organogenesis such as optic and otic vesicle growth.

2. Antisense Oligodeoxynucleotides

To interfere with the synthesis of embryonic (pro)insulin, we designed 15-base antisense ODNs complementary to two regions of the chicken preproinsulin

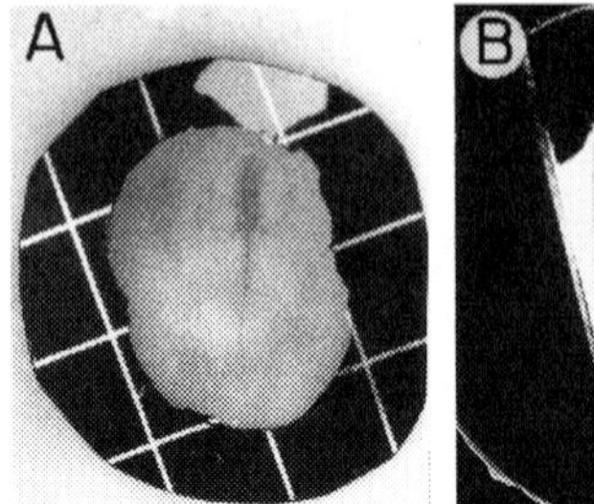

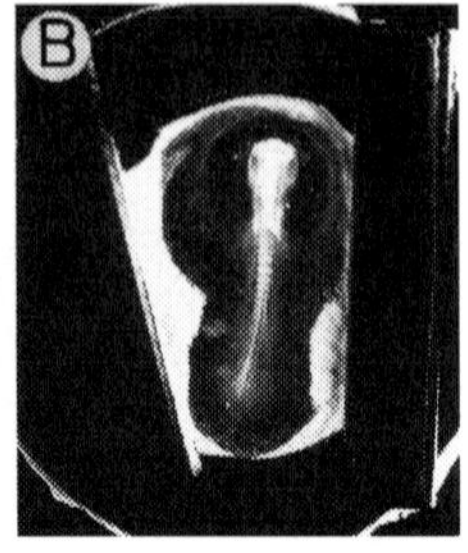

Fig. 3. Culture system for neurulating chick embryos. (A) E1.5 embryo inside the egg, stained with neutral red to distinguish the morphology of the embryo from the opaque yolk, and surrounded by a nitrocellulose ring. (B) The embryo has been removed from the egg and placed on a soft gel (see text) in a well. Two metal weights placed on the ring maintain the embryo and membrane tension. Scale bar, 1.5 mm.

mRNA, one of them complementary to the AUG region. Routinely, we use as controls the corresponding sense sequence and a 15-mer random-sequence ODN with the same nucleotide composition as one of the antisense ODNs. Phosphorothioate derivatives have worked effectively in our system (Morales *et al.*, 1997).

The ODNs are deposited in a 25-μl drop on the soft gel (12.5 nmole; 40 μ*M* final concentration considering the soft gel volume) before transferring the embryos on top. Different concentrations should be tested in pilot experiments to avoid toxicity. The route of administration of the ODNs was also tested using a [^{33}P]-labeled ODN. At this stage of development, application under the ventral region of the embryo in the culture dish provides the best uptake (see Fig. 2B). Effects on cell survival were specifically caused in this embryo model by antisense ODNs against (pro)insulin. This type of culture allows rescue experiments with the appropriate factor added exogenously. The processing of stage 10–11 embryo for evaluation of different cellular parameters is similar to that described for retina organocultures in Chapter 7 (De la Rosa *et al.*).

An alternative culture system for early chick embryos in which the antisense ODN strategy has been successfully used to inhibit a transcription factor mRNA, *slug*, has been described by Nieto *et al.* (1994) and Cooke (1995).

B. Application to the Chick Embryo *in Vivo*

Chick Embryos in Mid-neurulation

Although cultured embryos provide a well controlled system, they suffer from growth factors and nutrient deprivation, and only short-term survival of the embryo is possible. The more physiologic approach of *in vivo* interference with ODNs may, however, be useful for cell-autonomous gene transcripts and products because the yolk stores many growth factors. For example, we have applied ODNs against the insulin receptor mRNA to E1.5 chicken embryos *in ovo* (Morales *et al.*, 1997). After incubating the eggs for the desired time, 1 ml of albumen was removed through a lateral puncture. A window was cut laterally in the shell and the developmental stage of the embryo was assessed under magnifying binoculars. Thirty microliters of 250 μ*M* ODN (7.5 nmole in our case; the optimal concentration should be titrated) in the basal medium containing 0.7% (w/v) methylcellulose, as described earlier, was applied over the embryo, remaining largely localized (the lower concentration of methylcellulose did not keep the ODN localized). The eggs were then sealed with cellophane tape and incubated for 10–48 hr at 38.2°C until the cellular process of interest is evaluated—in our example, apoptotic cell death.

2. Intraocular Applications of Oligodeoxynucleotides

Oligodeoxynucleotides can be easily injected in the vitreous body of chick embryos between E4 and E8 of *in ovo* development. This period covers the proliferative and early differentiative stages of the neuroretina and other eye structures. The advantage of the vitreous injection of ODNs is that they diffuse widely within the retinal layers, while they remain contained in the injected eye at high concentration for several hours (Fig. 2E). The other eye provides a nonmanipulated control. We usually use eggs at E5 to study early retinal development. A lateral window is opened in the shell, and the membranes covering the embryo are carefully cut (avoiding hemorrhage) to get access to the eye. Using a tungsten needle, under binoculars, a puncture is made in the eye close to the lens. Through this hole, 1 μl of 2 m*M* ODN is injected into the vitreous using a sharpened capillary glass tube. The final ODN concentration in the vitreous in this case is approximately 40 μ*M* (the estimated volume of the vitreous in E5 is 50 μl). Then the egg is sealed with cellophane tape and the embryo can be incubated further. If necessary, a second injection can be performed 1 to several days after the initial one, but this increases the risk of microphthalmos from the procedure. A similar protocol has worked for several groups using excitatory amino acids and DNA vectors (Catsicas and Clarke, 1987; Fekete and Cepko, 1993).

V. Optimization of Oligodeoxynucleotide Delivery

As previously mentioned, a source of variability in their effects is the inability of antisense ODNs to cross cellular membranes efficiently. The studies that show the most potent antisense effect, using lesser amounts of ODNs, have relied on additional cell permeabilization tools to introduce ODNs directly into the cytoplasm. A variety of cationic lipids, such as lipofectin, lipofectamine, and lipofectace (Life Technologies, Gaithersburg, MD), dotap (Boehringer

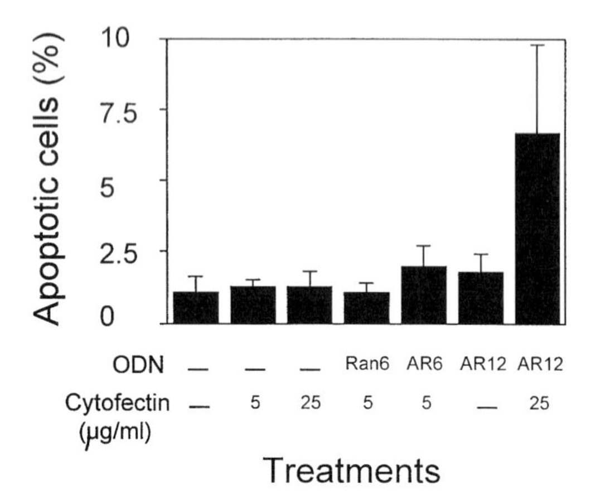

Fig. 4. Increased efficiency of antisense ODN in combination with cytofectin GS2888. Embryos at E1.5 were treated *in ovo* with 6 or 12 nmole of the ODN antisense to the insulin receptor mRNA (AR), or with 6 nmole of a random control ODN (Ran). When indicated, 5 or 25 μg/ml of cytofectin GS2888 was added alone, to test its toxicity, or in combination with the indicated ODN. After 10 hr of further incubation, the embryos were fixed and dissociated, and the cells were stained with DAPI to determine pyknotic nuclei. A minimum of 400 cells was counted per experimental point. The mean values of apoptotic cells found in three to four embryos in two independent experiments are expressed in absolute values. The standard deviation is represented over each bar.

Mannheim),transfectam (Promega), and the like, are available. An improved cytofectin, termed GS2888, has been described that overcomes many of the shortcomings of the other lipid formulations for delivering ODNs into cells (Lewis *et al.*, 1996).

We have tested whether cytofectin GS2888 (Gilead, CA) improves the efficiency (without increasing toxicity) of antisense ODNs against insulin receptor mRNA in neurulating chick embryos *in vivo* (system described in IV.B.1). Antisense ODNs designed complementary to a sequence in the tyrosine kinase region of the receptor (AR, 6 or 12 nmole) and a random control (Ran, 6 nmole) were applied alone or in combination with GS2888 cytofectin (5 or 25 μg/ml final concentration) as described in IV.B.1 (Fig. 4). Previously, to form the ODN–lipid complex, the ODN and the GS2888 cytofectin 4× solutions were mixed together (1:1) by pipetting up and down several times in a polystyrene tube (polypropylene tubes, i.e., Eppendorf tubes, may not work because the lipid–ODN formulation adheres to the tube) and let it settle for 10–15 min. Finally, methylcellulose 1.4% (w/v) was added at 1:1 proportion and thoroughly mixed. Embryos were incubated at 38.2°C for 10 hr, dissected out of the egg, and processed to analyze apoptotic cell death as described in Chapter 7 (De la Rosa *et al.*).

As shown in Figure 4, cytofectin at the doses tested, 5–25 μg/ml, is not toxic for neurulating chick embryos. Six nanomoles (AR6) of ODN in the presence of 5 μg/ml of cytofectin is slightly more effective in induction of apoptotic death than 12 nmole (AR12) ODN alone, both relative to the control ODN (Ran6). Remarkably, when a higher dose of antisense ODN (AR12) is applied in combination with a higher dose of cytofectin (25 μg/ml), the percentage of apoptotic cells increased 3.7 times over that caused by the antisense ODN alone (AR12, see Fig. 4 for treatment combinations). Thus, the use of GS2888 cytofectin to improve the penetration of oligonucleotides considerably increases the efficiency of the ODNs. In each biologic system, cells, tissue slices, or whole organs, appropriate pilot experiments should be done to determine if cationic lipids provide advantages over ODNs alone.

VI. Concluding Remarks

The use of antisense ODNs to produce a hybridization arrest of gene expression allows us to address the question of protein function by "reverse genetics." It is a technique that complements the use of antibodies, and its range of application is as wide as that of antibody application. For studies in rat, chicken, and other vertebrate species in which the gene knockout techniques are not yet developed, it is a good alternative applicable by single injection or infusion, in addition to administration in culture. ODN effects are usually slower to begin than the effects of blocking antibodies, and can be quite transient, thus producing effects during a short window of time that needs to be appropriately monitored. The use of antisense or dominant mutant DNA constructs injected in early embryos has been a particularly useful alternative tool in studies in *Xenopus* and is addressed in Chapter 4.

The antisense ODN approach has provided important information in understanding the control of cell proliferation, but also has helped in a diversity of areas, from early morphogenesis (Nieto *et al.*, 1994), axonal patterning (Rétaux *et al.*, 1996), to the problem of learning in adult animals (Castro-Alamancos and Torres-Aleman, 1994). It is expected that refining the chemical modifications of ODNs, synthesizing RNAs, improving the uptake by the target cell, and making ODNs less expensive reagents, will consolidate the reputation of antisense synthetic nucleic acids as a powerful pharmacologic tool.

Acknowledgments

We thank Enrique J. de la Rosa, who provided Figure 3, and Begoña Díaz and Belén Pimentel, who participated in the establishment of the protocols described here. We also wish to thank Angela Nieto for advice. Research in our laboratory is supported by grants from DGICYT (94-152), FIS (94/151), and CAM (AE 376/95).

References

Austin, C. P., Feldman, D. E., Ida, J. A., and Cepko, C. L. (1995). Vertebrate retinal ganglion cells are selected from competent progenitors by the action of *Notch*. *Development* **121,** 3637–3650.

Castro-Alamancos, M. A., and Torres-Aleman, I. (1994). Learning of the conditioned eye-blink response is impaired by an antisense insulin-like growth factor I oligonucleotide. *Proc. Natl. Acad. Sci. U.S.A.* **91,** 10203–10207.

Catsicas, S., and Clarke, P. G. H. (1987). Spatiotemporal gradients of kainate-sensitivity in the developing chicken retina. *J. Comp. Neurol.* **262,** 512–522.

Colman, A. (1990). Antisense strategies in cell and developmental biology. *J. Cell Sci.* **97,** 399–409.

Cooke, J. (1995). Improved *in vitro* development of the chick embryo using roller-tube culture. *Trends Genet.* **11,** 259–260.

Fekete, D. M., and Cepko, C. L. (1993). Replication-competent retroviral vectors encoding alkaline phosphatase reveal spatial restriction of viral gene expression/transduction in the chick embryo. *Mol. Cell. Biol.* **13,** 2604–2613.

Gewirtz, A. M., Stein, C. A., and Glazer, P. M. (1996). Facilitating oligonucleotides delivery: Helping antisense deliver on its promise. *Proc. Natl. Acad. Sci. U.S.A.* **93,** 3161–3163.

Hamburger, V., and Hamilton, H. L. (1951). A series of normal stages in the development of the chick embryo. *J. Morphol.* **88:**49–92.

Lewis, J. G., Lin, K. Y., Kothavale, A., Flanagan, W. M., Matteucci, M. D., DePrince, R. B., Mook, R. A. Jr., Hendren, R. W., and Wagner, R. W. (1996). A serum-resistant cytofectin for cellular delivery of antisense oligodeoxynucleotides and plasmid DNA. *Proc. Natl. Acad. Sci. U.S.A.* **93,** 3176–3181.

Morales, A. V., Serna, J., Alarcón, C., De la Rosa, E. J., and De Pablo, F. Role of prepancreatic (pro)insulin and the insulin receptor in prevention of embryonic apoptosis. *Endocrinology* (in press).

New, D. A. T. (1955). A new technique for the cultivation of the chick embryo in vitro. *Journal of Embryology and Experimental Morphology* **3,** 320–331.

Nieto, M. A., Sargent, M. G., Wilkinson, D. G., and Cooke, J. (1994). Control of cell behavior during vertebrate development by *slug*, a zinc finger gene. *Science* **264,** 835–839.

Osen-Sand, A., Catsicas, M., Staple, J. K., Jones, K. A., Ayala, G., Knowles, J., Grenningloh, G., and Catsicas, S. (1993). Inhibition of axonal growth by snap-25 antisense oligonucleotides *in vitro* and *in vivo*. *Nature* **364,** 445–448.

Pérez-Villamil, B., De la Rosa, E. J., Morales, A. V., and De Pablo, F. (1994). Developmentally regulated expression of the preproinsulin gene in the chicken embryo during gastrulation and neurulation. *Endocrinology* **135,** 2342–2350.

Prochiantz, A. (1996). Getting hydrophilic compounds into cells: Lessons from homeopeptides [Commentary]. *Curr. Opin. Neurobiol.* **6,** 629–634.

Rétoux, S., McNeill, L., and Harris, W. A. (1996). Engrailed, retinotectal targeting and axonal patterning in the midbrain during *Xenopus* development: An antisense study. *Neuron* **16,** 63–75.

Sambrook, J., Fritsch, E. F., and Maniatis, T. (1989). "Molecular Cloning: A Laboratory Manual," 2nd ed. Cold Spring Harbor Laboratory Press, Cold Spring Harbor, NY.

Sariola, H., Saarma, M., Sainio, K., Arumäe, U., Palgi, J., Vaahtokari, A., Thesleff, I., and Karavanov, A. (1991). Dependence of kidney morphogenesis on the expression of nerve growth factor receptor. *Science* **254,** 571–573.

Souza, P., Sedlackova, L., Kuliszewski, M., Wang, J., Liu, J., Tseu, I., Liu, M., Tanswell, A. K., and Post, M. (1994). Antisense oligodeoxynucleotides targeting PDGF-B mRNA inhibit cell proliferation during embryonic rat lung development. *Development* **120,** 2163–2173.

Toulmé, J. J. (1992). Artificial regulation of gene expression by complementary oligonucleotides: An overview. *In* "Antisense RNA and DNA" (J. A. H. Murray, ed.), pp. 175–194. Wiley-Liss, New York.

Van der Krol, A. R., Mol, J. N. M., and Stuitje, A. R. (1988). Modulation of eukaryotic gene expression by complementary RNA or DNA sequences. *Biotechniques* **6,** 958–976.

Wagner, R. W. (1994). Gene inhibition using antisense oligodeoxynucleotides. *Nature* **372,** 333–335.

3

Lineage Analysis Using Retroviral Vectors

Constance L. Cepko and Shawn Fields-Berry
Department of Genetics
Harvard Medical School and Howard Hughes Medical Institute
Boston, Massachusetts 02115

Elizabeth Ryder
Department of Biology and Biotechnology
Worcester Polytechnic Institute
Worcester, Massachusetts 01609

Christopher Austin
Merck Research Laboratories
West Point, Pennsylvania 19486

Jeffrey Golden
Department of Pathology
Children's Hospital of Philadelphia
Philadelphia, Pennsylvania 19104

The complexity and inaccessibility of many types of embryos have made lineage analysis through direct approaches, such as time-lapse microscopy and injection of tracers, almost impossible. A genetic and clonal solution to lineage mapping is through the use of retrovirus vectors. The basis for this technique is summarized, and the strategies and current methods in use in our laboratory are detailed.

I. Transduction of Genes via Retrovirus Vectors

A retrovirus vector is an infectious virus that transduces a nonviral gene into mitotic cells *in vivo* or *in vitro* (Weiss *et al.*, 1984–1985). These vectors use the

Current Topics in Developmental Biology, Vol. 36

0070-2153/98 $25.00

same efficient and precise integration machinery of naturally occurring retroviruses to produce a single copy of the viral genome stably integrated into the host chromosome. Those that are useful for lineage analysis have been modified so that they are replication incompetent and thus cannot spread from one infected cell to another. They are, however, faithfully passed on to all daughter cells of the originally infected progenitor cell, making them ideal for lineage analysis.

Retroviruses use RNA as their genome, which is packaged into a membrane-bound protein capsid. They produce a DNA copy of their genome immediately after infection using reverse transcriptase, a product of the viral *pol* gene that is included in the viral particle. The DNA copy is integrated into the host cell genome and is thereafter referred to as a "provirus." Integration of the genome of most retroviruses requires that the cell go through an M phase (Roe *et al.*, 1993), and thus only mitotic cells serve successfully as hosts for integration of most retroviruses. [There has been a recent generation of retrovirus vectors based on the human immunodeficiency virus (Naldini *et al.*, 1996) that can integrate into postmitotic cells. However, because lineage analysis is designed to ask about the fate of daughter cells, infection of postmitotic cells is not desirable.] Most vectors began as proviruses that were cloned from cells infected with a naturally occurring retrovirus. Although extensive deletions of proviruses were made, vectors retain the *cis*-acting viral sequences necessary for the viral life cycle. These include the Ψ packaging sequence (necessary for recognition of the viral RNA for encapsidation into the viral particle), reverse transcription signals, integration signals, viral promoter, enhancer, and polyadenylation sequences. A cDNA can thus be expressed in the vector using the transcription regulatory sequences provided by the virus. Because replication-incompetent retrovirus vectors usually do not encode the structural genes whose products comprise the viral particle, these proteins must be supplied through complementation. The structural proteins gag, pol, and env are typically supplied by "packaging" cell lines or co-transfection with packaging constructs into highly transfectable cell lines (for review, see Cepko and Pear in Ausubel *et al.*, 1997). Packaging cell lines are stable lines that contain the *gag*, *pol*, and *env* genes as a result of the introduction of these genes by transfection. However, these lines do not contain the packaging sequence, Ψ, on the viral RNA that encodes the structural proteins. Thus, the packaging lines make viral particles that do not contain the genes *gag*, *pol*, and *env*.

Retrovirus vector particles are essentially identical to naturally occurring retrovirus particles. They enter the host cell through interaction of a viral envelope glycoprotein (a product of the viral *env* gene) with a host cell receptor. The murine viruses have several classes of env glycoprotein that interact with different host cell receptors. The most useful class for lineage analysis of rodents is the ecotropic class. The ecotropic env glycoprotein allows entry only into rat and mouse cells via the ecotropic receptor on these species. It does not allow infection of humans and thus is considered relatively safe for gene transfer experiments. The first packaging line commonly in use was the Ψ2 line (Mann *et al.*, 1983). It encodes the ecotropic *env* gene and makes high titers of vectors. However, it can also lead to the production of helper virus (discussed later). A second generation of ecotopic packaging lines, ΨCRE (Mann *et al.*, 1983), GP+E-86 (Markowitz *et al.*, 1988), and ΩE (Morgenstern and Land, 1990), has not been reported to lead to production of helper virus to date. A third generation of "helper-free" packaging lines, exemplified by the ecotropic line Bosc23 (Pear *et al.*, 1993), was made in 293T cells and have the advantage over the earlier lines of giving high-titer stocks transiently after transfection. Similarly, cotransfection of 293T cells with packaging constructs and vectors can lead to the transient production of high-titer stocks (Soneoka *et al.*, 1995). The first two generations of packaging lines, which are based on mouse fibroblasts, require production of stably transduced lines for production of high-titer stocks.

For infection of nonrodent species, an envelope glycoprotein other than the ecotropic glycoprotein must be used for virion production. The one that endows the greatest host range is the vesicular stomatitis virus (VSV) G-glycoprotein, which allows infection of most species, including fish (Yee *et al.*, 1994). For lineage analysis of avian species, packaging lines and vectors based on avian retroviruses are available (Stoker and Bissell, 1988; Cosset *et al.*, 1990; Boerkoel *et al.*, 1993).

Any of these packaging systems can be used to produce vector stocks for lineage analysis. All stocks should be assayed for the presence of helper virus. A detailed description of protocols for making stocks, for titering and concentrating them, and for checking for helper virus contamination has been published and is given here (see Cepko and Pear in Ausubel *et al.*, 1997).

II. Production of Virus Stocks for Lineage Analysis

Replication-incompetent vectors that encode a histochemical reporter gene, such as *Escherichia coli lacZ* or human placental alkaline phosphatase (PLAP), are the most useful for lineage studies because they allow analysis of individual cells in tissue sections or whole mounts. When stable lines are used, it is best to obtain lines that make high-titered stocks of lineage vectors from the laboratories that have created them. We have placed Ψ2 and ΨCRE producers of BAG (Price *et al.*, 1987), a *lacZ* virus that we have used for lineage analysis, on deposit at the American Type Culture Collection (ATCC) in Rockville, Maryland. They can be obtained by anyone and are listed as ATCC CRL numbers 1858 (ΨCRE BAG) and 9560 (Ψ2 BAG). Similarly, Ψ2 producers of DAP (Fields-Berry *et al.*, 1992), a vector encoding PLAP (described further later), is available as CRL no. 1949. For reasons that are unclear, the DAP line is more reliable for production of high-titer stocks. Both of these vectors transcribe the reporter gene from the viral long-terminal repeat (LTR) promoter and are generally useful for expression of the reporter gene in most tissues. We compared the expression of *lacZ* driven by several different promoters (Cepko *et al.*, 1990; Turner *et al.*, 1990) and found that the LTR was in general the most reliable and non-cell-type-specific. This is an important consideration because it is desirable to see all of the cells descended from an infected progenitor, and thus a constitutive promoter is the most useful for lineage studies. However, even with constitutive promoters, it has been noted that some infected cells do not express detectable β-gal protein, even among clones of fibroblasts infected *in vitro*. Thus, it is important to restrict one's conclusions about lineage relationships to cells that are marked and not to make assumptions about their relationships to cells that are unmarked.

For lineage applications it is usually necessary to concentrate virus to achieve sufficient titer. This is typically due to a limitation in the volume that can be injected at any one site. Viruses can be concentrated fairly easily by a relatively short centrifugation step. Virions also can be precipitated using polyethylene glycol or ammonium sulfate, and the resulting precipitate collected by centrifugation. Finally, the viral supernatant can be concentrated by centrifugation through a filter that allows only small molecules to pass (e.g., Centricon filters). Regardless of the protocols used, it must be kept in mind that ecotropic and amphotropic retroviral particles are fragile, with short half-lives even under optimum conditions. In contrast, if the VSV G protein is used as the viral envelope protein, the virions are more stable. To prepare the highest-titered stock for multiple experiments, we usually concentrate several hundred milliliters of producer cells supernatant. These concentrated stocks are titered and tested for helper virus contamination, and can be stored indefinitely at −80°C or in liquid N_2 in small (10- to 50-μl) aliquots.

III. Replication-Competent Helper Virus

Replication-competent virus is sometimes referred to as helper virus because it can complement ("help") a replication-incompetent virus and thus allow it to spread from cell to cell. It can be present in an animal through exogenous infection (e.g., from a viremic animal in the mouse colony), through expression of an endogenous retroviral genome (e.g., the *akv* loci in AKR mice), or through recombination events in an infected cell that occurs between two viral RNAs encapsidated in retroviral virions produced during packaging. The presence of helper virus is an issue of concern when replication-incompetent viruses are used for lineage analysis because it can lead to horizontal spread of the marker virus, creating false lineage relationships. The most likely source of helper virus is the viral stock used for lineage analysis. The genome(s) that supplies the *gag*, *pol*, and *env* genes in packaging lines does not encode the Ψ sequence, but can still become packaged, although at a low frequency. If it is coencapsidated with a vector genome, recombination in the next cycle of reverse transcription can occur. If the recombination allows the Ψ genome to acquire the Ψ sequence from the vector genome, a recombinant that is capable of autonomous replication is the result. This recombinant can spread through the entire culture (although slowly because of envelope interference). Once this occurs, it is best to discard the producer clone or stock because there is no convenient way to eliminate the helper virus. As would be expected, recombination giving rise to helper virus occurs with greater frequency in stocks with high titer, with vectors that have retained more of the wild-type sequences (i.e., the more homology between the vector and packaging genomes, the more opportunity for recombination), and within stocks generated using *gagpol* and *env* on a complementing genome during transfection (as opposed to two separate genomes, one for *gagpol* and one for *env*; see previous discussion). Note that a murine helper genome itself does not encode a histochemical marker gene because apparently there is neither room nor flexibility within murine viruses that allows them to be both replication competent and capable of expressing another gene like *lacZ*. The way that spread would occur is by a cell being infected with both the *lacZ* virus and a helper virus. Such a doubly infected cell would then produce both viruses.

In lineage analysis, several signs can indicate the presence of helper virus within an individual animal. If an animal is allowed to survive for long periods of time after inoculation, particularly if embryos or neonates are infected, the animal is likely to acquire a tumor when helper virus is present. Most naturally occurring replication-competent viruses are leukemogenic, with the disease spectrum being at least in part a property of the viral LTR. Second, if analysis is done in either the short or the long term after inoculation, the clone size, clone number, and spectrum of labeled cells may be indicative of helper virus. For example, the eye of a newborn rat or mouse has mitotic progenitors for retinal neurons, as well as mitotic progenitors for astrocytes and endothelial cells. By targeting the infection to the area of progenitors for retinal neurons, we only rarely see infection of a few blood vessels or astrocytes because their progenitors are outside the immediate area that is inoculated and they become infected only by leakage of the viral inoculum from the targeted area. However, if helper virus were present, we would see infection of a high percentage of astrocytes, blood vessels, and, eventually, other eye tissues because virus spread would eventually lead to infection of cells outside the targeted area. One would expect to see a correlation between the percentage of such nontargeted cells that are infected and the degree to which their progenitors are mitotically active after inoculation, because infection requires a mitotic target cell. If tissues other than ocular tissues were examined, evidence would similarly be seen of virus spread to cells whose progenitors would be mitotically active during the period of virus spread. In addition, the size and number of "clones" may also appear to be too large for true "clonal" events if helper virus were present. This interpretation, of course, relies on some knowledge of the area under study.

IV. Determination of Sibling Relationships

In lineage analysis, it is critical to unambiguously define cells as descendents of the same progenitor. This can be relatively straightforward when sibling cells remain rather tightly, and reproducibly, grouped. An example of such a straightforward case is the rodent retina, where the descendents of a single progenitor migrate to form a coherent radial array (Turner and Cepko, 1987; Turner *et al.*, 1990). The two analyses described in the following were applied to the rodent retina and are applicable in any system where clones are arranged simply and reproducibly. The first assay is to perform a standard virologic titration, in which a particular viral inoculum is serially diluted and applied to tissue. In the retina, the number of radial arrays, their average size, and their cellular composition were analyzed in a series of animals infected with dilutions that covered a 3 log range. The number of arrays was found to be linearly related to the inoculum size, whereas the size and composition were unchanged. Such results indicate that the working definition of a clone, in this case a radial array, fulfilled the statistical criteria expected of a single-hit event.

The second assay is to perform a mixed infection using two different retroviruses in which the histochemical reporter genes are distinctive. Two such viruses might encode cytoplasmically localized vs nuclear-localized β-gal. This can work when the cytoplasmically localized β-gal is easily distinguished from the nuclear-localized β-gal (Galileo *et al.*, 1990; Hughes and Blau, 1990). We have found that this is not the case in rodent nervous system cells because the cytoplasmically localized β-gal quite often is restricted to neuronal cell bodies and is therefore difficult to distinguish from nuclear-localized β-gal. To overcome this problem, we created the aforementioned DAP virus (Fields-Berry *et al.*, 1992), which is distinct from the *lacZ*-encoding BAG virus. A stock containing BAG and DAP was produced by Ψ2 producer cells grown on the same dish. The resulting supernatant was concentrated and used to infect rodent retina. The tissue was then analyzed histochemically for the presence of blue (due to BAG infection) and purple (due to DAP infection) radial arrays. If radial arrays were truly clonal, then each one should be only one color. Analysis of approximately 1100 arrays indicated that most were clonal. However, five comprised both blue and purple cells. This value will be an underestimate of the true frequency of incorrect assignment of clonal boundaries because sometimes two BAG or two DAP virions infect adjacent cells and thus not lead to formation of bicolored arrays. A closer approximation of the true frequency can be obtained by using the formula (for derivation, see Fields-Berry *et al.*, 1992)

$$\frac{(\text{No. Bicolored arrays})\,\dfrac{(a+b)^2}{2ab}}{\text{No. total arrays}} = \%\ \text{errors},$$

where *a* and *b* are the relative titers comprising the virus stock. The relative titer of BAG and DAP used in the coinfection was 3:1, and thus the value for percentage of errors in clonal assignments was 1.2%.

The value of 1.2% for errors in assignment of clonal boundaries places an upper limit of 1.2% as the frequency of aggregation, because this figure includes errors due to both aggregation and independent virions (e.g., perhaps due to helper virus-mediated spread) infecting adjacent progenitors. The percentage of errors in other areas of an animal depends on the particular circumstances of the injection site, and on the multiplicity of infection (MOI, the ratio of infectious virions to target cells). Most of the time, MOI is quite low (e.g., in the retina it was approximately 0.01 at the highest concentration of virus injected). With regard to the injection site, injection into a lumen, such as the lateral ventricles, should not promote aggregation or high local MOI, but injection into solid tissue in which most of the inoculum has access to a limited number of cells at the inoculation site could present problems. By coinjecting BAG and DAP, one can monitor the frequency of these events and thus determine if clonal analysis is feasible.

An error rate as small as 1.2% does not affect the interpretation of "clones" that are frequently found in a large data set. However, as with any experimental procedure that relies in some way on statistical analysis, rare associations of cell types must be interpreted with some caution, and conclusions cannot be drawn independently of other data.

The preceding analysis was performed using viruses that were produced on the same dish and concentrated together. This was done because we thought that the most likely way that two adjacent progenitors might become infected would be through small aggregates of virions. Aggregation most likely occurs during the concentration step because macroscopic aggregates often can be seen after pellets of virions are resuspended. Thus, when the two-marker approach is used to analyze clonal relationships, it is best to coconcentrate the two vectors together in order for the assay to be sensitive to aggregation because of this aspect of the procedure. (Although aggregation of virions may frequently occur during concentration, it apparently does not frequently lead to problems in lineage analysis, presumably due to the high ratio of noninfectious particles to infectious particles found in most retrovirus stocks. It is estimated that only 0.1–1.0% of the particles will generate a successful infection. Moreover, most aggregates are probably not efficient as infectious units; it must be difficult for the rare infectious particles within such a clump to gain access to the viral receptors on a target cell.)

To determine the ratio of two genomes present in a mixed virus stock (e.g., BAG plus DAP), there are several methods that can be used. The first two methods are performed *in vitro* and are simply an extension of a titration assay. Any virus stock is normally titered on NIH 3T3 cells to determine the amount of virus to inject. The infected NIH 3T3 cells are then either selected for the expression of a selectable marker when the virus encodes such a gene (e.g., *neo* in BAG and DAP) or stained directly, histochemically, for β-gal or PLAP activity without prior selection in drugs. If no selection is used, the relative ratio of the two markers can be scored directly by evaluating the number of clones of each color on a dish. Alternatively, selected G418-resistant colonies can be stained histochemically for both enzyme activities and the relative ratio of blue vs purple G418-resistant colonies computed. A third method of evaluating the ratio of the two genomes is to use the values observed from *in vivo* infections. After animals are infected and processed for both histochemical stains, the ratio of the two genomes can be compared by counting the number of clones, or infected cells, of each color. When all these methods were applied to lineage analysis in mouse retina (Fields-Berry *et al.*, 1992) and rat striatum (Halliday and Cepko, 1992), the value obtained for the ratio of G418-resistant colonies scored histochemically was almost identical to the ratio observed *in vivo*. Directly scoring histochemically stained, non-G418-selected NIH 3T3 cells led to an underestimate of the number of BAG-infected colonies, presumably because such cells often are only a faint blue, whereas DAP-infected cells are usually an intense purple. *In vivo*, this is not usually the case because BAG-infected cells are usually deep blue.

The method of injecting two distinctive viruses is a straightforward and feasible one of assessing clonal boundaries when they are fairly easy to define. This method does not require circumstances in which there is a wide range of dilutions that can be injected to give countable numbers of events, which is required for a reliable dilution analysis. Moreover, it does not rely as critically on controlling the exact volume of injection, as is required in the dilution analysis. The use of a small number of vectors that encode distinct histochemical products for definition of sibling relationships is appropriate only when there is very little migration of sibling cells. In these cases, the arrangement of cells that will be used to identify clonal relationships can be defined, and then this definition can be tested as described previously. The fit of the definition with true clonal relationships revealed by the percentage of defined "clones" that are of more than one color. When the error is too great, one can reevaluate the criteria, make a new definition, and again test it by looking for "clones" that are more than one color. Through trial and error, an accurate definition of sibling relations should be possible when migration is not too great.

When clonal relationships cannot be accurately defined with a few distinctive viruses, a much greater number of vectors must be used. A library of retroviral vectors, each member of which is tagged with a unique small insert of an irrelevant DNA, can be used. Each vector is scored using the polymerase chain reaction (PCR). The library/PCR method is tedious but worthwhile for dealing with problematic areas.

Regardless of which method is used to score sibling relationships, one further recommendation to aid in the assignments is to choose an injection site that allows the inoculum to spread. If the injection is made into a packed tissue, the viral inoculum will most likely infect cells within the injection tract, and it will be very difficult to sort out sibling relationships. For example, a lumen, such as the neural tube, provides an ideal site for injection. Regardless of site, one must inject such that the virus has clear access to the target population; the virus binds to cells at the injection site and will not gain access to cells that are not directly adjoining that site.

The procedures described in the following sections are those that we have used for infection of rodents and chick embryos, histochemical processing of tissue for β-gal and PLAP, and preparation and use of a library for the PCR method.

V. Procedures

A. Infection of Rodents

1. Injection of Virus *in Utero* in Rodents

The following protocols may be used with rats or mice. Note that clean, but not aseptic, technique is used throughout. We routinely soak instruments in 70% ethanol before operations, use the sterile materials noted, and include penicillin/streptomycin (final concentrations of 100 units/ml each) in the lavage solution. We have not had difficulty with infection using these techniques.

Materials include the following:

Ketamine HCl injection (100 mg/ml ketamine)
Xylazine injection (20 mg/ml)
Animal support platform
Depilatory
Scalpel and disposable sterile blades
Cotton swabs and balls, sterile
Tissue retractors
Tissue scissors
Lactated Ringer's solution (LR) containing penicillin/streptomycin
Fiberoptic light source
Virus stock
Automated microinjector
1- to 5-μl micropipettes
3-O Dexon suture
Tissue stapler

1. Mix ketamine and xylazine 1:1 in a 1-ml syringe with 27-gauge needle; lift the animal's tail and hindquarters with one hand and with the other inject 0.05 ml (mice) or 0.18 ml (rats) of anesthetic mixture intraperitoneally. One or more additional doses of ketamine alone (0.05 ml for mice and 0.10 ml for rats) are usually required to induce or maintain anesthesia, particularly if the procedure

takes over 1 h. Respiratory arrest and spontaneous abortion appear to occur more often when a larger dose of the mixture is given initially or any additional doses of xylazine are given.

2. Remove hair over entire abdomen using depilatory agent (any commercially available formulation, such as Nair, works well); shaving of remaining hair with a razor may be necessary. Wash skin several times with water, then with 70% ethanol, and allow to dry.

3. Place animal on its back in support apparatus. For this purpose, we find that a slab of Styrofoam with two additional slabs glued on top to create a trough works well. With the trough appropriately narrow, no additional restraint is then needed to hold the anesthetized animal.

4. Make a midline incision in the skin from the xiphoid process to the pubis using scalpel, and retract; attaching retractors firmly to Styrofoam support creates a stable working field. Stop any bleeding with cotton swabs before carefully retracting fascia and peritoneum and incising them in the midline with scissors (care is required here not to incise underlying bowel). Continue incision cephalad along midline of fascia (where there are few blood vessels) to expose entire abdominal contents. If necessary to expose the uterus, gently pack the abdomen with cotton balls or swabs to remove the intestines from the operative field, being careful not to lacerate or obstruct the bowel. Fill the peritoneal cavity with LR, and lavage until clear if the solution turns at all turbid. Wide exposure is important to allow the later manipulations. During the remainder of the operation, keep the peritoneal cavity moist and free of blood; dehydration or blood around the uterus increases the rate of postoperative abortion.

5. Elevate the embryos one at a time out of the peritoneal cavity, and transilluminate with a fiberoptic light source to visualize the structure to be injected. For lateral cerebral ventricular injections, for example, the cerebral venous sinuses serve as landmarks. When deciding on a structure to inject, keep in mind that free diffusion of virus solution throughout a fluid-filled structure lined with mitotic cells is best for ensuring even distribution of viral infection events throughout the tissue being labeled. The neural tube is an example of such a structure; when virus is injected into one lateral ventricle, it is observed to diffuse quickly throughout the entire ventricular system.

6. Using a heat-drawn glass micropipette attached to an automatic microinjector, penetrate the uterine wall, extraembryonic membranes, and the structure to be infected in one rapid thrust; this minimizes trauma and improves survival. Once the pipette is in place, inject the desired volume of virus solution, usually 0.1 to 1.0 μl. Coinjection of a dye such as 0.005% (wt/vol) trypan blue or 0.025% fast green helps determination of the accuracy of injection and does not appear to impair viral infectivity; coinjection of the polycation polybrene (80 μg/ml) aids in viral attachment to the cells to be infected.

The type of instrument used to deliver the virus depends on the age of the animal and the tissue to be injected. At early embryonic stages, the small size and easy penetrability of the tissue make a pneumatic microinjector (such as the

Eppendorf 5242) best for delivering a constant amount of virus at a controlled rate with a minimum of trauma. Glass micropipettes should be made empirically to produce a bore size that allows penetration of the uterine wall and the tissue to be infected. At later ages (late embryonic and postnatal), a Hamilton syringe with a 33-gauge needle works best.

When injections are given through the uterine wall, all embryos may potentially be injected except those most proximal to the cervix on each side (injection of these greatly increases the rate of postoperative abortion). In practice, it is often not advisable to inject all possible embryos, if excessive uterine manipulation would be required. At the earliest stages at which this technique is feasible (E12 in the mouse or E13 in the rat), virtually any uterine manipulation may cause abortion, so any embryo that cannot be reached easily should not be injected.

7. Once all animals have been injected, lavage the peritoneal cavity until it is clear of all blood and clots, ensure that all cotton balls and swabs have been removed, and move retractors from the abdominal wall/fascia to the skin. Filling the peritoneal cavity with LR with penicillin/streptomycin before closing increases survival significantly, probably by preventing maternal dehydration during recovery from anesthesia as well as preventing infection.
8. Using 3-O Dexon or silk suture material on a curved needle, suture the peritoneum, abdominal musculature, and fascia from each side together, using a continuous locking stitch. After closing the fascia, again lavage using LR with penicillin/streptomycin.
9. Close skin using surgical staples (such as the Clay-Adams Autoclip) placed 0.5 cm apart. Sutures may also be used, but these require much more time (often necessitating further anesthesia, which increases abortion risk) and are frequently chewed off by the animal resulting in evisceration.
10. Place animal on its back in the cage and allow anesthesia to wear off. Ideally, the animal wakes up within 1 h of the end of the operation. Increasing time to awakening results in increasing abortion frequency. Food and water on the floor of the cage should be provided for the immediate postoperative period.
11. Allow mothers to deliver progeny vaginally, or harvest pups can be by cesarean section. Maternal and fetal survival is approximately 60% at early embryonic ages of injection and increases with gestational age to virtually 100% after postnatal injections.

2. Injection of Virus into Mice Using *exo Utero* Surgery

Injections into small or delicate structures (such as the eye) require micropipettes that are too fine to penetrate the uterine wall. In addition, it is impossible precisely to target many structures through the rather opaque uterine wall. These problems can be circumvented, although with a considerable increase in technical difficulty and decrease in survival, by use of the *exo utero* technique (Muneoka *et al.*, 1986). The procedure is similar to that detailed earlier, with the following modifications to free the embryos from the uterine cavity.

The technique works well in our hands only with outbred, virus-antigen-free CD-1 and Swiss-Webster mice, but even these strains may have different embryo survival rates when obtained from different suppliers, or different colonies of the same supplier. This variability presumably results from subclinical infections, which may render some animals unable to survive the stress of the operation. We have had no success with this technique in rats.

1. After exposing the uterus and before filling the peritoneum with LR, incise the uterus longitudinally along its ventral aspect with sharp microscissors. The uterine muscle will contract away from the embryos, causing them to be fully exposed, surrounded by their extraembryonic membranes.
2. Using a dry sterile cotton swab, scoop out all but two embryos, each with its placenta and extraembryonic membranes, and press firmly against the uterus where the placenta was removed for 30 to 40 sec to achieve hemostasis. Only two embryos in each uterine horn can be safely injected, apparently because of trauma induced by neighboring embryos touching each other. It is very important to stop all bleeding before proceeding. From this point on, the embryos must be handled extremely gently, because only the placenta is tethering an embryo to the uterus, and it tears easily.
3. Fill the peritoneal cavity with LR, and cushion each embryo to be injected with sterile cotton swabs soaked in LR. Keeping the embryos submerged throughout the remainder of the procedure is essential to their survival.
4. Perform the injection with a pneumatic microinjector and heat-pulled glass micropipette. This may usually be done by puncturing the extraembryonic membranes first and then the structure to be injected; for some very delicate injections it may be necessary to make an incision in the extraembryonic membranes, which is then closed with 10-O nylon suture after the injection.

Infection of Chick Embryos

The following description is an example of an infection protocol used for the chick neural tube. More details of infection protocols for chick embryos can be found in Morgan and Fekete (1997).

Fertilized, virus-free White Leghorn chicken eggs were obtained from SPAFAS (Norwich, CT) and kept at 4°C for 1 week or less until they were transferred to a high-humidity, rocking incubator (Petersime, Gettysburg, OH) at 38°C, which was designated Time 0. Line O eggs are the most desirable hosts because they do not encode any endogenous retroviruses that could lead to helper virus generation (Astrin *et al.*, 1979).

1. To prevent the embryo from sticking to the shell, lower the embryo by removing albumin at an early stage. To accomplish this, set the eggs on their sides and rinse with 70% ethanol. Poke a hole in both ends of the egg as it lies on its side using a sharp pair of forceps, scissors, or needle. Use a 5-ml syringe and

21-gauge needle and withdraw 1.5 ml of albumin from the pointed end of the egg. Angle the needle so as not to disrupt the yolk by pointing it down and by not putting it too deep into the egg. Cover both holes with clear tape and return to the incubator. Alternatively, the embryo locate and stage by cutting a hole on the top (side facing up as it lies on its side) with curved scissors. Remove shell to make a hole 0.5 to 1 inch in diameter. Locate the embryo and then enlarge the hole to allow easy access to the embryo. If the embryo is to be used later, cover the hole with clear tape and return to the incubator.

2. At approximately 18–42 h incubation [Hamburger and Hamilton stage 10–17 (Hamburger and Hamilton, 1951)], inject embryos with 0.1–1.0 μl of viral inoculum including 0.25% fast green dye (Fekete and Cepko, 1993). Deliver the inoculum by injection directly into the ventricular system, which is easily accessed at these early times. For delivery, load the inoculum into a glass micropipette, made as described previously for rodent injections. A micromanipulator and Stoelting microsyringe pump (Catalogue No. 51219, Wood Dale, Illinois) are convenient and deliver the maximum volume in about 1 min. After injection, cover the hole with clear tape and return the embryo to the incubator.

Infected chick or rodent embryos can be incubated to any desired point. Chicks can be allowed to hatch and rodents can be delivered by cesarean section and reared to maturity. Embryonic brains are dissected in PBS followed by overnight fixation in 4% paraformaldehyde in phosphate-buffered saline (PBS, pH 7.4) at 4°C. Posthatch or postnatal animals are perfused with the same fixative and are incubated overnight in the fixative. They are then washed overnight in three changes of PBS and cryoprotected in 30% sucrose. After cryoprotection, the brains are embedded in OCT media and cut on a Reichart-Jung 3000 cryostat at 60–90 μm. Sections are histochemically stained for the appropriate marker and are mounted with gelvatol. For details of the histochemical reaction for lacZ or PLAP, see Cepko *et al.* in Morgan and Fekete (1997).

Solutions

PBS (10×)
80 g NaCl
2 g KCl
11.5 g Na_2HPO_4
2 g KH_2PO_4
in 1 liter H_2O

Twenty-five percent stock (Sigma Chemical Company, St. Louis, MO) can be stored at −20°C and frozen/thawed many times. Make dilution immediately before use.

4 g solid paraformaldehyde
2 m*M* $MgCl_2$
1.25 m*M* EGTA (0.25 ml of a 0.5 *M* EGTA stock, pH 8.0)
in 100 ml PBS, pH 7.2–7.4

Heat H_2O to 60°C. Add paraformaldehyde. Add NaOH to get paraformaldehyde in solution. Cool to room temperature, add 1/10 volume 10× PBS, and adjust pH with HCl. Can be stored at 4°C several weeks.

C. Preparation and Use of a Retroviral Library for Lineage Analysis Using PCR and Sequencing

We developed a more direct approach to address lumping and splitting errors (Walsh and Cepko, 1992) by constructing a library of viruses. In our first libraries, each virus of the library carried one member from a pool of approximately 100 DNA fragments from *Arabidopsis thalliana* DNA, in addition to the *lacZ* or *PLAP* gene. Infected cells, recognized by their enzyme activity, were mapped and the positive cells cut from cryosections. The *A. thalliana* DNA was amplified by PCR and characterized by size and restriction enzyme digestion patterns. If the sizes and restriction digestion patterns of the PCR product from two or more cells were the same, they were considered siblings with a probability calculated on the basis of the number of infections in that brain and the complexity of the library (Walsh and Cepko, 1992; Walsh *et al.*, 1992). Lineage analysis using such libraries revealed novel lineal relationships in the rat cerebral cortex (Walsh and Cepko, 1992, 1993; Reid *et al.*, 1995) and chick diencephalon (Arnold-Aldea and Cepko, 1996). However, the limited number of unique members in the library made from *A. thalliana* DNA restrained the analysis to tissues with low infection rates.

More data could be acquired with each experiment and additional questions could be addressed in the central nervous system and other tissues with a more complex library containing a greater number of DNA tags. We therefore constructed several retroviral vectors, of which CHAPOL (*ch*ick *a*lkaline *p*hosphatase with *o*ligonucleotide *l*ibrary) is the prototype (Golden *et al.*, 1995), that include degenerate oligonucleotides with a theoretical complexity of 1.7×10^7. Studies in the developing nervous system of the chick have been successfully completed using CHAPOL (Golden and Cepko, 1996; Szele and Cepko, 1996).

A summary of the production of CHAPOL and BOLAP (*B*abe-derived *o*ligonucleotide *l*ibrary with *a*lkaline *p*hosphatase, an oligo library in a murine vector) is given here; a detailed description of the construction of CHAPOL can be found elsewhere (Golden *et al.*, 1995). For either avian or murine retroviruses, the overall strategy is the same. A population of double-stranded DNA molecules that includes a short degenerate region, $[(G \text{ or } C)(A \text{ or } T)]_{12}$, is generated by PCR amplification of a chemically constructed single-stranded oligonucleotide population of the same sequence. The oligo preparation is ligated into a retrovirus vector and a preparation of highly competent *E. coli* is transformed. The library is then grown as a pool and a preparation of plasmids from the pool is produced. The DNA of the pool is transfected into an avian or mammalian packaging cell line to produce a library of virus particles. The library is injected into an area to be mapped. Infected cells are detected histochemically and each infected cell is recovered for PCR amplification. Each PCR product is then sequenced. Two cells with the same sequence are considered siblings.

1. Preparation of CHAPOL

The avian replication-incompetent virus CHAP (Fields-Berry, 1990; Ryder and Cepko, 1994), encoding the human *PLAP* gene, was modified to accept the oligo inserts. CHAP was linearized, purified, and mixed with the degenerate oligonucleotides in the presence of ligase, and aliquots of the resulting ligation products were used to transform *E. coli* DH5α. Following transformation, all aliquots were pooled. One hundred microliters of the pool was plated at varying dilutions on plates containing ampicillin. The remainder of the pool was divided and added to eight 2-liter flasks containing 1 liter of LB medium with 50 μg/ml ampicillin. The cultures were shaken overnight at 37°C. Plasmid DNA was extracted from these cultures by the Triton lysis procedure and purified on CsCl gradients (Sambrook *et al.*, 1989).

CHAPOL DNA was transfected into the avian virus packaging line Q_2bn (Stoker and Bissell, 1988), and the transiently produced virus was collected and concentrated. Aliquots of $CaPO_4$ precipitates of 100 μg CHAPOL DNA were made in 10 ml of HBS (Hepes buffered saline). The precipitate in each aliquot was then distributed equally on ten 10-cm plates of Q_2bn, and glycerol shock was carried out for 90 s at room temperature 4 h later. At 24 h postglycerol shock, the supernatants were collected and pooled. This was repeated at 48 h. The supernatants from the 24- and 48-h harvests were pooled and the titer was calculated by infection of QT6 cell and assay of the PLAP activity as described (Cepko, 1992). The stock was filtered through a 0.45-mm filter and concentrated by centrifugation in an SW27 rotor at 4°C, 20,000 rpm, for 2 h. The concentrated stock was titered on QT6 and tested for helper virus, which proved negative. The titer of CHAPOL was determined to be 1.1×10^7 colony-forming units (cfu)/ml. The same stock was used for all experiments conducted over a 3-year period, and many aliquots remain. We recommend making large stocks and storing them as small aliquots.

2. Use of CHAPOL

CHAPOL was used to infect the developing brain of chick embryos using the procedures outlined previously. At various times later, the tissue was harvested and stained for alkaline phosphatase (AP) activity. The outline of each section was drawn by camera lucida and the location and type of cells labeled on each section were recorded. A single cell or cluster of cells with a small group of surrounding cells was removed using a heat-pulled glass micropipette (Fig. 1)

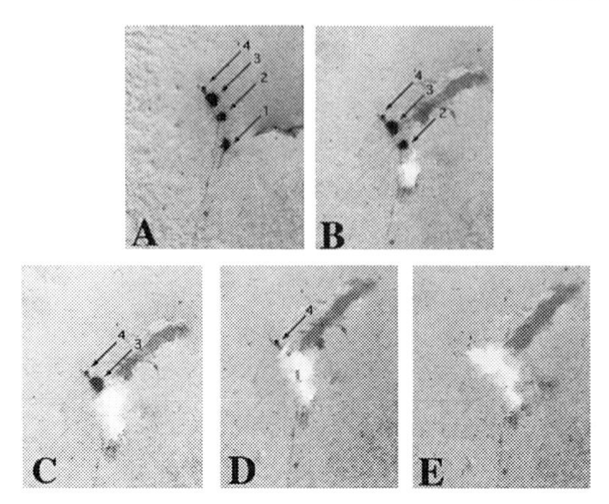

Fig. 1. Picking of AP+ cells from cryosections of CHAPOL-infected chick diencephalon. (A) Four cells (labeled 1–4) are present in the section. Note the long processes projecting toward the bottom of the panel. These processes were joining a white matter tract and allowed the cells to be defined as neurons. (B) After removal of cell 1. (C) After removal of cell 2. (D) After removal of cell 3. (E) All cells have been removed.

and transferred to a 96-well PCR plate for proteinase K digestion (as described later). Following digestion, nested PCR was performed (see later). The product of each PCR was run on a 1.5% agarose gel to determine if a product of the appropriate size had been amplified (Fig. 2). The recovery of a PCR product of the proper size occurred from PCR of 50–85% of the picks (the frequency varied depending on the batch of PCRs and the tissue being studied) using CHAPOL. Sequencing of the oligonucleotide insert (see later) was performed on all reactions, which gave the expected product on the agarose gel analysis (Fig. 3) and was successful approximately 75% of the time. All sequences were stored in the software program GCG (1991). All common sequences were pulled from the database created in GCG and the corresponding cells labeled. Sections were then aligned to determine the three-dimensional boundaries of clonal expansion. Each type of cell (e.g., neuron, glia) was also recorded to determine the variety of cells that can arise from a single progenitor.

The value of using a complex library of vectors is illustrated by the view of more heavily infected brains (Fig. 4), where closely aggregated AP+ cells would have been lumped into a single clone based on proximity. In addition, even lightly infected brains could give rise to lumping errors (also see Walsh and Cepko, 1992).

Two issues are important in determining the value of this type of library of DNA markers: the number of unique members in the library, and the distribution of the library members (Walsh *et al.*, 1992). If only two members exist in the library, for example, then there is a one in two chance that the same tag will be

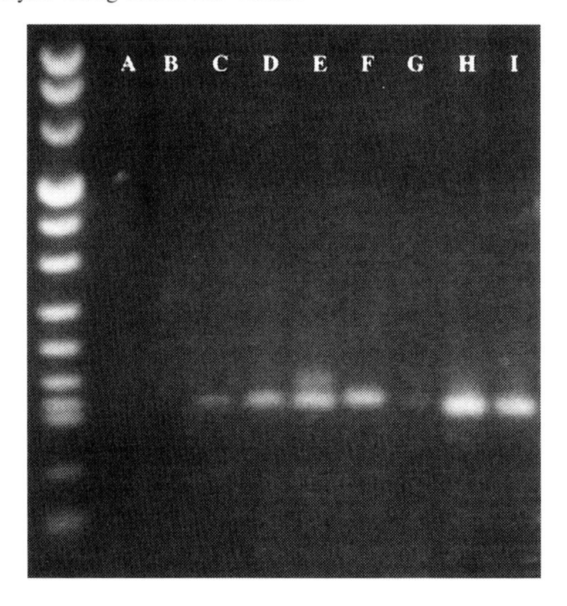

Fig. 2. The products of nine nested PCRs from CHAPOL-infected brain tissue were analyzed on a 1.5% agarose gel. The first lane is a molecular weight marker (MWM VIII, Boehringer Mannheim). Lane A is a control lane of the product of a PCR without DNA. Lane B is a control lane with the PCR products from tissue picked from an infected brain in a region having no purple precipitate. Lanes C–I are all the PCR products from tissue fragments with 1–5 AP+ cells. A single band at the expected size of 121 bp is found in lanes C, D, F, H, and I. A suggestion of a faint band is present in lane G, but no definite band is present. Lane E shows the expected band and a slightly larger band. The lighter extra band represents some priming in the second PCR from primer carried over from the first reaction.

selected in two consecutive picks. If 100 members exist in the library, the chance that two picks come up with the same member is reduced to 10^{-2}. The second important variable determining the quality of the library is the distribution of the members within the library. This can be illustrated as follows. Consider a library composed of 10^6 members, with 50% of the library composed of one member. If two neighboring or distant cells are found to carry the overrepresented insert, the probability that the two cells arose from separate clones is still 0.5. CHAPOL was found to have an equal distribution in that each of the inserts picked to date ($n > 500$) has occurred independently only once. One further issue to consider is the level of difficulty in using the library. We have found in practice that this method of tag identification is in fact easier than our previous method based on the analysis of the size and restriction digestion pattern of *A. thalliana* DNA.

These libraries should be useful for application in a wide range of tissues and species. The host range has been expanded to previously uninfectable hosts (Yee

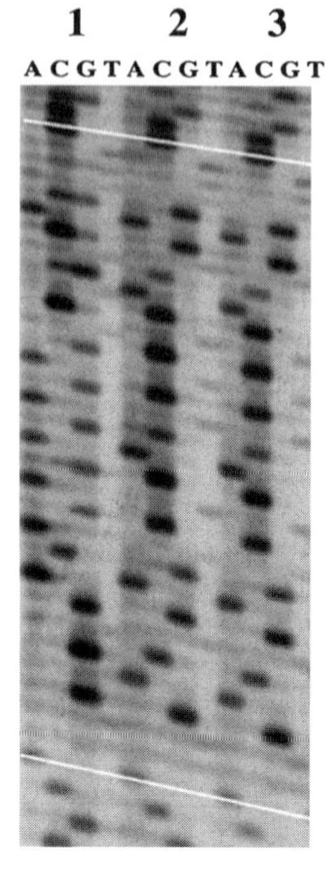

Fig. 3. The sequencing reactions from the PCR products of CHAPOL-infected tissue of three different samples are shown (1–3). The white lines delineate the beginning and end of the degenerate oligonucleotide sequence. PCR product 1 reveals a unique sequence. The sequence of PCR products 2 and 3 are the same, indicating that they are from siblings.

et al., 1994), and infection of nonneural tissue with CHAPOL has been observed, as would be expected based on experience with avian and murine retroviruses.

D. PCR and Sequencing from CHAPOL

Proteinase K Digestion

The coverslips were removed from slides by immersion in sterile H_2O. Single cells or small clusters of cells containing purple nitroblue tetrazolium (NBT) precipitate with surrounding unlabeled tissue (approximately 0.5- to 2-mm tissue fragments) were scraped from the slide using a heat-pulled glass micropipette (Fig. 1). The cells were transferred to a 96-well PCR (Hybaid) plate with 10 μl of a proteinase K solution (50 m*M* KCl, 10 m*M* Tris–HCl, pH 7.5, 2.5 m*M* $MgCl_2$, 0.02% Tween-20, 200 μg/ml proteinase K). Each well was overlaid with one drop of light mineral oil (Sigma) and the plates were heated to 60°C for 2 h, 85°C for 20 min, and 95°C for 10 min in a Hybaid OmniGene thermocycler.

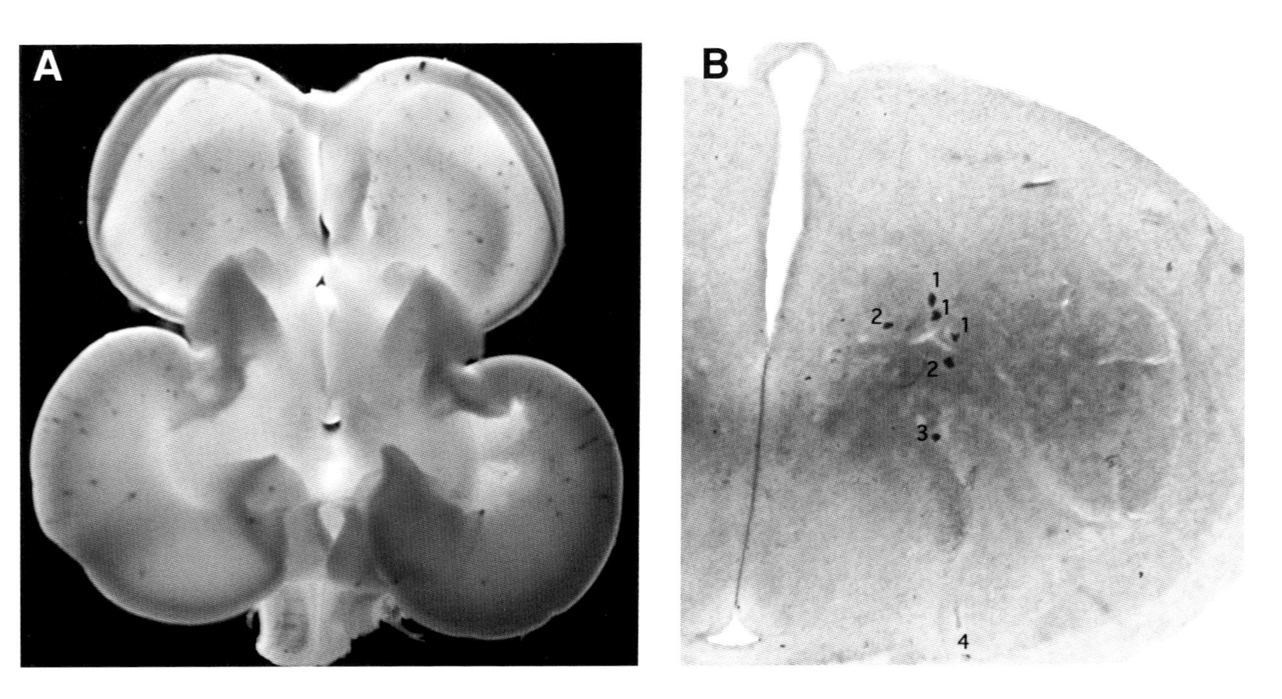

Fig. 4. CHAPOL-infected brains at E8 and E18. (A) Dorsal view of an E8 brain fixed and reacted with X-Phos and NBT as a whole mount. The ventricles have been opened to expose >100 AP+ columns of cells. (B) A representative 60-μm section of the diencephalon from a brain infected with CHAPOL and analyzed at E18. The AP+ cells were removed and the inserts were sequenced following PCR. The AP+ cells labeled "1" were each found to carry the same insert. The two adjacent cells labeled "2" each carried the same insert, which was different than those in clone "1." Two other cells, "3" and "4," in the same section each carried unique inserts that were also different from either "1" or "2." The proximity of the cells labeled "1" and "2" would have resulted in a lumping error if clonal definitions were based on geometric boundaries.

2. Nested PCR

The first PCR was accomplished by adding 0.15 μl *Taq* polymerase (Boehringer Mannheim), 0.15 μl deoxynucleotide triphosphate (dNTP) mix (Boehringer Mannheim), 0.75 μl each of 10 μ*M* oligonucleotide 0 (5′TGTGGCTGC-CTGCACCCCAGGAAAG3′) and 10 μ*M* oligonucleotide 5 (5′GTGTGCTG-TCGAGCCGCCTTCAATG3′), 2 μl PCR buffer with $MgCl_2$ (Boehringer-Mannheim), and 16.2 μl of H_2O to each well of the 10-μl proteinase K solution (final volume, 30 μl). This was cycled at 93°C × 2.5 min; [(94°C × 45 s)(72°C × 2 min)] × 40 cycles; 72°C × 5 min.

The second PCR was performed with 1 μl of reaction product from the first PCR added to 0.25 μl *Taq* polymerase, 0.25 μl dNTP mix, 1 μl each 10 μ*M* oligonucleotide 2 (5′GCCACCACCTACAGCCCAGTGG3′) and 10 μ*M* oligonucleotide 3 (5′GAGAGAGTGCCGCGGTAATGGG3′), 2 μl PCR buffer with $MgCl_2$, and 14.5 μl of H_2O (final volume, 30 μl). The reaction was thermocycled at 93°C × 2.5 min; [(94°C × 45 s)(70°C × 2 min)] × 30 cycles; 72°C × 5 min. An aliquot of DNA was run on a 1.5% agarose gel (0.75% SeaKem, 0.75% NuSeive) to ensure that the appropriate insert was present (Fig. 2).

3. Sequencing

Sequencing was performed using the Cyclist Exo-Pfu DNA sequencing kit from Stratagene. Briefly, 5 μl of each deoxy/dideoxyNTP mix was added to 4 wells of a 96-well Hybaid plate. To each of these wells 25% of the following mixture was added: 1 μl of the nested PCR product, 1 μl of 10 μ*M* Oligo 3, 3 μl 10× sequencing buffer, 1 μl Exo-Pfu, 0.75 μl ^{35}S (10 μCi), 4 μl dimethyl sulfoxide and 11.25 μl H_2O. This was cycled at 95°C × 5 min; [(95°C × 30)(60°C × 30 s)(72°C × 1 min)] × 30 cycles. Sequencing reactions were analyzed on a 6% acrylamide denaturing gel (Fig. 3).

For all procedures, reagents, instruments, and glass microscope slides should be handled with scrupulous technique, and ultraviolet-irradiated when needed to destroy contaminating DNA.

E. Oligo Libraries for Other Species

1. Creation of BOLAP

The BOLAP library was created in a murine retrovirus vector, pBABE (Morgenstern and Land, 1990), into which the *PLAP* gene was inserted to create pBABE-AP (Fields-Berry and Cepko, unpublished) (Fig. 5). The BOLAP library was generated by inserting PCR-amplified DPL2 into the *Asc*I and *Bgl*III sites of BABE-AP. After the ligation, BOLAP was digested with *Asc*I and *Xho*I, phenol/chloroform extracted, and ethanol precipitated.

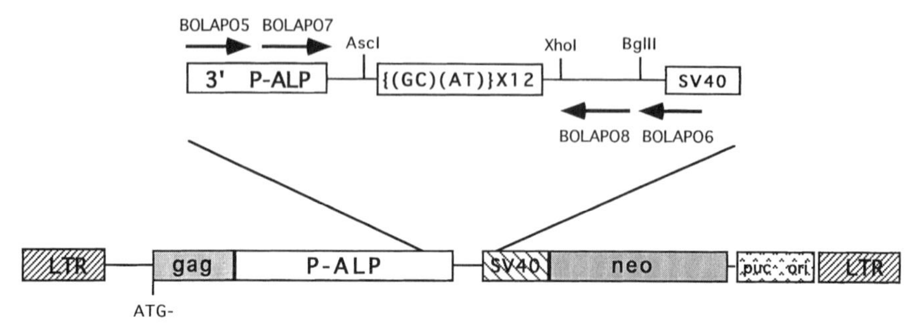

Fig. 5. BOLAP, a murine library encoding an oligonucleotide library. A degenerate oligonucleotide pool (DPL1) was inserted into the BABE-Neo plasmid encoding PLAP. The positions and orientations of the oligonucleotides BOLAPO 5–8 used to amplify each insert and for sequencing are indicated by the arrows. LTR, long terminal repeat, gag, 5′ terminus of MMLV (Maloney Murine Leukemia Virus) *gag* gene; SV40, simian virus 40 early promoter; neo, Tn5 neomycin resistance gene; puc ori, bacterial origin of replication.

Double-stranded DPL was prepared by PCR amplification of the following reaction mix (for all oligos referred to in this section, see below):

1 μl 1/100 diluted DPL1 (0.6 mg.ml)
1 μl DPLP (0.44 mg/ml)
1 μl DPLP5 (0.2 mg/ml)
5 μl 2.5 m*M* dNTP mix
10 μl 10× PCR buffer (BM)
79 μl H_2O
2 μl *Taq* DNA polymerase (Boehringer-Mannheim)

The amplification program was:

93°C for 2.5 min
[94°C for 45 s, 70°C for 2 min] × 30
72°C for 5 min

The product was phenol/chloroform extracted, ethanol precipitated, digested with *Asc*I and *Xho*I, gel purified, and ethanol precipitated. Approximately 3 μg of *Asc*I/*Xho*I-digested BABE-AP-X was ligated to 25 ng of *Asc*I/*Xho*I-digested DPL1 in a 100-μl ligation reaction at 16°C overnight. The product was phenol/chloroform extracted, ethanol precipitated, and resuspended in 50 μl TRIS-EPTA. Twenty microliters of ligation product was used to transform ElectroMax DH10B (Gibco/BRL; Grand Island, New York) competent cells according to the manufacturer's instructions. The transformed cells were pooled. One microliter of the pooled cells was serially diluted and plated on LB/amp plates. The remainder of the pool was split into five 1-l cultures of TB (Terrific Broth) containing ampicillin and kanamycin sulfate (each at 50 μg/ml) and shaken overnight at

37°C. BOLAP plasmid, 3.7 mg, was extracted from the overnight cultures by Triton lysis, followed by purification via CsCl gradient. The number of colonies on the LB/amp plates projects the total number of transformants to be 1.28×10^7. A control ligation containing no DPL1 projects the background of vector without DPL1 insert to be 0.9%.

2. BOLAP Virus Production and Evaluation

Eleven confluent 10-cm dishes of Bosc23 cells (Pear *et al.*, 1993) were split into fifty dishes. The next morning, 350 μg of BOLAP DNA was combined with 1.5 ml of Lipofectamine (Gibco/BRL) in 25 ml of Optimem (Gibco/BRL). The mixture was incubated for 20 min at room temperature and then added to 200 ml of DME (Dulbecco's Modified Eagle's Medium). The 50 plates were washed with DME, and 5 ml of the DNA/Lipofectamine/DME mixture was added to each plate. The plates were incubated for 5 h at 37°C. Five milliliters of 20% fetal bovine serum in DME was then added to each plate and the plates were returned to the incubator overnight. The next morning, the supernatant was harvested and replaced with 5 ml of 10% fetal calf serum in DME. The next morning, this supernatant was harvested. The supernatants were pooled, filtered through a 0.45-μm filter unit, and concentrated by centrifugation at 20,000 rpm for 2 h at 4°C. The final concentration of the viral supernatant was 1×10^8 cfu/ml.

To test the complexity of the library, the protocols of Golden *et al.* (1995) were followed. One microliter of viral supernatant was diluted into 30 ml of DME with 10% calf serum. One to two microliters of this diluted viral stock was used to infect a 30–50% confluent 6-cm dish of NIH 3T3 cells. Five to six hours later, the infected cells were trypsinized, diluted 10-fold, and plated on 96-well dishes. Three days later, the plates were washed with PBS, fixed with 4% paraformaldehyde for 10 min, washed thrice with PBS for 5 min, heated to 65°C for 25 min, and then stained overnight for AP activity with X-Phos and NBT. The following morning, each well was examined for the presence of a single discrete grouping of AP+ cells. The cells in chosen wells were washed with PBS. Ten microliters of 400 μg/ml proteinase K solution (50 m*M* KCl, 10 m*M* Tris–HCl, pH 7.5, 2.5 m*M* $MgCl_2$, 0.02% Tween-20) was added and the cells and solution were scraped/suctioned off and placed in individual wells in a 96-well dish. A drop of mineral oil was placed over each well and the plate was heated to 65°C for 2 h, 85°C for 20 min, and 95°C for 10 min.

A nested PCR was performed as follows. To each well was added 20 μl of the following reaction mix:

2 μl 10× PCR buffer with Mg (Boehringer-Mannheim)
0.15 μl BOLAPO 5 and 6 (0.6 mg/ml)
0.15 μl dNTP mixture (25 m*M*)
17.4 μl H_2O
0.4 μl *Taq* DNA polymerase (Boehringer-Mannheim)

The reactions were cycled as follows:

93°C for 2.5 min
[94°C for 45 s, 67°C for 2 min, 72°C for 2 min] × 33
72°C for 5 min

One microliter of the preceding reaction was added to 20 μl of the same reaction mix, substituting BOLAPO 7 and 8 for BOLAPO 5 and 6. The amplification program was

93°C for 2.5 min
[94°C for 45 s, 72°C for 2 min] × 30
72°C for 5 min

Eight microliters of the second reaction mix was added to 2.5 μl of gel loading buffer and then fractionated on a 3% NuSieve GTG/1% SeaKem ME agarose gel. Amplifications that yielded a DNA product of the appropriate molecular weight (bp) were sequenced using the Exo-Pfu Cyclist kit. One microliter of nested PCR product was added to the following reaction mix:

0.15 μl BOLAPO 7 (0.6 mg/ml)
3 μl 10× sequencing buffer
0.75 μl ^{35}S-(alpha)dATP
4 μl Dimethyl sulfoxide
12.1 μl H_2O
1 μl Exo-Pfu DNA polymerase

This reaction was mixed and 5-μl portions were added separately to 5 μl of the four dNTP/ddNTP mixtures. The reactions were overlaid with oil and cycled as follows:

95°C for 5 min
[95°C for 30 s, 60°C for 30 s, 72°C for 1 min] × 30

The reactions were terminated by the addition of 5 μl of stop buffer and then fractionated by 6% acrylamide gel electrophoresis.

At this point, 98 individual clones have been sequenced. All carry a unique DPL1 insert. Of those, the degenerate oligo region was shorter than 24 bases in 7 clones and longer in 1 clone. Occasional variations in the (GC)(AT) sequence were also seen.

Sequences of Oligonucleotides Referenced

DPL2—5′-TAGGAGGCGCGCCTTT-[(GC)(AT)]12-GTTCTCGAGGACAC-CTGACTGGGCTGAGGGCTTCCGCGACCCGAGATCTCAGCTTCC-3′
DPL1—5′-TAGGAGGCGCGCCTTT-[(GC)(AT)]12-GTTACGCGTTA-ATTAACTCGAGATCTCAGCTTC-3′
DPLP—5′-GAAGCTGAGATCTCGAGTTA-3′
DPLP5—5′-TAGGAGGCGCGCCTTT-3′

BOLAPO 5—5′-CCAGGGACTGCAGGTTGTGCCCTGT-3′
BOLAPO 6—5′-AGACACACATTCCACAGGGTCGAAG-3′
BOLAPO 7—5′-GGCTGCCTGCACCCCAGGAAAGGAG-3′
BOLAPO 8—5′-GGTCTCGGAAGCCCTCAGCCCAGTC-3′

References

Arnold-Aldea, S., and Cepko, C. (1996). Dispersion patterns of clonally related cells during development of the hypothalamus. *Dev. Biol.* **173,** 148–161.

Astrin, S. M., Buss, E. G., and Hayward, W. S. (1979). Endogenous viral genes are non-essential in the chicken. *Nature* **282,** 339–341.

Ausubel, F. M., Brent, R., Kingston, R. E., Moore, D. D., Seidman, J. G., Smith, J. A., and Struhl, K. (1997). "Current Protocols in Molecular Biology." Greene Publishing Associates, New York.

Boerkoel, C. F., Federspiel, M. J., Salter, D. W., Payne, W., Crittenden, L. B., Kung, S.-J., and Hughes, S. H. (1993). A new defective retroviral vector system based on the Bryan strain of Rous sarcoma virus. *Virology* **195,** 669–679.

Cepko, C. L. (1992). "Transduction of Genes Using Retrovirus Vectors." Green Publishing Associates and Wiley-Interscience, New York.

Cepko, C. L., Austin, C. P., Walsh, C., Ryder, E. F., Halliday, A., and Fields-Berry, S. (1990). Studies of cortical development using retrovirus vectors. *Cold Spring Harb. Symp. Quant. Biol.* **55,** 265–278.

Cosset, F. L., Legras, C., Chebloune, Y., Savatier, P., Thoraval, P., Thomas, J. L., Samarut, J., Nigon, V. M., and Verdier, G. (1990). A new avian leukosis virus-based packaging cell line that uses two separate transcomplementing helper genomes. *J. Virol.* **64,** 1070–1078.

Fekete, D., and Cepko, C. (1993). Replication-competent retroviral vectors encoding alkaline phosphatase reveal spatial restriction of viral gene expression/transduction in the chick embryo. *Mol. Cell. Biol.* **13,** 2604–2613.

Fields-Berry, S. C. (1990). Termination of DNA replication and construction of retrovirus vectors encoding histochemical reporter genes. Ph.D. thesis, Harvard University, Cambridge, MA.

Fields-Berry, S. C., Halliday, A., and Cepko, C. L. (1992). A recombinant retrovirus encoding alkaline phosphatase confirms clonal boundary assignment in lineage analysis of murine retina. *Proc. Natl. Acad. Sci. U.S.A.* **89,** 693–697.

Galileo, D. S., Gray, G. E., Owens, G. C., Majors, J., and Sanes, J. R. (1990). Neurons and glia arise from a common progenitor in chicken optic tectum: demonstration with two retroviruses and cell type-specific antibodies. *Proc. Natl. Acad. Sci. U.S.A.* **87,** 458–462.

Golden, J., Fields-Berry, S., and Cepko, C. (1995). Construction and characterization of a highly complex retroviral library for lineage analysis. *Proc. Natl. Acad. Sci. U.S.A.* **92,** 5704–5708.

Golden, J. A., and Cepko, C. L. (1996). Clones in the chick diencephalon contain multiple cell types and siblings are widely dispersed. *Development* **122,** 65–78.

Halliday, A. L., and Cepko, C. L. (1992). Generation and migration of cells in the developing striatum. *Neuron* **9,** 15–26.

Hamburger, V., and Hamilton, H. L. (1951). A series of normal stages in the development of the chick embryo. *Journal of Experimental Morphology* **88,** 49–92.

Hughes, S., and Blau, H. (1990). Migration of myoblasts across lamina during skeletal muscle development. *Nature* **345,** 350–352.

Mann, R., Mulligan, R. C., and Baltimore, D. (1983). Construction of a retrovirus packaging mutant and its use to produce helper-free defective retroviruses. *Cell* **33,** 153–159.

Markowitz, D., Goff, S., and Bank, A. (1988). A safe packaging line for gene transfer: separating viral genes on two different plasmids. *J. Virol.* **62,** 1120–1124.

Morgan, and Fekete. (1997). "Methods in Cell Biology." Cold Spring Harbor Laboratory Press, Cold Spring Harbor, NY.

Morgenstern, J. P., and Land, H. (1990). Advanced mammalian gene transfer: high titer retroviral vectors with multiple drug selection markers and a complementary helper-free packaging cell line. *Nucleic Acids Res.* **18,** 3587–3596.

Muneoka, K., Wanek, N., and Bryant, S. V. (1986). Mouse embryos develop normally exo utero. *J. Exp. Zool.* **239,** 289–293.

Naldini, L., Blomer, U., Gallay, P., Ory, D., Mulligan, R., Gage, F. H., Verma, I. M., and Trono, D. (1996). *In vivo* gene delivery and stable transduction of nondividing cells by a lentiviral vector. *Science* **272,** 263–267.

Pear, W. S., Nolan, G. P., Scott, M. L., and Baltimore, D. (1993). Production of high titer helper-free retroviruses by transient transfection. *Proc. Natl. Acad. Sci. U.S.A.* **90,** 8392–8396.

Price, J., Turner, D., and Cepko, C. (1987). Lineage analysis in the vertebrate nervous system by retrovirus-mediated gene transfer. *Proc. Natl. Acad. Sci. U.S.A.* **84,** 156–160.

Reid, C. B., Liang, I., and Walsh, C. (1995). Systematic widespread clonal organization in cerebral cortex. *Neuron* **15,** 299–310.

Roe, T., Reynolds, T., Yu, G., and Brown, P. (1993). Integration of murine leukemia virus DNA depends on mitosis. *EMBO J.* **12,** 2099–2108.

Ryder, E. F., and Cepko, C. L. (1994). Migration patterns of clonally related granule cells and their progenitors in the developing chick cerebellum. *Neuron* **12,** 1011–1028.

Sambrook, J., Fritsch, E., and Maniatis, T. (1989). "Molecular Cloning: A Laboratory Manual." Cold Spring Harbor Laboratory Press, Cold Spring Harbor, NY.

Soneoka, Y., Cannon, P. M., Ramsdale, E. E., Griffiths, J. C., Ramano, G., Kingsman, S. M., and Kingsman, A. J. (1995). A transient three-plasmic expression system for the production of high titer retroviral vectors. *Nucleic Acids Res.* **23,** 628–633.

Stoker, A. W., and Bissell, M. J. (1988). Development of avian sarcoma and leukosis virus-based vector-packaging cell lines. *J. Virol.* **62,** 1008–1015.

Szele, F. G., and Cepko, C. L. (1996). A subset of clones in the chick telencephalon are arranged in rostrocaudal arrays. *Cur. Biol.* **6,** 1685–1690.

Turner, D. L., and Cepko, C. L. (1987). A common progenitor for neurons and glia persists in rat retina late in development. *Nature* **328,** 131–136.

Turner, D. L., Snyder, E. Y., and Cepko, C. L. (1990). Lineage-independent determination of cell type in the embryonic mouse retina. *Neuron* **4,** 833–845.

Walsh, C., and Cepko, C. L. (1992). Widespread dispersion of neuronal clones across functional regions of the cerebral cortex. *Science* **255,** 434–440.

Walsh, C., and Cepko, C. L. (1993). Clonal dispersion in proliferative layers of developing cerebral cortex. *Nature* **362,** 632–635.

Walsh, C., Cepko, C. L., Ryder, E. F., Church, G. M., and Tabin, C. (1992). The dispersion of neuronal clones across the cerebral cortex [letter]. *Science* **258,** 317–320.

Weiss, R., Teich, N., Varmus, H., and Coffin, J. (1984–1985). "RNA Tumor Viruses." Cold Spring Harbor Laboratory, Cold Spring Harbor, NY.

Yee, J. K., Friedmann, T., and Burns, J. C. (1994). Generation of high-titer pseudotyped retroviral vectors with very broad host range. *Methods Cell Biol.* **43,** 99–112.

4

Use of Dominant Negative Constructs to Modulate Gene Expression

Giorgio Lagna and Ali Hemmati-Brivanlou
Laboratory of Molecular Embryology
The Rockefeller University
New York, New York 10021-6399

I. Introduction

Molecular techniques have revolutionized the study of developmental problems in the latest years, allowing the dissection of signaling and genetic pathways in several developmental systems. Among these new tools are mutant proteins that inhibit in a dominant fashion the function of their normal counterpart *in vivo*. Compared to the now standard approach of gene inactivation by homologous recombination ("knockout"), which is highly specific but time-consuming and limited to the yeast and mouse systems, dominant negative strategies have the advantage of being very rapid and versatile. Instead of inactivating one by one the genes thought to participate in a function, the researcher has the alternative of

Current Topics in Developmental Biology, Vol. 36

0070-2153/98 $25.00

gaining rapid insights into the role of entire gene families through the "dominant negative" approach. This shortcut, however, cannot be taken without caution: the experimenter has to provide stringent controls of specificity, lack of toxicity, rescue by downstream factors, and so on. In this chapter, we review the many different ways to generate dominant negative mutations in a protein, and discuss their mechanisms of action. We also review the potential interpretations of the biologic effects observed and the most stringent controls. Finally, we provide specific examples of dominant negative proteins that have been engineered or occur naturally, and refer to them to illustrate specific advantages and caveats of their use.

A. Assay for Gene Function in Higher Eukaryotes

Diverse techniques allow modern embryologists to isolate novel genes from a variety of organisms at an amazing pace. With the various genome sequencing projects promising to characterize virtually *all* the genes of a selected number of organisms, the challenge for the present and future has become the assignment of specific functions to each gene product. This task appears particularly formidable, as we realize that the concept of "one gene, one function," advanced in the early 1960s, does not reflect the degree of complexity encountered in the study of genes involved in the development of higher eukaryotes.

B. Embryologic Function

The functions of a gene can evidently be studied at several levels. Firstly, there is a *biochemical function*; for example, kinases are involved in the phosphorylation of substrates, transcription factors activate or repress transcription, and so on. The *cellular function* of a gene product, then, derives directly from the integration of its biochemical functions; for example, tyrosine kinases such as Src can modulate cell growth by binding and phosphorylating a variety of polypeptides, whereas specialized cytoskeleton proteins can determine the shape and motion of cells. Finally, there is an *embryologic function*, which is the integration of cellular functions. The secretion of a given ligand from a subset of cells, for example, instructs the differentiation of other cells in a different region of the embryo. Although a variety of *in vitro* assays usually allows a direct assignment of the biochemical functions of a gene product, the assignment of cellular and embryologic functions has proved more challenging and has relied on either powerful mutational screens, such as the ones performed in *Caenorhabditis elegans*, *Drosophila*, zebrafish, or mouse, or the generation of molecular tools that allow a reverse genetic approach for the study of organisms that are not amenable to genetic screens.

C. Tools Available to the Molecular Embryologist

The molecular tools used to infer gene function can be classified into two main groups, those employed by *gain-of-function* studies and those used for *loss-of-function* analyses.

1. Gain-of-Function Approach

In a typical scheme, the ectopic expression of a polypeptide is targeted to cells that do not usually express the protein, or is forced at a time when the factor is not usually present. Alternatively, the protein of interest can be overexpressed in cells that are already producing it. Any cellular or embryonic phenotype derived from such studies points to the potential functions of the gene (for a review of this type of approach, see Vize *et al.*, 1991).

2. Loss-of-Function Approach

In the second class are methods that aim at eliminating the activity of a factor or class of factors. Many molecular strategies are used toward this aim. In organisms in which ectopic DNA can undergo homologous recombination, such as yeast and mouse, the gene targeting technique (or "knockout") has provided precious insights into the function of numerous genes. Unfortunately, this procedure is restricted to only a few systems, requires long periods of time, and does not fare well against gene families (see Section VI.D). Furthermore, until the recent development of the *flp* recombinase system, gene inactivation could not be induced or expressed at the desired stage and tissue. It is also possible to interfere with gene activity at the RNA level by using antisense constructs or ribozymes. Alternatively, an activity can be inhibited at the protein level by using neutralizing antibodies that block specific polypeptides. Although these techniques have been used to assign cellular or embryologic functions with varying degrees of success, they have recently been joined by a very powerful method: the use of dominant negative mutants. In this chapter, we focus on the different ways of generating dominant negative mutants, the most stringent controls, the limitations of the approach, and the interpretation of the observed effects. In detailing the use of dominant negative constructs, we focus primarily on developmental systems, drawing examples from the study of vertebrate embryogenesis and in particular of the amphibian system, *Xenopus laevis*, where this strategy has been extremely useful in providing molecular solutions to classical embryologic problems.

II. Dominant Negative Mutations

A dominant negative mutation in a gene generates a mutant protein that, when coexpressed with the wild-type polypeptide, can inhibit its activity. Mutants of this type were noticed by Muller (1932), who named them "antimorphs." Muller defined them as mutants having an effect contrary to that of the genes from which they are derived (Muller, 1932). In genetic screens, heterozygote antimorphic mutations often exhibit a phenotype more severe than homozygote null mutations, probably because dominant negative mutants are often able to inhibit entire gene families, not only their wild-type counterparts. Very elegant mutagenesis studies involving dominant negative alleles were later performed in prokaryotes, such as those on the *lacI* gene (Muller-Hill *et al.*, 1968) of the β-galactosidase operon and the *cI* gene (Oppenheim and Salomon, 1970, 1972) of the λ phage. Eventually, the usefulness of mutants isolated from genetic screens paved the way for the molecular design of such variants for reverse genetic use. In 1987, Ira Herskowitz proposed the use of dominant negative mutations for gene inactivation in higher eukaryotes. He predicted with remarkable accuracy not only how they could be generated and how they might act, but how to interpret phenotypes derived from the expression of such mutants (Herskowitz, 1987). It was not long after Herskowitz's pioneering proposal that his ideas were tested experimentally and proven to work as predicted. These approaches are now standard practice for the elimination of gene activity.

III. Generation of Dominant Negative Mutants

A large number of successful attempts to produce dominant negative derivatives has recently been reported in the scientific literature. In the next sections, we state some of the fundamental concepts that have emerged from these studies, regarding both the strategies of mutagenesis and the mechanisms of action of the inhibitory mutants.

A. Strategies of Mutagenesis

Generation of a dominant negative product *in vitro* requires the manipulation of the protein coding region of a gene so that an inactivating mutation is introduced. As defined previously, this mutation should be such that the protein product is not only nonfunctional, but able to inhibit the function of its wild-type counterpart when coexpressed. The simplest and most widely used mutation consists of a more-or-less extended *deletion* of the N-terminal or C-terminal end of the protein of interest. This should also be the approach followed in the first attempt to produce a dominant negative variant of a protein of unknown function. In cases in which the structural and functional studies of the target gene are more advanced, site-directed aminoacid *substitutions* allow investigators to produce a dominant negative construct without affecting the overall structure of the protein. It is important to remember, however, that in some cases deletions or point mutations can generate *constitutively active* derivatives, for example by inactivating intramolecular inhibitory domains; appropriate controls should be devised to address this possibility. Finally, in a few studies dealing with transcription factors, the *addition* of an inhibitory domain swapped from a different protein has resulted in the creation of a dominant negative variant.

B. Mechanisms of Action

Three general mechanisms of action of dominant negative gene products can be envisioned. The first, to which we refer as *inhibition by multimerization*, is specifically suited for homo- or heteromultimeric proteins. The second, designated *inhibition by titration of upstream or downstream targets*, can be used to suppress monomeric proteins as well. The third, *inhibition by active repression*, is also the newest and has so far been applied only to transcription factors.

1. Inhibition by Multimerization

As has been shown in numerous studies of multimeric proteins, a functionally defective mutant polypeptide that retains its ability to form multimeric complexes is often inhibitory. This effect can be achieved through various mechanisms. For example, the lack of an enzymatic function in the mutated subunit can result in the inactivation of the entire complex. This is the case of membrane-bound tyrosine and serine/threonine kinase receptors (see Section VI.B). Alternatively, the defective subunit might prevent the whole complex from interacting with other polypeptides. In a typical example, subunits of multimeric transcription factors with a mutated activation domain render the entire complex unable to interact with RNA polymerase and its ancillary machinery, resulting in a repression of transcription (see Section VI.D).

Most proteins act through interactions, more or less stable, with other polypeptides. Therefore, unless there is a specific reason for pursuing a different strategy, the procedure of choice for generating a dominant negative derivative of most polypeptides consists of the deletion or inactivation of the functional domain(s), leaving intact the homo- or heteromultimerization domain(s). This method possesses the advantage of producing detectable effects on expression of physiologic doses of inhibitory derivative. Consider the case of a homodimer complex, for example, such as the one formed by tyrosine or serine/threonine kinase receptors in response to ligand binding (Fig. 1). If one of the two subunits lacks the kinase domain, as a result of a targeted deletion, the complex is unable to phosphorylate and activate the downstream components of the signal transduction pathway. Therefore, one expects to observe an inhibitory effect (about 50% inhibition) at stoichiometric ratios of wild-type to dominant negative variant close to equimolar. This is in fact what has been reported for the dominant negative fibroblast

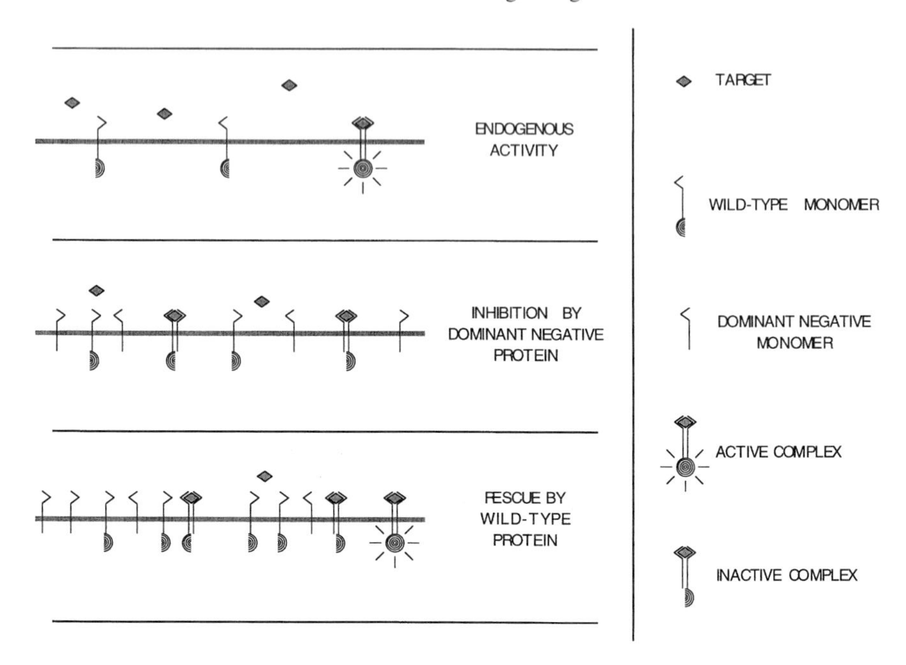

Fig. 1. Inhibition by multimerization. In this example, a hypothetical transmembrane receptor requires dimerization to transduce a signal. (Top) A receptor molecule dimerizes in a ligand- (gray diamond) dependent manner and generates a functional intracellular domain (concentric circles) that transduces the signal. (Middle) A truncated version of the receptor containing only the extracellular, ligand-binding domain, is still able to dimerize with the wild-type receptor. The overexpression of this construct dominantly inhibits the function of the endogenous receptor. (Bottom) Coexpression of an ectopic wild-type receptor with the dominant negative form rescues signaling by sequestering the dominant negative mutants.

growth factor (FGF) (Amaya *et al.*, 1991) and activin type II (Hemmati-Brivanlou and Melton, 1992) receptors in *Xenopus*.

2. Inhibition by Titration of Upstream or Downstream Targets

Enzymes, as well as regulatory and structural proteins, generally require interaction with a target molecule to carry out their function. For metabolic enzymes, the target molecule can be an upstream modulator, the substrate, the product of the catalysis, or a cofactor. If a nonmodifiable substrate is available, this often allows investigators to inhibit the enzymatic function *in vivo*; most kinases, for example, are inhibited by γ-S-ATP, a nonhydrolyzable form of ATP that irreversibly docks into the ATP-binding domain. The target of regulatory factors is often another protein, which may be recruited, anchored, transported, or modified in response to the interaction. The target could also be a specific sequence on the DNA, as in the case of transcription factors. Regardless of the nature of the

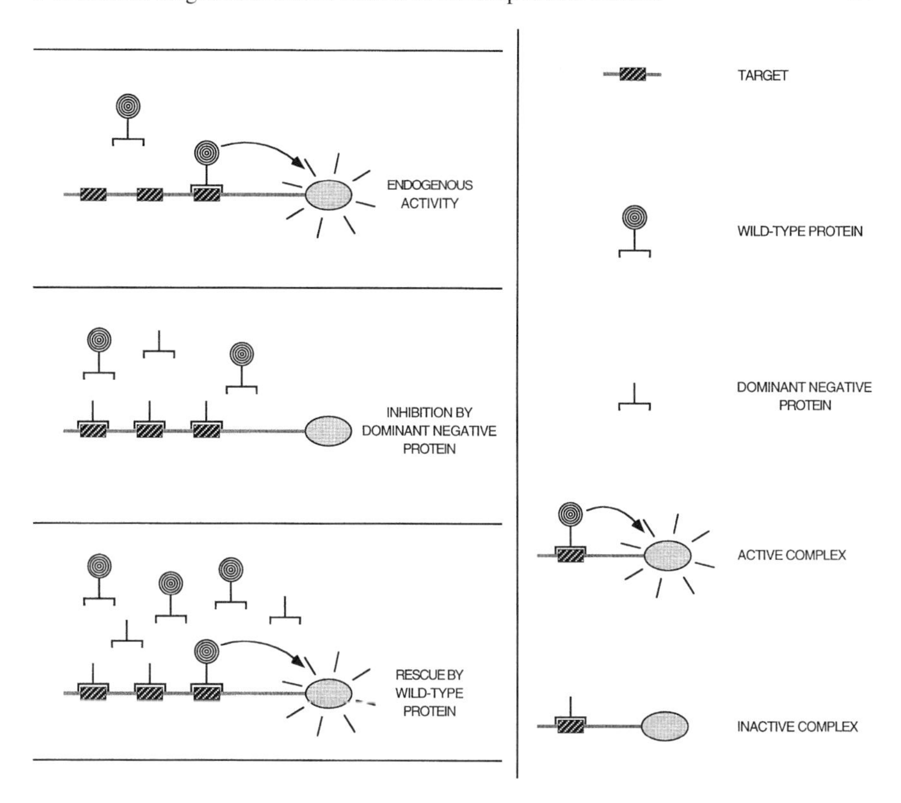

Fig. 2. Inhibition by titration of target. A hypothetical transcription factor is shown, composed of a DNA-binding domain (bracket) and an activation domain (concentric circles). This factor binds the target (stippled rectangle) on a promoter element and activates the basal transcription machinery (gray oval). (Top) Activation of transcription under normal (wild-type) conditions. (Middle) A dominant negative derivative of the same transcription factor saturates the target, interfering with the activity of the endogenous factor. (Bottom) Ectopic expression of wild-type factor rescues the inhibition by the dominant negative construct by competing for binding sites.

target, however, a mutant protein still able to bind it, but otherwise defective, could be inhibitory. As a typical example of downstream target inhibition, consider a monomeric transcription factor in which the activation domain has been deleted, leaving only the sequence-specific DNA-binding domain (Fig. 2). When overexpressed *in vivo*, such a derivative could saturate the target sites of the endogenous wild-type protein on the chromosomes, resulting in a dominant negative phenotype (see Section VI.D). In an example of titration of upstream targets, this method has been used in *Drosophila* to saturate the ligand of the tyrosine kinase receptor *torso*, by overexpressing a mutant nonsignaling variant of *torso* that is still able to bind the ligand (Sprenger and Nusslein-Volhard, 1992).

Because it does not require multimerization, the "titration of target" approach is more versatile than the "inhibition by multimerization" method described in Section III.B.1. On the other hand, it often entails the expression of higher doses of mutant derivative, generally proportional to the concentration of the target (when this is not limiting) rather than to that of the wild-type counterpart. As a consequence, this method works best in the cases in which the target is the limiting component in the interaction.

Importantly, special care should be taken in the choice of the controls, because the titration of a specific target might affect the activity of all the different proteins that use the same target.

Finally, it should be noted that, at least with transcription factors *in vitro*, an inhibitory effect due to titration of the target has been observed with the overexpression of the *wild-type* protein, a phenomenon often referred to as *squelching* (Ptashne, 1988). The explanation for this effect is provided in Section IV.A.3.

3. Inhibition by Active Repression

In studies that are treated in detail in Section VI.D, dominant negative derivatives of transcription factors have been obtained by fusion of the DNA-binding domain of the protein of interest with the repressor domain of the *Drosophila* Engrailed (En) homeodomain transcription factor. These constructs *actively* repress transcription from a distance, presumably interfering with the capacity of other positive factors bound to the promoter to contact the basal machinery and activate transcription (Fig. 3).

Accordingly, these repressors are more effective than simple deletion mutants containing only the DNA-binding domain (Han and Steinberg, 1991; Jaynes and O'Farrell, 1991). Although the applications of this approach to developmental systems remain few, it is tempting to speculate that its strength and specificity will lead to a flourishing of "active dominant negative" studies in the immediate future. The target proteins need not necessarily be transcription factors: for example, a kinase domain could be substituted by a phosphatase domain, or a glycosylase domain by a glycosydase domain, and so on.

IV. Interpretation of Dominant Negative Effects

A. Detecting Inhibitory Effects and Avoiding Artifacts

In the next sections, we mention some of the possible causes of misinterpretation and the controls that may help to avoid them. It is important to realize, however, that even the best controls may fail to unveil unexpected features of the biologic process under study, so that caution in drawing final conclusions from dominant negative experiments, as from any other genetics experiment, is strongly recommended.

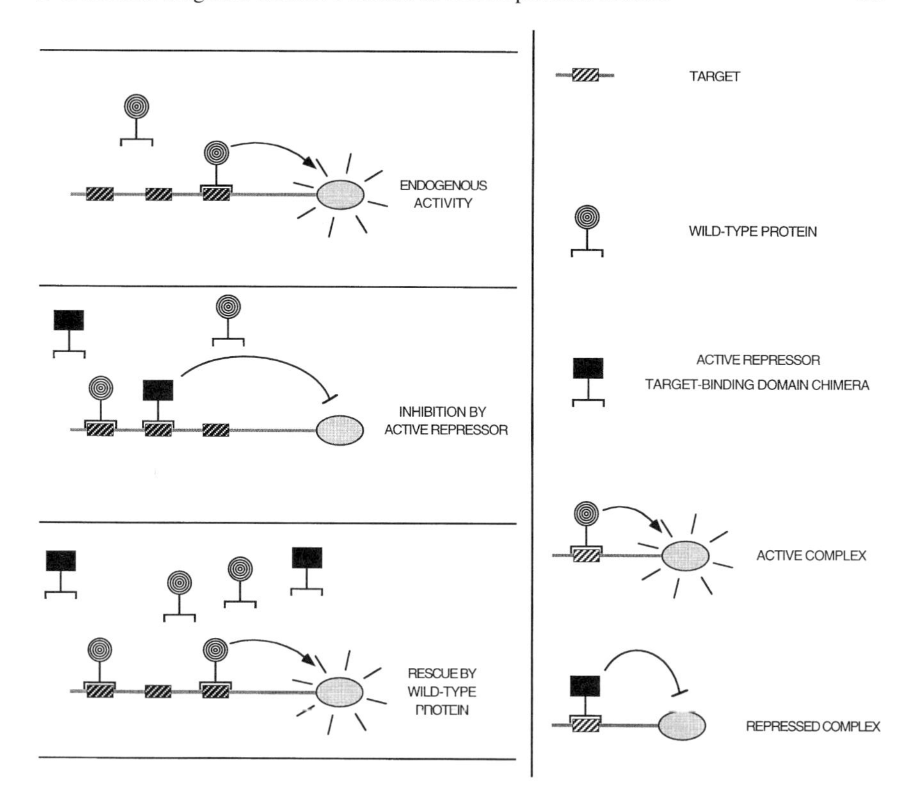

Fig. 3. Inhibition by active repression. This example depicts a variation on the theme displayed in Figure 2. (Top) Activation of transcription under normal (wild-type) conditions. (Middle) A chimeric protein in which the activation domain of the transcription factor has been substituted by a strong inhibitory domain (black square), binds the same binding site and actively represses transcription. (Bottom) The negative effect of the active repressor is eliminated by overexpression of the wild-type protein, which now competes for binding sites.

1. Toxicity

Whenever the expression in embryos of a dominant negative mutant protein results in the *lack* or *disappearance* of an organ, a tissue or a cell type, an immediate concern should be the potential toxicity of the inhibitory derivative.[1] The deletion or mutation of a protein domain, in fact, could alter the specificity of the variant for its target, rendering it more promiscuous with nonphysiologic partners. These atypical targets, in turn, could be involved in housekeeping cellular functions, so that their inhibition would result in nonspecific lethality or toxicity. A mutation could also affect the chemical–physical properties of a

[1]We are assuming that the molecule of interest is not being investigated for a potential role in apoptosis or cell growth, in which cases different criteria might be required.

polypeptide, for example causing it to become insoluble, as in the case of the partially dominant point mutation in the β-globin gene that causes sickle cell anemia (Ingram, 1956).

To eliminate these potential causes of artifacts, a control that has often proved useful is the *labeling* of the cells that express the dominant negative derivative. This can be achieved by coexpressing a reporter gene (such as β-galactosydase or green fluorescent protein) with the inhibitory mutant. The researcher can then monitor the amount of tracer present in samples treated or untreated with the dominant negative product when the effect of the mutant is scored. If the treatment has not killed or arrested the growth of the cells, comparable measurements should be obtained. As we shall see, using a tracer to follow cells that received the inhibitory derivative is also essential in cases of mosaic expression.

Finally, there might be toxic effects resulting solely in a block of differentiation, without altering the capacity of cells to survive or grow. To rule out these possibilities, rescue (see Sections IV.B.1 and IV.B.2) and specificity (see Section IV.B.3) experiments are often necessary.

2. Derepression

As our understanding of the processes that control development and differentiation progresses, we realize how many of these pathways are under negative control. In fact, proteins that act as repressors via direct interaction with target molecules have been described among secreted factors (e.g., see Nakamura *et al.*, 1990; Piccolo *et al.*, 1996; Zimmerman *et al.*, 1996), membrane-bound receptors (e.g., Eide *et al.*, 1996), as well as cytoplasmic and nuclear components of transduction systems (e.g., Benezra *et al.*, 1990). Overexpression of a protein derivative, therefore, could produce a phenotype not by inhibiting the effect of the wild-type counterpart, but rather by titrating away an inhibitor of the native protein. To illustrate this problem, we borrow an example first used by Herskowitz (1987). Suppose that the overexpression of a mutant derivative of a transcription factor causes the activation of a target gene. If the wild-type protein binds to the promoter of this gene, we might be tempted to surmise that it acts as a transcriptional repressor. But if the native protein was under negative regulation, the effect of the mutant derivative might have been the result of saturation of the interacting repressor. This could have unveiled the transcriptional activation potential of the wild-type protein, now free to pursue its regular function. Note that the observed result is perfectly consistent with both models, although the role inferred for the wild-type protein is opposite. Herskowitz (1987) concluded that in this or similar cases, additional experiments are required. We recommend a control that generally allows for obtaining a direct answer in these and other circumstances: *rescue* by the wild-type protein. As we shall see in a broader context (see Section IV.B.1), this experiment invokes the coexpression of dominant negative and wild-type proteins. In the example supplied previously, if the transcription factor were actually a repressor, then expression of the dominant negative protein would activate the target gene, whereas coexpression of both the mutated and the wild-type proteins would restore the repressed state of the promoter. Conversely, if the transcription factor were an activator whose function was blocked by a repressor, then the overproduction of the mutated derivative would also activate the target gene (by derepression of the endogenous native polypeptide), but coexpression of wild-type and mutant proteins would only *further increase* the amount of transcripts driven from the target promoter. Therefore, rescuing with the unmodified protein should supply a convenient tool to discriminate true inhibition by a dominant negative variant from derepression of the endogenous factor.[2]

3. Squelching

In 1988, Ptashne suggested that transcription activators act at a distance by binding and recruiting basal factors on the minimal promoter, looping out the DNA in between. Because this mechanism entails a direct protein–protein interaction between activator and basal factors, he predicted that, in a dose–response experiment, overexpression of an activator will first proportionally increase the rate of transcription of the target gene. At higher doses, however, overexpression will eventually result in a proportional repression by engaging the basal factors in sterile complexes away from the promoter, which is already saturated with activator molecules (Ptashne, 1988). These predictions have since been verified, at least *in vitro*. Here, we can extend this concept to any inhibitory effect resulting from expression of a gene (wild-type or mutated) to levels that titrate interacting factors into inactive complexes by stoichiometric overloading. This mechanism of repression differs from the "inhibition by titration of target," described in Section III.B.2, insofar as it cannot be rescued by coexpression of the dominant negative variant together with the wild-type protein.

To avoid this problem, overexpression of the inhibitory mutant should be carefully calibrated, so that only the amount sufficient to inhibit the wild-type counterpart is routinely used. This usually allows the investigator a wide margin of concentrations at which a rescue by coexpression of wild-type protein can be attempted.

4. Inhibition of Gene Families

As more and more genes are cloned and sequenced, it is becoming clear that factors that do *not* belong to gene families are exceptions to the rule. In attempt-

[2]Note that in this case, the simple overexpression of wild-type protein alone would also suffice in distinguishing between these two hypotheses, but it would fail to control for other artifacts, such as nonspecific derepression of *another* activator, in which case the rescue by the wild-type would not *further* increase transcription.

ing to inhibit the function of the protein of interest, then, it is impossible to neglect the existence (potential or ascertained) of several other closely related molecules in the same cell. Members of gene families, probably the result of ancient duplications and subsequent divergence, can fulfill still redundant or overlapping functions, or can have drifted toward very distinct roles in the organism. As studies of single gene "knockouts" have indicated in the mouse, often the lack of a phenotype is due to complementation of the missing gene by other structurally or functionally related factors (see Section VI.D for an example). In these instances, the use of a reagent that can inhibit a whole family of genes is certainly advantageous. Dominant negative derivatives have frequently filled this role. For example, a mutant transcription factor retaining the ability to bind DNA, but failing to activate transcription, can interfere with the function of all the factors that specifically recognize the same sequence (see Section VI.D). Once an entire family of factors has been inhibited by a dominant negative construct, the task becomes discerning which of its members are more directly involved in the loss of activity. Whenever this is essential, and most of the related genes have been already isolated, the investigator can attempt a rescue of the missing function with each wild-type family member, or combinations thereof. However, in cases in which this approach is not practical, the experimenter should at least provide evidence that the inhibition observed is specific to the gene family targeted and does not originate from toxicity.

B. The Best Controls

The previous sections have shown that some supplementary analyses are particularly useful in the interpretation of results obtained with dominant negative constructs. Following are three sections dealing with some of the most stringent controls: *rescue by the wild-type protein*, *rescue by a downstream factor*, and *assessment of specificity*.

1. Rescue by the Wild-Type Protein

The examples detailed previously have already illustrated the usefulness of eliminating artifactual explanations by a rescue experiment with the wild-type protein. Whenever this control works, it excludes simultaneously several arguments against the specificity of the effect triggered by the dominant negative derivative. For instance, toxicity, derepression, and squelching artifacts are eliminated. Specificity within a conserved family is also strongly suggested.

Because this experiment can be very beneficial, it is worth investing some time in its optimization. A suggested simple strategy is the following: first calibrate the minimum amount of dominant negative derivative required to obtain an appreciable inhibitory effect; second, coexpress this fixed amount of mutant with increasing doses of wild-type protein, until the repressed phenotype is completely reversed.

Note that there are cases in which rescue, even in the absence of artifacts, cannot be obtained. Consider, for example, a case in which the mutation introduced in a gene to produce the dominant negative derivative results in an increase of the protein's affinity (i.e., its association constant) for its partner–substrate. This phenomenon could occur as a result of deletion of an intramolecular regulatory domain, as in the case of the transcription factor Ets1. A derivative of Ets1 containing only the DNA-binding domain, in fact, interacts with its target sequence on the DNA much more avidly than the native counterpart (Petersen *et al.*, 1995). Therefore, when an amount of truncated derivative sufficient to inhibit endogenous functions is expressed in *Xenopus* embryos, it is impossible to revert the observed phenotype by coexpression of the wild-type protein, at least at concentrations of injected RNA that are tolerated by the embryo (G. Lagna and A. Hemmati-Brivanlou, unpublished observations). Another circumstance in which rescue by the wild-type protein might not be attainable is the described inhibition of an entire family of related factors (see Section IV.A.4). In these specific instances, controls of specificity and toxicity, wherever applicable, should be pursued.

2. Rescue by a Downstream Factor

Whenever the signal transduction pathway or the molecular cascade upstream and downstream of the mutagenized gene is known, the type of rescue described here is a viable approach to the elimination of artifacts. This control can be achieved by coexpression of the inhibitory construct together with a wild-type (or constitutively active) protein acting downstream of the mutated gene *in vivo*. For example, Yamauchi and colleagues (1994) showed that microinjection of fertilized mouse eggs with constructs expressing dominant negative *ras* inhibited subsequent development through the two-cell stage. The effect of the inhibitory *ras* derivative was overcome by simultaneous injection of a plasmid expressing an active *raf* oncogene, indicating that it resulted from interference with the Ras/Raf signaling pathway (Yamauchi *et al.*, 1994).

3. Assessment of Specificity

Depending on the different systems and processes studied, the investigator will recognize which experiments might more appropriately control for the specificity of the inhibitory reagent used. In general, the aim is to identify another developmental process, distinct from the one we want to inhibit, but also characteristic of the cells targeted, and verify that expression of the dominant negative derivative does not interfere with it. Ideally, the test and the control processes should be phenotypically closely related, but molecularly distinct. In the case of the described dominant negative activin type II receptor, dn-ActR-IIB, the use of this inhibitory derivative blocked mesoderm induction in *Xenopus* ectodermal explants by activin (Hemmati-Brivanlou and Melton, 1992). As control of specificity, it was shown that dn-ActR-IIB does not block FGF-mediated mesoderm induction (Hemmati-Brivanlou and Melton, 1994), which occurs through a *ras*/*raf*/MAPK pathway (Whitman and Melton, 1992).

V. How to Overproduce a Dominant Negative Mutant

The methods used for delivery of the mutagenized construct to the target cells are the same as those used for ectopic expression of any gene: they range from transfection (in cell culture), to lipofection and viral infection of DNA, to microinjection of RNA, DNA, and proteins in embryos. The transgenic technique has also been widely used in mice and flies. Here we stress only one aspect of the method of expression that can affect the final outcome: the choice of transgenic versus episomal expression.

If the aim is inhibiting an endogenous function, the investigator will wish to expand the expression of the dominant negative derivative to the maximum number of target cells. This purpose often can be achieved optimally by use of a transgenic technique, which usually takes advantage of strong, ubiquitous promoters. In some cases, tissue- and stage-specific promoters are also available, allowing a more directly targeted expression. For systems in which transgenic animals are not yet widely available, several alternatives exist. Virus-deriven expression has been used in chickens, whereas DNA or RNA injections have prevailed in *Xenopus* studies. The advantage of these methods of ectopic expression resides in their extreme rapidity. In the case of RNA injection in *Xenopus*, its further convenience is that the injected transcripts are added to the maternal pool of mRNA, when transcription of zygotic genes has not yet initiated. Its shortcoming lies in the restriction of its usefulness to the early developmental stages. As for viral or plasmid DNA vectors, their main disadvantage is mosaicism, because episomal DNA does not segregate properly in dividing cells. When only a fraction of cells express the construct, the investigator should attempt a mosaic analysis by labeling the cells that receive the inhibitory gene.

VI. Examples of Dominant Negative Mutants

A bewildering array of dominant negative studies has surfaced in the literature in recent years. Because commenting exhaustively on this large number of examples would be impossible, we have selected a few cases organized by *classes* of molecules. In the following paragraphs, these examples are arranged in the order of an ideally linear signal transduction pathway.

A. Soluble Ligands

In 1990, Mercola and coworkers reported the trans-dominant inactivation of wild-type platelet-derived growth factor A (PDGF-A) by coexpression of mutant derivatives incorporating either a processing lesion or a conserved cystein substitution (Mercola *et al.*, 1990). Because other growth factors, like PDGF, function as dimers, this strategy proved useful later for revealing the function of secreted ligands during early embryonic development. Members of the transforming growth factor β (TGF-β) superfamily have been the object of studies, beginning with the assessment of the role of activin in the induction of mesoderm in vertebrates (Wittbrodt and Rosa, 1994). For this purpose, two activin dominant negative variants were generated: one behaving as an inhibitor of pre-existing maternal activin protein, the second being able to inhibit only when cotranslated with wild-type activin. Injection of RNA encoding these variants in the embryo of the Japanese medaka fish (*Oryzias latipes*) enabled the authors to demonstrate that only the maternally provided activin protein is required for mesoderm formation *in vivo* (Wittbrodt and Rosa, 1994). Subsequent studies focused on another class of TGF-β-like molecules, the bone morphogenic proteins (BMPs). By altering the cleavage sites required for maturation of the active dimeric forms of BMP-7 and BMP-4, Hawley and collaborators (1995) generated inhibitory mutants that were able to dorsalize ventral mesoderm and induce neural tissue from naive ectoderm in *X. laevis* (Hawley *et al.*, 1995). This confirmed results obtained with the truncated BMP type IA receptor (Graff *et al.*, 1994; Suzuki *et al.*, 1994; Xu *et al.*, 1995b), lending further support to the default model of neural induction (Hemmati-Brivanlou and Melton, 1994).

B. Receptors

A vast numbers of studies have dealt with the cell-autonomous inhibition of signal transduction pathways by repression of their primary effectors: the receptors.

1. Tyrosine Kinase Receptors

The transmembrane tyrosine kinase receptors offered an easy target because their mechanism of action requires a ligand-induced homodimerization step, which results in the activation of the kinase domains and the subsequent chain of intracellular phosphorylation events (Kazlauskas, 1994). The first hints in favor of the dominant inhibitory strategy, however, came from the study of naturally occurring mutants. Consider the proto-oncogene *c-kit*, for example, encoding a tyrosine kinase receptor that is allelic with the murine white-spotting locus (W). The W42 allele of this gene has severe effects, even in the heterozygous state, on melanogenesis, gametogenesis, and hematopoiesis. Knowing that the W42 variant contains a missense mutation that abolishes its *in vitro* kinase activity, Ray and colleagues (1991) surmised that *c-kit*^{W42} could act as a dominant negative derivative. To test this hypothesis, they generated transgenic mice in which ectopic expression was driven by an ubiquitous promoter (human β-actin). The effects they observed on pigmentation and the number of tissue mast cells in the resulting mouse lines confirmed the dominant negative role of the W42 mutation (Ray *et al.*, 1991).

More recently, a truncated form of *Sek-1* (a member of the Eph-related receptor tyrosine kinase family) lacking the entire cytoplasmic kinase domain has been used in *Xenopus* and zebrafish to gain insights into the role of its wild-type counterpart in the establishment of rhombomeric identity (Xu *et al.*, 1995a). For the analysis of a PDGF receptor in *Xenopus*, instead, Ataliotis and collaborators (1995) have taken advantage of the lesson learned from the *c-kit* study described previously and have generated a point mutant of PDGFRα analogous to the W37 mutation isolated in mice. In embryos injected with RNA encoding this variant, the involuting mesodermal cells fail to adhere to the overlying ectoderm, pointing to a potential requirement of PDGF for mesodermal cell–substratum interaction (Ataliotis *et al.*, 1995).

Finally, among the tyrosine kinase receptors, the dominant negative FGF receptor construct has seen the most widespread use across the spectrum of vertebrate experimental systems. Thank to its use, the FGF pathway has been implicated in the formation of posterior mesoderm (Amaya *et al.*, 1991) and neural tissue (Launay *et al.*, 1996) in *Xenopus*, in terminal differentiation of limb muscle in chickens (Itoh *et al.*, 1996), and in differentiation of the fiber cells in the mouse lens (Chow *et al.*, 1995; Robinson *et al.*, 1995).

2. Serine–Threonine Kinase Receptors

Serine–threonine kinase receptors are also prone to inhibition by multimerization, as type I and type II receptors form hetero-oligomers in response to ligand binding (Massague, 1996). The dominant negative activin type II receptor dn-ActR-IIB, constructed by deletion of the cytoplasmic kinase domain, inhibits both the mesoderm-inducing activin pathway and the epidermis-inducing BMP pathway in *Xenopus* embryos (Hemmati-Brivanlou and Melton, 1992; Wilson and Hemmati-Brivanlou, 1995). This promiscuity, although usually inconvenient, was in this case instrumental in discovering that the formation of neuroectoderm in vertebrates is under inhibitory control, and ensues by default once TGF-β-like signals are removed (Hemmati-Brivanlou and Melton, 1994). Later studies with type I serine–threonine receptors took advantage of a higher specificity to pinpoint the role of BMPs to the inhibition of neural and dorsal mesodermal fates [BMPR-IA receptor; also known as Alk 3; (Graff *et al.*, 1994; Suzuki *et al*., 1994; Xu *et al*., 1995b)], and the role of activin-like signals to the induction of mesoderm (ActR-IB receptor, also known as Alk 4; Chang *et al*., 1997).

The use of a truncated BMPR-IB receptor (also known as Alk 6) allowed assessment of the requirement for BMP signaling in apoptosis of the interdigital membrane cells in the chick limb (Zou and Niswander, 1996). Decisive evidence for the involvement of BMPs in this process came from the correlation of two observations: first, that chick embryos infected with a virus encoding the mutated receptor displayed webbed feet; second, that duck embryos differed from chicken in that they did not express BMP-4 in the interdigital cells (Zou and Niswander, 1996).

2. Hormone Receptors

The retinoic acid receptors (RARs) have also been the object of dominant negative approaches to uncover the role of retinoids in embryogenesis. In 1994, Tsai and coworkers were able to immortalize murine lymphohematopoietic progenitor cells by a retroviral vector harboring a dominant negative RAR (Tsai *et al*., 1994). This study and a previous one (Tsai *et al*., 1992) also implicated RARs at multiple different stages of hematopoiesis. The involvement of RARs in skin development was revealed by a subsequent report (Saitou *et al*., 1995), in which the authors generated a dominant inhibitory RAR by a single aminoacid substitution and targeted its expression to the mouse epidermis.

C. Cytoplasmic Components of Signal Transduction Pathways

Dominant negative derivatives of this category of molecules have been widely used to assess whether a given pathway is involved in the phenomenon under study. Therefore, a dominant inhibitory *ras* mutant helped assign a role to the *ras/raf*/MAPK pathway in progression through the two-cell stage in mouse (Yamauchi *et al*., 1994), in the induction of mesoderm in *Xenopus* (Whitman and Melton, 1992), in *torso* signaling in *Drosophila* (Lu *et al*., 1993), and in formation of the vulva in *C. elegans* (Han and Sternberg, 1991), among others.

Studies on the role of the Wnt pathway in vertebrates have also benefited from the generation of an inhibitory mutant: a dominant negative glycogen synthase kinase 3 (GSK-3) was constructed by two different groups (He *et al*., 1995; Pierce and Kimelman, 1995). This gene is the vertebrate homologue of *Drosophila shaggy* (*zeste-white 3*), a serine–threonine kinase required for signaling by *wingless*. When RNA encoding dominant negative GSK-3 was injected into *Xenopus* embryos, it mimicked the ability of Wnt to induce a secondary dorsal axis. Conversely, wild-type GSK-3 induced ventralization, suggesting that dorsal differentiation may involve the suppression of GSK-3 activity by a wingless–Wnt-related signal (He *et al*., 1995; Pierce and Kimelman, 1995).

The use of a dominant negative DPC4, an antioncogene deleted in pancreatic carcinomas, has allowed a role to be assigned to this member of the Smad family of TGF-β signal transducers in pathways that signal mesoderm induction and patterning in *Xenopus* embryos, as well as antimitogenic and transcriptional responses in mammalian cells (Lagna *et al.*, 1996).

D. Transcription Factors

A multitude of studies have used dominant negative transcription factors. Only a few representative examples are mentioned here.

In 1996, Mead and colleagues reported that injection of mRNA encoding Mix.1, a paired class homeoprotein, results in ventralization of mesoderm in *Xenopus* embryos. Knowing that Mix.1 forms homodimers, the authors generated a dominant inhibitory variant by substituting a single aminoacid in the turn connecting α-helixes II and III of the homeodomain. The resulting mutant binds DNA with approximately 100-fold lower affinity than the wild-type, but is still able to dimerize with the native protein and decrease its ability to bind DNA. When overexpressed in embryos, the inhibitory mutant produced the opposite phenotype of the wild-type, leading to dorsalization of mesoderm and blocking the activity of BMP-4, a ventralizing growth factor. As a result of coexpression, the effects of the mutant and the native proteins canceled each other (Mead *et al.*, 1996). In this example of "inhibition by multimerization," the mutant derivative interacts directly with the wild-type protein and interferes with its ability to bind DNA. An alternative approach was used by Chong and collaborators (1995) to study REST, a zinc finger protein that binds to the silencer element of the type II voltage-dependent sodium channel gene. The silencer element, which is active in non-neuronal cells, is held responsible for restricting expression of the channel gene to neurons. Misexpression of recombinant REST confers the ability to silence type II reporter genes to neurons that normally lack REST. Conversely, the expression of a dominant negative REST, constructed by deleting all but the DNA-binding zinc finger domain relieved the silencing of the type II reporter by the wild-type REST in non-neuronal cells (Chong *et al.*, 1995). This example illustrates the "inhibition by titration of target" approach: dominant negative REST competes with the wild-type protein for the DNA site in the target promoter. Note that, because the REST mutant *induces* the expression of the target gene, a control to investigate a potential "derepression" phenomenon (see Section IV.A.2) is essential. This is supplied by the expression of wild-type REST in neuronal cells, with the resulting *repression* of the target gene.

We mentioned in Section IV.A.4 that dominant negative derivatives can often inhibit a whole family of genes. In this respect, they can occasionally outperform targeted mutagenesis approaches. Consider, for example, the case of the cAMP response element binding protein (CREB). Homozygous null mice for this basic–

leucine zipper transcription factor appear healthy and exhibit no impairment of growth or development (Hummler *et al.*, 1994). The authors of this study noted, however, that a compensatory change had occurred in the level of at least one related protein, the cAMP response element modulation factor. Previous and later reports, therefore, made use of transgenic mice expressing a dominant negative CREB to assess the role of its *family* of related factors in specific tissue contexts (Barton *et al.*, 1996; Struthers *et al.*, 1991). Both groups used a CREB derivative that cannot be phosphorylated (because of the substitution of an alanine for a serine), and is therefore transcriptionally inactive. Expressed under the control of anterior pituitary-specific (Struthers *et al.*, 1991) and T-cell-specific (Barton *et al.*, 1996) promoters, this mutant produced transgenic mice exhibiting a dwarf phenotype with atrophied pituitary glands (Struthers *et al.*, 1991) or animals with T cells displaying profound proliferative defects in response to different activation signals (Barton *et al.*, 1996), respectively. Both studies relate observations that testify to the specificity of the effects detected: pituitary glands expressing dominant negative CREB are deficient in somatotrophic but not other cell types (Struthers *et al.*, 1991), whereas T cells expressing the CREB mutant develop normally, but fail to respond properly to stimulation (Barton *et al.*, 1996).

In Section III.B.3, we first met the concept of "inhibition by active repression." To date, it has mainly used the described Engrailed (En) repressor domain fused to the transcription factor of interest. Like repression domains identified in the *Drosophila* repressors Even-skipped and Krüppel, the En repression domain is rich in alanine residues (26%) and can be transferred to a heterologous DNA-binding domain (Jaynes and O'Farrell, 1991). Importantly, En was found to act as an active repressor in cell culture studies, blocking activation by mammalian and yeast activators that bind to sites some distance away from those bound by En (Han and Sternberg, 1991; Jaynes and O'Farrell, 1991). In 1994, Badiani and colleagues took advantage of these features to investigate the role of the proto-oncogene c-Myb in T-cell development. Using a T-cell-specific promoter to drive the expression of the constructs, the authors generated transgenic mice with two dominant interfering Myb alleles. The first encoded a "classic" competitive inhibitor of DNA binding, consisting only of the DNA-binding domain of Myb. The second was an active repressor comprising the Myb DNA-binding domain linked to the En repressor domain. Both alleles partially blocked thymopoiesis and inhibited proliferation of mature T cells, but the Myb–En chimera was 50- to 100-fold more potent than the competitive inhibitor construct in generating a strong phenotype (Badiani *et al.*, 1994).

E. Adhesion Molecules

Several of the gene products involved in cell–cell communication and adhesion are transmembrane multimeric proteins with interacting partners inside and outside the cell. Given their characteristics, these polypeptides are optimal targets for dominant inhibitory strategies. Among the embryologic studies on cadherins, which are primary cell–cell adhesion molecules, we mention only a few examples, starting from the work of Levine and colleagues (1994). These authors disrupted E-cadherin function in *Xenopus* embryos through the expression of a truncated E-cadherin mutant lacking the cytoplasmic domain. Overexpression of the dominant inhibitory version caused lesions to develop in the ectoderm during gastrulation, suggesting a potential role of the wild-type protein in maintaining the integrity of the ectoderm during epiboly in the gastrulating embryo. This defect was rescued by coexpression of wild-type E-cadherin, but not by C-cadherin, a close homologue. Furthermore, a similarly truncated N-cadherin failed to cause lesions, pointing to the specificity of the effect scored (Levine *et al.*, 1994). This same group later focused on the role of C-cadherin during tissue morphogenesis, using a similarly constructed deletion mutant (Lee and Gumbiner, 1995). Finally, a role for N-cadherin in cell communication in early *Xenopus* embryos was indicated by the study of Holt and coworkers (1994). This group had previously suggested that muscle progenitors in *Xenopus* interact in a community of 100 or more cells to activate myogenic genes and the muscle differentiation pathway. To study the involvement of cadherins in this process, dominant negative N-cadherin RNA was injected into the region of *Xenopus* embryos that gives rise to muscle. While a pan-mesodermal marker was still present, the treatment suppressed expression of MyoD, a master regulator of muscle differentiation, in muscle progenitor cells. Because MyoD inhibition in embryos injected with the dominant negative cadherin mRNA was rescued by coinjection of full-length cadherin RNA, the authors concluded that the inhibition of MyoD occurs through the cadherin pathway and that cadherin-mediated cell interactions play a role in the signaling events required for muscle progenitor cells to differentiate (Holt *et al.*, 1994).

N-cadherin was also the subject of a subsequent study performed in transgenic mice. Hermiston and Gordon (1995) expressed a dominant negative N-cadherin mutant in adult chimeric mice under the control of an intestinal-specific promoter. The phenotype observed was an inflammatory bowel disease resembling a human condition, Crohn's disease, that eventually lead to the formation of adenomas. Thus, this work established a model system to investigate the factors involved in inflammatory disease and intestinal neoplasia (Hermiston and Gordon, 1995).

G. Naturally Occurring Dominant Negative Mutants

As mentioned previously, several dominant inhibitory variants have been discovered through genetic screens. Occasionally, hereditary human diseases have also been linked to mutation producing dominant negative protein derivatives. Radovick and collaborators (1992), for example, in a study on Pit-1, a pituitary-specific POU-homeodomain transcription factor responsible for pituitary development and hormone expression in mammals, reported that mutations in the *Pit-1* gene were found in two dwarf mouse strains displaying reduction of hormone-secreting cells in the anterior pituitary. Furthermore, a point mutation in *Pit-1* was identified on only one allele in a patient with combined pituitary hormone deficiency. Consistently, mutant Pit-1 was still able to bind DNA normally, but it acted as a dominant inhibitor of Pit-1 action in the pituitary (Radovick *et al.*, 1992).

In some cases, it appears that naturally occurring dominant negative mutants have been adopted by the cell to modulate the function of its fully functional counterparts. For example, C-terminally truncated forms of *trkB*, the gene encoding the brain-derived neurotrophic factor receptor, have been identified *in vivo* and shown to act as dominant negative variants (Eide *et al.*, 1996).

VII. Concluding Remarks

Throughout this chapter, we have seen that the molecular embryologist is often faced with the necessity to inactivate a specific gene function *in vivo*. To this end, two methods have been consistently used in the most recent years: gene inactivation by homologous recombination ("knockout") and dominant negative approaches. The first is highly specific, but slow and applicable only to a limited number of systems. The second is very rapid and versatile, but requires appropriate controls to avoid misinterpretations. In general, a phenotype caused by a dominant inhibitory mutant should be rescued by the wild-type protein. Under these circumstances, the conclusions drawn from the inhibition experiments are usually reliable. We mention here only one exception to the "rescue" rule: the inhibition of entire gene families. The ability of dominant inhibitory variants to interfere with *classes* of related factors, in fact, can be seen as an advantage, rather than a disadvantage of this technique over the knockout strategy. Instead of inactivating one by one the genes thought to participate in a function, a process that at the current speed of knockout generation can take several years, the researcher has the alternative of gaining rapid insights into the role of entire gene families through the dominant negative approach. This shortcut, however, cannot be taken without caution: the experimenter has to provide stringent controls of specificity, lack of toxicity, rescue by downstream factors, and so on. If these criteria are met, the already vast collection of studies using dominant negative constructs *in vivo* is bound to increase exponentially in the near future, thanks to the foreseeable development of novel tools (such as active repressors) and the application of the proven ones to new tasks.

Acknowledgments

We would like to thank Akiko Hata, Paul Wilson, and Peter Model for critical reading of the manuscript. We are also grateful to Chenbei Chang, Atsushi Suzuki, Daniel Weinstein, Curtis Altman, and Francesca Carnevali for many engaging discussions. G. L., a doctoral student on leave of absence from the Università "La Sapienza" of Rome, is partially supported by the Rockefeller University. A. H. B. is supported by an NIH grant (ROI-HD32105-01A2). A. H. B. also gratefully acknowledges funding provided by the Searle Scholar Awards and the Klingestein and the McKnight foundations.

References

Amaya, E., Musci, T. J., and Kirschner, M. W. (1991). Expression of a dominant negative mutant of the FGF receptor disrupts mesoderm formation in *Xenopus* embryos. *Cell* **66,** 257–270.

Ataliotis, P., Symes, K., Chou, M. M., Ho, L., and Mercola, M. (1995). PDGF signalling is required for gastrulation of *Xenopus laevis*. *Development* **121,** 3099–3110.

Badiani, P., Corbella, P., Kioussis, D., Marvel, J., and Weston, K. (1994). Dominant interfering alleles define a role for c-Myb in T-cell development. *Genes Dev.* **8,** 770–782.

Barton, K., Muthusamy, N., Chanyangam, M., Fischer, C., Clendenin, C., and Leiden, J. M. (1996). Defective thymocyte proliferation and IL-2 production in transgenic mice expressing a dominant-negative form of CREB. *Nature* **379,** 81–85.

Benezra R., Davis R. L., Lockshon D., Turner D. L., and Weintraub H. (1990). The protein Id: a negative regulator of helix-loop-helix DNA binding proteins. *Cell* **61,** 49–59.

Chang, C., Wilson, P. A., Mathews, L. S. and Hemmati-Brivanlou, A. (1997). A *Xenopus* type I activin receptor mediates mesodermal but not neural specification during embryogenesis. *Development* **124,** 827–837.

Chong, J. A., Tapia-Ramirez, J., Kim, S., Toledo-Aral, J. J., Zheng, Y., Boutros, M. C., Altshuller, Y. M., Frohman, M. A., Kraner, S. D., and Mandel, G. (1995). REST: A mammalian silencer protein that restricts sodium channel gene expression to neurons. *Cell* **80,** 949–957.

Chow, R. L., Roux, G. D., Roghani, M., Palmer, M. A., Rifkin, D. B., Moscatelli, D. A., and Lang, R. A. (1995). FGF suppresses apoptosis and induces differentiation of fibre cells in the mouse lens. *Development* **121,** 4383–4393.

Eide, F. F., Vining, E. R., Eide, B. L., Zang, K., Wang, X. Y., and Reichardt, L. F. (1996). Naturally occurring truncated trkB receptors have dominant inhibitory effects on brain-derived neurotrophic factor signaling. *J. Neurosci.* **16,** 3123–3129.

Graff, J. M., Thies, R. S., Song, J. J., Celeste, A. J., and Melton, D. A. (1994). Studies with a *Xenopus* BMP receptor suggest that ventral mesoderm-inducing signals override dorsal signals in vivo. *Cell* **79,** 169–179.

Han, M., and Sternberg, P. W. (1991). Analysis of dominant-negative mutations of the *Caenorhabditis elegans* let-60 ras gene. *Genes Dev.* **5,** 2188–2198.

Hawley, S. H., Wunnenberg-Stapleton, K., Hashimoto, C., Laurent, M. N., Watabe, T., Blumberg, B. W., and Cho, K. W. (1995). Disruption of BMP signals in embryonic *Xenopus* ectoderm leads to direct neural induction. *Genes Dev.* **9,** 2923–2935.

He, X., Saint-Jeannet, J. P., Woodgett, J. R., Varmus, H. E., and Dawid, I. B. (1995). Glycogen synthase kinase-3 and dorsoventral patterning in *Xenopus* embryos [published erratum appears in *Nature* 1995 May 18;375(6528):253]. *Nature* **374,** 617–622.

Hemmati-Brivanlou, A., and Melton, D. A. (1992). A truncated activin receptor inhibits mesoderm induction and formation of axial structures in *Xenopus* embryos [see comments]. *Nature* **359,** 609–614.

Hemmati-Brivanlou, A., and Melton, D. A. (1994). Inhibition of activin receptor signaling promotes neuralization in *Xenopus*. *Cell* **77,** 273–281.

Hermiston, M. L., and Gordon, J. I. (1995). Inflammatory bowel disease and adenomas in mice expressing a dominant negative N-cadherin. *Science* **270,** 1203–1207.

Herskowitz, I. (1987). Functional inactivation of genes by dominant negative mutations. *Nature* **329,** 219–222.

Holt, C. E., Lemaire, P., and Gurdon, J. B. (1994). Cadherin-mediated cell interactions are necessary for the activation of MyoD in *Xenopus* mesoderm. *Proc Natl Acad Sci USA* **91,** 10844–10848.

Hummler, E., Cole, T. J., Blendy, J. A., Ganss, R., Aguzzi, A., Schmid, W., Beermann, F., and Schutz, G. (1994). Targeted mutation of the CREB gene: Compensation within the CREB/ATF family of transcription factors. *Proc. Natl. Acad. Sci. U.S.A.* **91,** 5647–5651.

Ingram, V. M. (1956). A specific chemical difference between the globins of normal human and sickle-cell anemia hemoglobin. *Nature* **178,** 792–794.

Itoh, N., Mima, T., and Mikawa, T. (1996). Loss of fibroblast growth factor receptors is necessary for terminal differentiation of embryonic limb muscle. *Development* **122,** 291–300.

Jaynes, J. B., and O'Farrell, P. H. (1991). Active repression of transcription by the engrailed homeodomain protein. *EMBO J.* **10,** 1427–1433.

Kazlauskas, A. (1994). Receptor tyrosine kinases and their targets. *Curr. Opin. Genet. Dev.* **4,** 5–14.

Lagna, G., Hata, A., Hemmati-Brivanlou, A., and Massagué, J. (1996). Partnership between DPC4 and SMAD proteins in TGFβ signalling pathways. *Nature* **383,** 832–836.

Launay, C., Fromentoux, V., Shi, D. L., and Boucaut, J. C. (1996). A truncated FGF receptor blocks neural induction by endogenous *Xenopus* inducers. *Development* **122,** 869–880.

Lee, C. H., and Gumbiner, B. M. (1995). Disruption of gastrulation movements in *Xenopus* by a dominant-negative mutant for C-cadherin. *Dev. Biol.* **171,** 363–373.

Levine, E., Lee, C. H., Kintner, C., and Gumbiner, B. M. (1994). Selective disruption of E-cadherin function in early *Xenopus* embryos by a dominant mutant. *Development* **120,** 901–990.

Lu, X., Chou, T. B., Williams, N. G., Roberts, T., and Perrimon, N. (1993). Control of cell fate determination by p21ras/Ras1, an essential component of torso signaling in *Drosophila. Genes Dev.* **7,** 621–632.

Massague, J. (1996). TGFβ signaling: Receptors, transducers, and Mad proteins. *Cell* **85,** 947–950.

Mead, P. E., Brivanlou, I. H., Kelley, C. M., and Zon, L. I. (1996). BMP-4-responsive regulation of dorsal-ventral patterning by the homeobox protein Mix.1. *Nature* **382,** 357–360.

Mercola, M., Deininger, P. L., Shamah, S. M., Porter, J., Wang, C. Y., and Stiles, C. D. (1990). Dominant-negative mutants of a platelet-derived growth factor gene. *Genes Dev.* **4,** 2333–2341.

Muller, H. J. (1932). Further studies on the nature and causes of gene mutations. *In* "Proceedings of the Sixth International Congress of Genetics" (D. F. Jones, ed.), pp. 213–255. Brooklyn Botanic Gardens, Menasha, Wisconsin.

Muller-Hill, B., Crapo, L., and Gilbert, W. (1968). Mutants that make more lac repressor. *Proc. Natl. Acad. Sci. U.S.A.* **59,** 1259–1264.

Nakamura, T., Takio, K., Eto, Y., Shibai, H., Titani, K., and Sugino, H. (1990). Activin-binding protein from rat ovary is follistatin. *Science* **247,** 836–838.

Oppenheim, A. B., and Salomon, D. (1970). Studies on partially virulent mutants of lambda bacteriophage: I. Isolation and general characterization. *Virology* **41,** 151–159.

Oppenheim, A. B., and Salomon, D. (1972). Studies can partially virulent mutants of lambda bacteriophage: II. The mechanism of overcoming repression. *Mol. Gen. Genet.* **115,** 101–114.

Petersen, J. M., Skalicky, J. J., Donaldson, L. W., McIntosh, L. P., Alber, T., and Graves, B. J. (1995). Modulation of transcription factor Ets-1 DNA binding: DNA-induced unfolding of an alpha helix. *Science* **269,** 1866–1869.

Piccolo, S., Sasai, Y., Lu, B., and De Robertis, E. M. (1996). Dorsoventral patterning in *Xenopus* : Inhibition of ventral signals by direct binding of chordin to BMP-4. *Cell* **86,** 589–598.

Pierce, S. B., and Kimelman, D. (1995). Regulation of Spemann organizer formation by the intracellular kinase Xgsk-3. *Development* **121,** 755–765.

Ptashne, M. (1988). How eukaryotic transcriptional activators work. *Nature* **335,** 683–689.

Radovick, S., Nations, M., Du, Y., Berg, L. A., Weintraub, B. D., and Wondisford, F. E. (1992). A mutation in the POU-homeodomain of Pit-1 responsible for combined pituitary hormone deficiency. *Science* **257,** 1115–1118.

Ray, P., Higgins, K. M., Tan, J. C., Chu, T. Y., Yee, N. S., Nguyen, H., Lacy, E., and Besmer, P. (1991). Ectopic expression of a c-kitW42 minigene in transgenic mice: Recapitulation of W phenotypes and evidence for c-kit function in melanoblast progenitors. *Genes Dev.* **5,** 2265–2273.

Robinson, M. L., MacMillan-Crow, L. A., Thompson, J. A., and Overbeek, P. A. (1995). Expression of a truncated FGF receptor results in defective lens development in transgenic mice. *Development* **121,** 3959–3967.

Saitou, M., Sugai, S., Tanaka, T., Shimouchi, K., Fuchs, E., Narumiya, S., and Kakizuka, A. (1995). Inhibition of skin development by targeted expression of a dominant-negative retinoic acid receptor [see comments]. *Nature* **374,** 159–162.

Sprenger, F., and Nusslein-Volhard, C. (1992). Torso receptor activity is regulated by a diffusible ligand produced at the extracellular terminal regions of the *Drosophila* egg. *Cell* **71,** 987–1001.

Struthers, R. S., Vale, W. W., Arias, C., Sawchenko, P. E., and Montminy, M. R. (1991). Somatotroph hypoplasia and dwarfism in transgenic mice expressing a non-phosphorylatable CREB mutant. *Nature* **350,** 622–624.

Suzuki, A., Theis, R. S., Yamaji, N., Song, J. J., Wozney, J., Murakami, K., and Ueno, N. (1994). A truncated BMP receptor affects dorsal-ventral patterning in the early *Xenopus* embryo. *Proc. Natl. Acad. Sci. U.S.A.* **91,** 10255–10259.

Tsai, S., Bartelmez, S., Heyman, R., Damm, K., Evans, R., and Collins, S. J. (1992). A mutated retinoic acid receptor-alpha exhibiting dominant-negative activity alters the lineage development of a multipotent hematopoietic cell line. *Genes Dev.* **6,** 2258–2269.

Tsai, S., Bartelmez, S., Sitnicka, E., and Collins, S. (1994). Lymphohematopoietic progenitors immortalized by a retroviral vector harboring a dominant-negative retinoic acid receptor can recapitulate lymphoid, myeloid, and erythroid development. *Genes Dev.* **8,** 2831–2841.

Vize, P. D., Melton, D. A., Hemmati-Brivanlou, A., and Harland, R. M. (1991). Assays for gene function in developing *Xenopus* embryos. *Methods Cell Biol.* **36,** 367–387.

Whitman, M., and Melton, D. A. (1992). Involvement of p21ras in *Xenopas* mesoderm induction. *Nature* **357,** 252–254.

Wilson, P. A., and Hemmati-Brivanlou, A. (1995). Induction of epidermis and inhibition of neural fate by Bmp-4. *Nature* **376,** 331–333.

Wittbrodt, J., and Rosa, F. M. (1994). Disruption of mesoderm and axis formation in fish by ectopic expression of activin variants: the role of maternal activin. *Genes Dev.* **8,** 1448–1462.

Xu, Q., Alldus, G., Holder, N., and Wilkinson, D. G. (1995a). Expression of truncated Sek-1 receptor tyrosine kinase disrupts the segmental restriction of gene expression in the *Xenopus* and zebrafish hindbrain. *Development* **121,** 4005–4016.

Xu, R. H., Kim, J., Taira, M., Zhan, S., Sredni, D., and Kung, H. F. (1995b). A dominant negative bone morphogenetic protein 4 receptor causes neuralization in *Xenopus* ectoderm. *Biochem. Biophys. Res. Commun.* **212,** 212–219.

Yamauchi, N., Kiessling, A. A., and Cooper, G. M. (1994). The Ras/Raf signaling pathway is required for progression of mouse embryos through the two-cell stage. *Mol. Cell Biol.* **14,** 6655–6662.

Zimmerman, L. B., De Jesus-Escobar, J. M., and Harland, R. M. (1996). The Spemann organizer signal noggin binds and inactivates bone morphogenetic protein 4. *Cell* **86,** 599–606.

Zou, H., and Niswander, L. (1996). Requirement for BMP signaling in interdigital apoptosis and scale formation. *Science* **272,** 738–741.

5

The Use of Embryonic Stem Cells for the Genetic Manipulation of the Mouse

Miguel Torres
Departamento de Inmunología y Oncología
Centro Nacional de Biotecnología
Campus Universidad Autónoma
Madrid 28049, España

I. Introduction

Embryonic stem (ES) cells are undifferentiated totipotent cells. They derive from the inner cell mass of the mouse blastocyst and can be perpetuated in culture (Robertson, 1987). Chimeric embryos can be generated by injecting ES cells into blastocysts or by aggregating them with morulae. Adult mice develop normally from these embryos and ES cells contribute to all tissues, including the germline (Robertson, 1986; Gossler *et al.*, 19886). These unique characteristics allow the introduction of foreign DNA, the screening for rare integration events in ES cells, and the generation of ES cell-derived mouse strains. This strategy has been successfully used for gene targeting by homologous recombination, which has become a powerful tool to study gene function. By this procedure, designed mutations are introduced into particular genes, and mouse lines carrying the mutated allele are generated (Capecchi, 1989). In addition, ES cells have been used for the random mutagenesis of the mouse genome by the insertion of gene and enhancer trap vectors (Gossler *et al.*, 1989). Gene trap allows the isolation, mutation, and monitoring of the expression of randomly targeted genes (Skarnes *et al.*, 1992). In this chapter, methods are described for the design and isolation of gene-targeted and gene-trapped ES cell lines. In addition, morula aggregation, a procedure alternative to the classic microinjection of ES cells into the blasto-

Current Topics in Developmental Biology, Vol. 36

0070-2153/98 $25.00

cysts, is described, and its performance is discussed in comparison with microinjection methods.

II. Materials and Instrumentation

Dulbecco's modified Eagle's medium (DMEM, 4.5 g glucose/l, cat. No. 041-01965H), nonessential amino acids (cat. No. 043-01140H), sodium pyruvate (cat. no. 043-01360H), fetal calf serum (FCS, cat. No. 011-06290M), geneticin (G418, cat. No. 066-1811), and leukaemia inhibitory factor (LIF, cat. No. 6203275SA or SB) are purchased from Gibco BRL (Life Technologies GmbH, Eggenstein, Germany). Ethylenediaminetetraacetic acid (EDTA, cat. No. 4040), NaCl (cat. No. 0278), KCl (cat. No. 0509), KH_2PO_4 (cat. No. 4008), $Na_2HPO_4 \cdot 2H_2O$ (cat. No. 0326), Trizma base (cat. No. 1414), and D^+glucose (cat. No. 0114) are purchased from J. T. Baker (Baker Chemikalien, Groβ-Gerau, Germany). Gelatin (cat. No. G1890), dimethyl sulfoxide (DMSO, cat. No. D 8414), mitomycin C (cat. No. M 0503), sodium dodecylsulfate (SDS, cat. No. L4509), and phenol red (cat. No. P 5530) are purchased from Sigma (Sigma Chemie GmbH, Deisenhofen, Germany). Trypsin (cat. No. L 2103) is purchased from Biochrom KG (Berlin, Germany). Proteinase K (cat. No. 1092766) is purchased from Böhringer Mannheim (Mannheim, Germany). Gene Pulser and electroporation cuvettes (0.4 cm and 0.8-ml volume) are purchased from Bio-Rad Laboratories GmbH (München, Germany). Sterile scissors and forceps, 50-ml sterilized glass beads (3-mm diameter), stirring bar in 500-ml Erlenmeyer flask, tissue culture dishes are also necessary.

III. Protocols

A. Solutions

1. Phosphate-buffered saline (PBS): To prepare 1 liter, dissolve 8.0 g of NaCl, 0.2 g of KC1, 1.15 g of $Na_2HPO_4 \cdot 2H_2O$ and 0.2 g of KH_2PO_4 in distilled water, adjust pH to 7.2, and bring to a total volume of 1 liter. Sterilize by autoclaving and store at room temperature.
2. Trypsin–EDTA: To prepare 1 liter, dissolve 8.0 g NaCl, 0.40 g KCl, 0.10 g $Na_2HPO_4 \cdot 2H_2O$, 1.0 g glucose, 3.0 g Trizma base, 0.01 g phenol red, and 2.50 g Trypsin (Difco 1:250) in distilled water, adjust pH to 7.6, and bring a total volume of 1 liter. Filter sterilize and store in aliquots at −20°C. This stock is diluted 1:4 in saline–EDTA for use as 0.05% Trypsin–EDTA solution. Store at −20°C in aliquots of 5–10 ml.
3. Saline–EDTA for dilution of Trypsin: To prepare 1 liter, dissolve 0.2 g EDTA (disodium salt), 8.0 g NaCl, 0.2 g KCl, 1.15 g $Na_2HPO_4 \cdot 2H_2O$, and 0.2 g

KH_2PO_4 in distilled water. Check pH (7.2) and bring to a total volume of 1 liter, filter sterilize or autoclave. Store at room temperature.

4. Gelatin: To prepare 1 liter, dissolve 1 g in 1 liter of distilled water and autoclave. Mix after autoclaving and store at room temperature.

5. Mitomycin C: Dissolve 2 mg of mitomycin C in 2 ml of PBS and filter sterilize. Keep in the dark at 4°C and use within a week.

6. Medium for embryonic fibroblasts: DMEM + 10% (v/v) FCS. To make 1 liter, dissolve 100 ml of FCS (the serum is always heat inactivated at 56°C for 30 min before use for all media) to 900 ml of DMEM. Store at 4°C.

7. Freezing medium for fibroblasts and ES cells: To prepare 10 ml of freezing medium, dissolve 1 ml DMSO (cell culture grade) and 1 ml FCS to 8 ml of ES cell or fibroblast medium. After adding freezing medium, the cells are kept for 1 hr at −20°C and transferred to −70°C overnight before being stored in liquid nitrogen.

8. ES cell medium: DMEM (4.5 g/l glucose); β-mercaptoethanol, 10^{-4} *M*; glutamine, 2 m*M*; nonessential amino acids, 1% of stock solution; Na-pyruvate, 1 m*M*; FCS, 15% (v/v) (tested batches); LIF, 1000 units/ml. To prepare 500 ml, add 5 ml of glutamine, 5 ml of nonessential amino acids, 5 ml of sodium pyruvate, 5×10^5 units of LIF, and 0.5 ml of β-mercaptoethanol to DMEM. Adjust to 500 ml with DMEM and add 89 ml of FCS. Store at 4°C.

9. β-Mercaptoethanol stock: To prepare 20 ml, add 140 μl of β-mercapto-ethanol (14 *M*) to 20 ml of PBS, mix well, and filter sterilize. Store in 1-ml stocks at −20°C. When thawed, store at 4°C in the dark and use within 1 week.

10. Selection media

a. Neomycin phosphotransferase positive selection medium: Dissolve G418 (Gibco) in PBS at a concentration of 250 mg/ml, filter sterilize, and store in small aliquots (1 ml) at −20°C. Thaw before use and add to the selection medium. G418 potency varies between batches and should be tested empirically with untransfected ES cells. The appropriate G418 concentration is usually between 200 and 350 μg/ml medium (100–175 μg active G418/ml medium). Ideally, control ES cells should die after 7 days and G418-resistant colonies will form by day 8–10, depending on the promoter driving neoexpression.

b. Thymidine kinase negative selection medium: Dissolve gancyclovir at 2 m*M* in PBS, aliquot and freeze at −20°C. Dilute 1:1000 into selection medium to achieve a final concentration of 2 μ*M*.

11. Lysis buffer for genomic DNA preparation: 100 m*M* Tris-HCl pH 8.5, 5 m*M* EDTA, 0.2% SDS, 200 m*M* NaCl, 100 μg/ml proteinase K.

12. Proteinase K stock solution: Dissolve powder at 10 mg/ml in H_2O, aliquot, and store at −20°C.

13. Tris-EDTA: 10 m*M* Tris.HCl pH 8.0, 1 m*M* EDTA, autoclaved.

14. Cell-fixing solution (LacZ staining): 0.2% glutaraldehyde, 5 m*M* EGTA, 2m*M* $MgCl_2$, 0.1M K_2HPO_4 pH 7.4. Glutaradlehyde stock solution at 25% is stored in aliquots at −20°C and used immediately after thawing.

15. Washing solution (LacZ staining): 0.01% desoxycholate, 0.02% Nonidet P-40 5 m*M*, EGTA, 2 m*M* $MgCl_2$, 0.1M K_2HPO_4 pH 7.4.

16. Staining solution: 0.5 mg/ml X-gal, 10 m*M* $K_3[Fe(CN)_6]$, 10 m*M* $K_4[Fe(CN)_6]$0.01% desoxycholate, 0.02% Nonidet P-40 5 m*M*, EGTA, 2 m*M* $MgCl_2$, 0.1 *M* K_2HPO_4 pH 7.4

B. Preparation, Culture, and Inactivation of Embryonic Fibroblasts

Embryonic stem cells should be cultured on a feeder layer of inactivated embryonic fibroblasts (Robertson, 1987). Embryonic fibroblasts produce LIF, which is necessary to maintain ES cells undifferentiated during culture. In this section, protocols for preparation, routine culture, and inactivation of the feeder layer are detailed.

1. Preparation and Routine Culture

1. Dissect ten 13- to 15-day embryos (we normally use NMRI or CD1 mice) in PBS. Discard liver and internal organs. Wash carcasses several times in PBS to remove blood.
2. Transfer carcasses to 5 ml of trypsin–EDTA and mince in small pieces.
3. Using a wide pipette, transfer the dissected embryo pieces to the Erlenmeyer flask containing glass beads and add 50 ml of trypsin–EDTA.
4. Incubate for 30 min at 34°C with gentle agitation on a magnet stirrer.
5. Using a pipette, remove the cell suspension from the flask and transfer to sterile 50-ml tube. Leave the cell clumps in the flask.
6. Add 50 ml trypsin–EDTA to the remaining cell clumps in the flask and repeat step 4.
7. Spin down the cell suspension from step 5 at 1000 rpm for 10 min, and resuspend pellet in embryonic fibroblast medium.
8. Remove the second cell suspension from the flask and proceed as in step 7.
9. Pool cells from steps 5 and 6 and plate on 14.5-cm tissue culture dishes (about two embryos/dish).
10. Change medium after 24 hr.
11. Once confluent (after 2–3 days), freeze cells by making five vials from each plate.
12. For routine culture, one vial is thawed on an 8.5-cm plate. When confluent, the plate is passaged on 4 × 8.5-cm plates, which are again passaged (when confluent) to 4 × 14.5-cm plates. When confluent, the 14.5-cm plates are treated for 2.5 hr with mitomycin C to inactivate the fibroblasts.

2. Fibroblast Inactivation

1. Prepare inactivation medium by adding 100 ml of mitomycin C solution to 10 ml culture medium (calculate 10 ml of medium for each 14.5-cm plate).
2. Change medium to inactivation medium and put plates back in incubator for 2.5 hr.
3. Stop inactivation by washing the cells twice with 10 ml of PBS, and add 20 ml of fresh medium to each plate.
4. Trypsinize cells and seed on gelatinized plates at a density of 8×10^4 cells/cm^2. Gelatinized plates are prepared by covering the culture surface with 0.1% gelatin for at least 15 min at room temperature and removing it before using.

C. Routine Culture of Embryonic Stem Cells

1. General Remarks

Embryonic stem cells are very sensitive to culture conditions, and therefore care should be taken at all steps to maintain them under constant parameters. They grow forming compact, three-dimensional, island-like colonies within which cells are densely packed (Fig. 1). They are round and show a prominent nucleus and a thin cytoplasmic region. Changes in cell or colony morphology might indicate differentiation of the cells and, consequently, the loss of totipotency and the ability to generate germline chimeras. ES cells grow fast, dividing every 8–10 hr, and tend to differentiate if colony size is too big or if they reach confluence. Therefore, splitting and feeding should be done frequently and, when splitting, cells should be trypsinized to a single-cell suspension to avoid differentiation. Totipotency of ES cells tends to be lost with passage number; therefore, it is advisable to subclone the cells every 20 passages to recover full potency.

2. Protocol

1. Day 0: Wash the confluent ES cells with PBS.
2. Add trypsin–EDTA (2 ml for a 8.5-cm culture dish, 1 ml for 6-cm dish, and 0.5 ml for 3.5-cm culture dish). Put back into the incubator for 5 min.
3. Pipette the cells up and down using a plugged Pasteur pipette until a single-cell suspension is obtained.
4. Add an excess of medium to stop trypsinization and mix well.
5. Spin down the cells at 1000 rpm for 5 min.
6. Remove the supernatant and resuspend the cells in new ES cell medium.

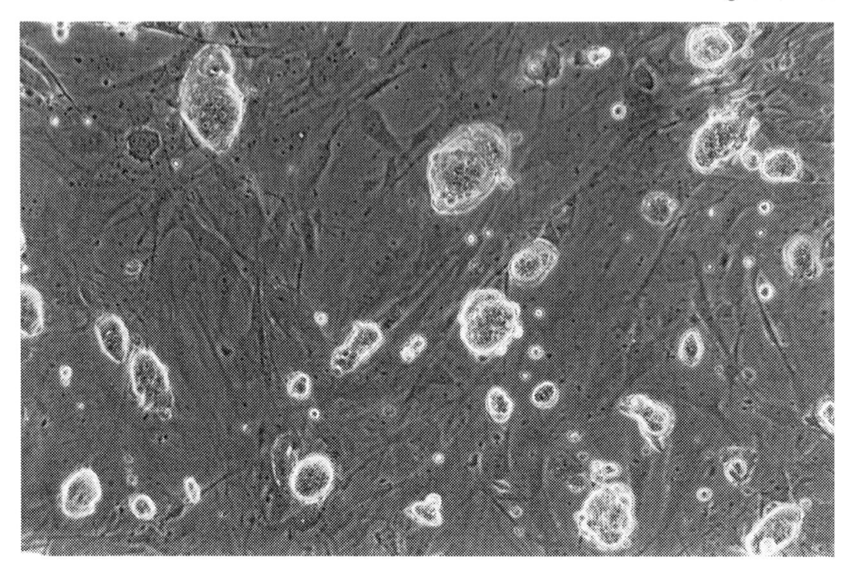

Fig. 1. Embryonic stem cell colonies growing on a feeder layer of mitotically inactive primary embryonic fibroblasts.

7. Split the cells 1/5 to 1/8 (depending on the growth rate, which can vary between serum batches and particular cell lines) on newly inactivated feeder plates.
8. Day 1: Change medium.
9. Day 2: Split cells on new feeder as in day 0.

D. Gene Targeting in Embryonic Stem Cells

1. Theoretical Background and Targeting Vector

Mammalian cells are able to incorporate introduced DNA into their genome by either heterologous or homologous recombination. Transformation of ES cells with constructs containing modified DNA sequences from a chosen locus allows targeting to the endogenous location of the modifications introduced in the construction. The most commonly introduced modification is deletion of an essential region of the targeted gene to achieve a loss-of-function mutation. For this purpose, the targeting construct is composed of a genomic fragment from which regions coding for essential parts of the gene have been removed and substituted for by a neomycin resistance casette (Fig. 2). Two events of homologous recombination, at the regions flanking the modification introduced, will replace the endogenous sequences by the mutated ones (Fig. 2). Expression of the *Neo* gene confers resistance to the drug G418. This will be used to select for ES cell clones

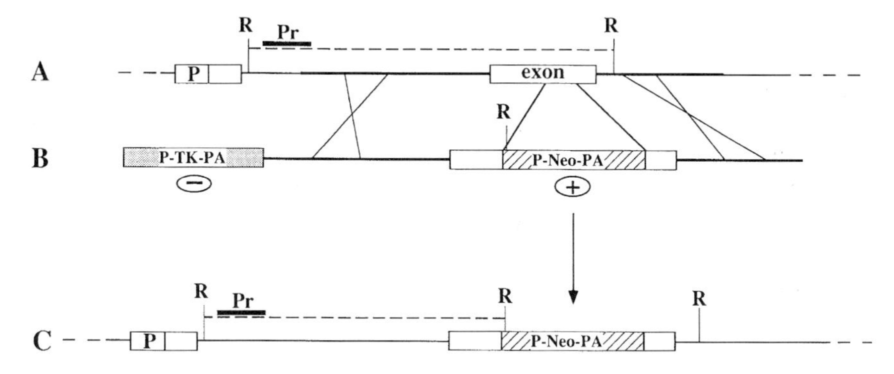

Fig. 2. Gene targeting strategy. A model genomic organization of the endogenous gene to be targeted is shown in (A). The targeting vector is represented in (B), and the resulting targeted allele after homologous recombination is shown in (C). The black segment indicated as Pr represents the location of a suitable "outside probe," which will not detect any changes in clones that incorporated the vector by nonhomologous recombination, but will detect a change in the mobility of a restriction fragment (dotted line) after the desired targeting event. P, endogenous promoter; P-TK-PA, promoter, thymidine kinase gene, and polyadenylation signal; P-TK-PA, promoter, neomycin resistance gene, and polyadenylation signal; R, restriction endonuclease site.

with stable integrations of the construct. Two factors have been shown to affect targeting efficiency, the size of the genomic fragment included in the construct (Deng and Capecchi, 1992) and the source of the DNA (Riele *et al.*, 1992). Good targeting efficiencies are often obtained with constructs including more than 10 kb of homology. The length of homologous DNA on each side of the *Neo* casette should not be under 1.5 kb. Increased targeting efficiencies have been observed when the source of the targeting vector DNA is the same mouse strain from which the ES cells derive, usually 129sv.

Once G418-resistant colonies are isolated, the next step is to discriminate between nonhomologous and homologous recombinants by genomic DNA analysis. The relative frequency between these two events is variable depending on the locus and the targeting vector. Enrichment strategies have been devised to increase the relative frequency of homologous recombination events. The most widely used strategy is based on the use of the negative selectable gene for thymidine kinase (*TK*) (Mansour *et al.*, 1988), placed at one of the ends of the homologous sequences of the targeting construct (Fig. 2). Insertional nonhomologous integrations, in most cases, keep and express the *TK* gene and do not survive in the presence of the drug gancyclovir. However, homologous recombinants lose the *TK* gene by recombination in the homology region next to it and survive in the presence of gancyclovir. The construct should be linearized and introduced into ES cells by electroporation.

2. Electroporation of Embryonic Stem Cells

1. Prepare DNA for electroporation by CsCl centrifugation or by using a Quiagen column. The targeting construct should be linearized outside the domain of homology by the appropriate restriction enzyme. After digestion, the DNA is treated once with phenol–chloroform, precipitated, and resuspended in TE at 0.5 μ/μl. Check DNA on agarose gel before electroporation.

2. Day 0: Split ES cells on 8.5-cm-diameter feeder plates as described in III.C and use freshly inactivated feeder. The culture of the feeder for inactivation should be started 1 week before electroporation. The fibroblasts should be prepared from embryos carrying the *Neo* gene, so that they are resistant to G418.

3. Day 1: Change medium.

4. Day 2: Change medium. Remove medium from the 8 × 8.5-cm feeder plates and add 6 ml of ES cell medium to each plate.

5. Four hours after changing medium, trypsinize the cells as described in III.C and add ES cell medium to 10 ml, mix well and spin down for 5 min at 1000 rpm.

6. Aspirate the supernatant and resuspend the cells in 30 ml of PBS. Determine the cell number. There should be around 1.5×10^7 cells.

7. Spin down for 5 min at 1000 rpm.

8. Resuspend approximately 10^7 ES cells in 0.8 ml PBS containing 25 μg/ml of linearized DNA. Let suspension stand for 5 min at room temperature.

9. Pipette the cells up and down and transfer 0.8 ml of the suspension to one electroporation cuvette, omitting air bubbles. Electroporate with one pulse of 500 μF and 250 V at room temperature. Let stand for 5 min at room temperature.

10. Using a plugged Pasteur pipette, transfer the cells directly to ES medium (28 ml) and plate the cells on 7 × 8.5-cm feeder plates. Plate control ES cells (electroporated without DNA) at the same call density.

11. Twenty-four hours after electroporation, change medium to selection medium with G418 and gancyclovir if using a positive–negative selection.

12. Change medium every day.

13. After 5 days of selection, remove gancyclovir from the selection medium.

14. After 8 days of selection, check control plate (there should be no ES cells left) and transfection plates for G418-resistant colonies. Large ES cell colonies (about 1000 cells) can already be seen. Do not let them grow too large because they will differentiate.

3. Picking Resistant Clones

1. Using a microscope marker, mark the clones to pick.

2. Prepare a 96-well plate with 30 μl trypsin–EDTA in each well.

3. Under a dissecting microscope, inside a laminar-flow sterile hood, pick every marked clone individually and transfer to one well in the 96-well plate. Pick the colonies with a yellow tip on a 100-μl micropipette, until 12–24 colonies are accumulated individually in the trypsin wells.

4. After the colonies are trypsinized (variable time), add 50 μl of warm medium with a 12-channel multipipette (20–200-μl capacity) and pipette up and down several times, until obtaining a single-cell suspension. Transfer to 96 wells covered with an embryonic fibroblast feeder layer and add up to 200 μl of ES cell medium.

5. After 3–4 days, the clones will be ready to be frozen and split. Trypsinize the cells to a single-cell suspension as described in section III.B.2, but using the multipipette. Use 30 μl of trypsin per well and stop trypsinization with 70 μl of ES cell medium.

6. Remove 50 μl from each well and transfer to a 24-well gelatinized plate containing 1 ml of ES cell medium. This operation is best done by inserting alternate tips into the multipipette and loading pair and even numbers alternatively. The alternate tips fit into the 24-well plate, so that clones from single wells of the 96-well plate can be transferred to corresponding single wells in the 24-well plate. During this procedure, the 96- and the 24-well plates should be numbered with correlative numbers. The 96-well plate will be used for freezing the clones and the 24-well plate to obtain DNA for screening.

7. After trypsinizing the whole 96-well plate and having removed the aliquot for the 24-well plate, add one volume of 2× freezing medium (double concentration of FCS and DMSO), mix well, wrap the plate with paper, and freeze at −80°C inside a Styrofoam box. The cells will remain viable for several months.

4. Analysis of the Picked Clones by Southern Blot

The isolated clones should be checked individually to distinguish between random integrations and homologous recombination. We recommend genomic Southern analysis for this purpose. Genomic DNA is digested with an enzyme that generates bands of different sizes from the wild type and the targeted allele, and probed with a genomic fragment not included in the targeting construct (see Fig. 2).

1. When the cells in the gelatinized plates are confluent, remove medium and add 500 μl of lysis buffer to each well.
2. Leave overnight in the incubator at 37°C.
3. Transfer the DNA solution from each well to an Eppendorf tube and add 500 μl of isopropanol to precipitate DNA.
4. Spin down for 10 min and remove supernatant.
5. Add 500 μl 70% ethanol to wash pellet and spin down for 5 min.
6. Remove ethanol and dry pellet. Resuspend DNA in 50 μl of 1/10 TE.
7. Use 15 μl from each sample for screening and digest with the appropriate restriction enzyme. Positive clones will be expanded and rechecked. They can be used now to generate chimeras by blastocyst injection or by morula aggregation (Bradley, 1987; see section F).

E. Gene Trap in Embryonic Stem Cells

1. Theoretical Background and Gene Trap Constructs

Gene trap strategies have been used in ES cells to identify developmentally regulated genes (Gossler *et al.*, 1989; Skarnes *et al.*, 1992). The approach is based on the introduction of a promoterless reporter gene that integrates randomly in the genome. A first screening is done in ES cells by selecting for successful gene targeting events (Fig. 3A). By introducing the respective ES cell clones into developing embryos, it is possible to perform a second screening based on *in vivo* embryonic expression (Fig. 3B–D). This approach allows the characterization of the genes trapped: the expression pattern analysis is facilitated because the expression of the reporter gene reflects the activity of the promoter of the trapped gene, and molecular analysis is straightforward because transcriptional fusions are produced between the endogenous and reporter genes. In addition, the screening is not biased by sequence constraints, allowing the isolation of relevant genes not necessarily conserved in other organisms. More important, the gene trap is a highly mutagenic screening, because the expression of the reporter involves, in most of cases, the interruption of the coding region of the endogenous gene.

Two basic types of constructs have been used in gene trap. One of these vectors (Fig. 4A) carries a fusion gene (β-*geo*) between the neomycin resistance gene and *LacZ* (Friedrich and Soriano, 1991), which serves both as a selectable gene and as a reporter. Because *Neo* expression depends on the effective fusion of the β-*geo* gene to an endogenous gene, the frequency of blue clones among the resistant ones is very high. The β-*geo* gene is promoterless; therefore, a transcriptional and translational fusion to an endogenous gene must occur for its expression to be detected. A splice acceptor is placed 5′ of the *LacZ* to increase the frequency with which it is incorporated into the mRNA.

In the second type of vector (Fig. 4B), the *Neo* gene is under the control of a constitutive promoter. When using this vector, all integrations render a resistant clone, but the frequency of blue clones among the resistant ones varies between 1% and 5%, depending on the particular vector used. This represents a disadvantage with respect to the first construct, because it renders the screening step more difficult. Yet, the β-*geo* trap selects only for those genes that are expressed in ES cells, whereas with the *LacZ-P-Neo* trap, many of the resistant white clones represent trapping of genes silent in ES cells. Differentiation procedures may then be applied to these white resistant clones to reveal the trapped genes. Furthermore, differentiation procedures can be directed to detect expression in a particular cell type of interest.

In this chapter, we describe those ES cell manipulation procedures specific for

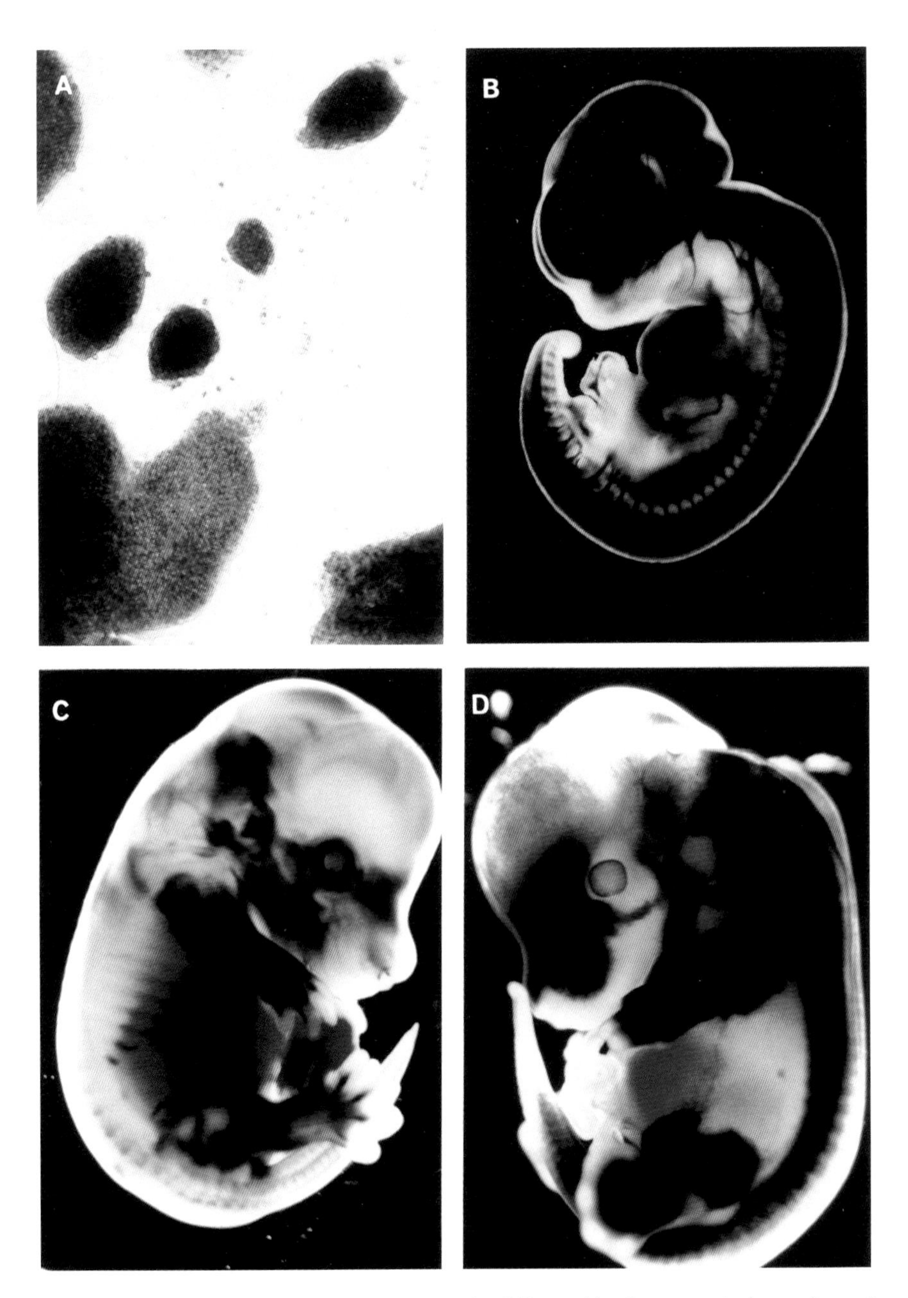

Fig. 3. Gene trap strategy. (A) Colonies from an ES cell line positive for a gene trap integration and stained for *LacZ* activity. (B–D) Three examples of expression patterns in mouse lines carrying gene trap insertions.

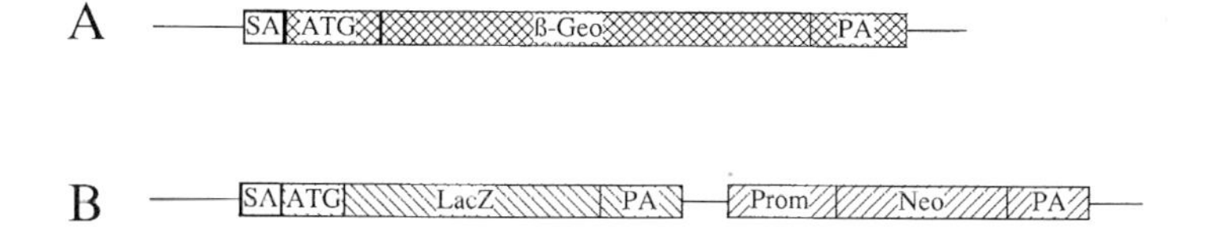

Fig. 4. Gene trap vectors. (A) The structure of a β-*geo*-based gene trap vector and (B) of a *LacZ*-promoter-*Neo*-based vector. SA, splice acceptor; Prom, constitutive promoter; PA, polyadenylation signal.

the gene trap approach and refer to other sections for general ES handling procedures.

2. Electroporation and Selection

1. Grow ES cells in standard conditions to obtain 10^8 for β-*geo* electroporation and 2×10^7 for the *PGK-Neo* construct. Maintain cells in underconfluent conditions and electroporate the cells about 1.5 days after the last splitting, when they are in the exponential growth phase.
2. Electroporate as described in section III.D.2 at a cell density of 10^7 cells/ml and a DNA concentration of 40 μg/ml. Plate the cells at a density of about 6×10^4/cm^2 in standard ES cell medium on neomycin-resistant feeder layers. Replace the medium with selection medium after 24 hr of culture.
3. Pick clones into 96-well with feeder layer as described in section III.D.3.
4. Once confluent, freeze as described in section III.D.3, but instead of preparing a 24-well plate for DNA analysis, prepare a 96-well replica plate for β-galactosidase staining.
5. Comments

Linearization of the vector: If possible, choose an enzyme that cuts at least 1 kb away from any essential motif in the construct, because in most of cases, sequences at the ends of the molecule will be lost on integration of the trap.

Selection: The concentration of G418 in the selection medium should be determined empirically by establishing a curve of lethality for different concentrations of G418. The optimal concentration depends on the particular ES cell line used and on culture conditions. Choose the G418 concentration that kills all untransfected colonies after 8 days for experiments with β-*geo* and after 6 days for experiments with *PGK-Neo*. The lower concentration of G418 used for β-*geo* experiments avoids the death of clones in which the trapped gene is expressed at low levels and therefore show low *Neo* resistance.

Freezing β-geo *lines*: Because the frequency of blue clones is very high when using the β-*geo* construct, we recommend not freezing a whole 96-well plate for experiments with this vector, because too many clones will have to be handled at once when thawing the plate.

3. Screening for *LacZ*-Positive Clones

For detection of *LacZ* activity in the selected clones, it is very important to achieve low background levels because this allows development of the reaction for longer times, and clones with very low levels of expression can be identified. Use the following protocol for best signal-to-background ratios:

1. Rinse cells in PBS. Do it quickly, because if living cells stay in PBS even for a short time, a significant background will appear.
2. Fix for 3 min at room temperature. Be strict with fixation times because both signal and background are sensitive to fixation conditions.
3. Wash 3 × 5 min in washing solution at room temperature.
4. Incubate in staining solution at 37°C in the dark until signal develops (for up to 3 days).
5. Stop the reaction by rinsing in PBS and fix for 5 min in 2% formaldehyde in PBS.
6. Wash 3 × 5 min in PBS and store at 4°C.

Because in many cases a fusion occurs with an endogenous protein, there is variable subcellular localization of the staining. In addition, for many clones expression appears restricted to a few cells in each colony or to a few colonies in the well; therefore, check very carefully all the cells in each well.

In experiments with β-*geo*, we have observed that some clones show G418-dependent blue staining, and therefore we recommend maintaining G418 selection after picking until the plate is stained. Between 30% and 50% of the resistant clones should be positive. In experiments with *LacZ-P-Neo*, if the staining is going to be scored only in undifferentiated ES cells, proceed as for the β-*geo* construct. If differentiation protocols are to be used, split the plate for staining using the multipipette into as many 96 wells as differentiation protocols are going to be applied. After the differentiation protocol is completed, stain the plates as described previously.

4. Analysis of the Positive Clones

Once the positive clones are identified, thaw the plates and grow and make stocks out of the positive clones. Culture the cells for at least 1 week before using them for generating chimeras. Two alternative ways are possible for analysis of the expression pattern in the embryo, either a transient expression analysis in chimeric embryos or the establishment of mouse lines corresponding to the ES cell clones and subsequent analysis of the expression pattern in these lines. In either case, chimeric embryos or adults can be generated by either injection of the cells into a blastocyst (Bradley, 1987) or by aggregation with an eight-cell-stage embryo (Nagy *et al*., 1993; see section F).

F. Production of Mouse Embryonic Stem Cell Chimeras by Morula Aggregation

1. Overview

The classic method for generation of germline mouse chimeras has been the microinjection of ES cells into the blastocyst. However, the establishment of ES cell lines with especially good performance in aggregation experiments, the use of aggregation wells to hold the aggregates, and the modification of culture media (Nagy *et al.*, 1993), have shown that aggregation is an efficient and versatile method for the generation of germline chimeras in gene targeting and gene trap experiments (Yamada *et al.*, 1995; Torres *et al.*, 1995; Subramanian *et al.*, 1995; Mansouri *et al.*, 1996; and unpublished results). In addition, aggregation chimeras present the following advantages with respect to injection chimeras:

No requirement for micromanipulation equipment
Less time required to learn the technique
Processing of a larger number of embryos per experiment
No need to use inbred mouse strains to obtain germline chimeras
Possibility of generating completely ES-cell-derived embryos

The following protocols are used for aggregation of ES cells with normal morula to obtain chimeric animals. These protocols have been established for the R1 cell line (Nagy *et al.*, 1993) and have been shown not to work for a variety of other traditionally used ES cell lines. With this method, R1 genetically modified cells efficiently produce chimeras with both C57B1/6 (inbred) and CD1 or NMRI (outbred) embryos. The efficiencies vary from clone to clone, but on average two of three clones produce chimeras that transmit the ES cells genome through the germline. The average number of embryos necessary to transfer to obtain a germline transmitter is around 30, and each person can easily transfer 60 embryos each day. Those procedures specific for the morula aggregation technique are described in this section, whereas other sources are quoted for more general procedures. In particular, for embryo donor superovulation, preparation of foster mothers, embryo recovery, culture, and retransfer, as well as preparation of M2 and M16 media and acid Tyrode's (AT) solution, the procedures described in the Cold Spring Harbor manual (Hogan *et al.*, 1994) were followed.

2. Procedure

1. Two days before the day of the experiment, plate ES cells onto a 3.5-cm plate; cell density should be kept low enough so that cells will form medium-sized colonies after 2 days, but will not reach confluence. When starting the experiment from a frozen stock, cells should be split at least twice before aggregation. Change the medium to the ES cells the day before the experiment.

2. The morning of the experiment, dissect 2.5 dpc pregnant superovulated females and obtain eight-cell-stage embryos by oviduct flushing. Collect them in microdrops of M2 medium under mineral oil using a mouth-controlled pulled Pasteur pipette.

3. Zona removal: Prepare a 3.5-cm tissue culture plate with several AT solutions and at least four M2 drops. Transfer the embryos in groups of about 20 into the AT, minimizing M2 carryover. Observe zona dissolution under binoculars, which takes from a few seconds to a minute, depending on the amount of medium carried with the embryos. Use a fresh AT drop every time. As soon as the zona disappears, transfer the embryos to the M2 drops and wash them by successive changes in the different M16 drops. Too long a time in the AT solution results in lysis of the embryos.

4. Preparation of the aggregation plate: Prepare a 3.5-cm plate with 2 normal M16 drops and 10 small drops. A small percentage of ES cell medium should be added to the M16 medium to obtain good ES cell contribution; we routinely use 2% for inbred embryos and 4% for outbred embryos. Make six conical depressions in each small drop by pressing vertically with a darning needle. Place one embryo in each depression and leave the plate in a 37°C, 5% CO_2 humidified incubator until the ES cells are ready for the aggregation.

5. Preparation of the cells: Wash the ES cell dish two times with PBS (without Ca and Mg). Place 0.5 ml of trypsin solution and leave the dish in the incubator for a 4 and 5 min, depending on the colony size. Next, add 2 ml of ES cell medium and slowly pipette up and down several times with a 2-ml pipette. Check the cells periodically under the microscope and continue pipetting until cell clumps of the desired size are obtained. Place the cells in a sterile Falcon tube for transportation to the animal house, and dilute up to 4–5 ml with ES cell medium.

6. Aggregation: Aggregation can be done late in the morning or early in the afternoon. Place 2 ml of the ES cells in a 3.5-cm plate and, under binoculars and using the mouth-controlled pipette, choose a large number of cell clumps of approximately the required size and transfer them to the aggregation plate inside the two drops without depressions. Choose individual clumps from these two plates and sandwich them between two embryos inside each depression; the shape of the depression will hold the aggregate together (Fig. 5A). Alternatively, use only one embryo in each aggregation, because there is no detectable difference in the quality of the chimeras and the requirement for embryos decreases. After aggregating all the embryos, transfer the plate for overnight culture inside the incubator. The number of cells that are aggregated with each embryo is crucial for the success of the experiment: too many cells will kill the embryos, whereas too few cells will give poor chimerism. We have found that the optimal number of cells is between 10 and 15 for aggregations with outbred strains and between 6 and 8 with the C57B1/6 strain; however, it might be necessary to adjust the cell number and percentage of ES cell medium to particular experimental conditions.

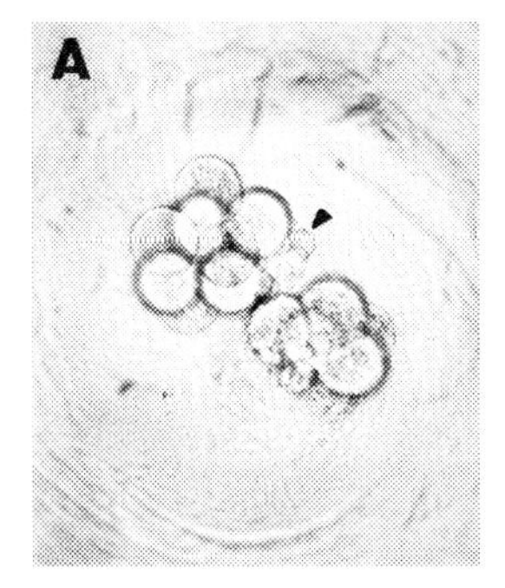

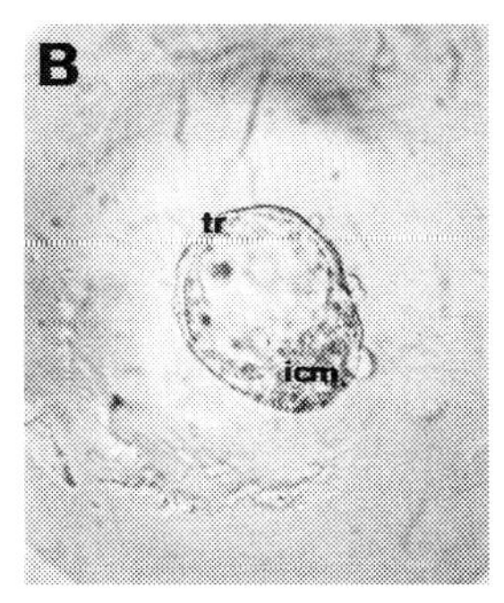

Fig. 5. Aggregation procedure. (A) A sandwich between two eight-cell-stage embryos and a clump of ES cells; the arrowhead points to the ES cells in between the two embryos. (B) A blastocyst that developed after overnight culture of an aggregate such as the one shown in (A). tr, trophoblast; icm, inner cell mass.

7. On the afternoon of the next day, choose the embryos that have developed to the blastocyst stage (Fig. 5B; no more than 5% should be underdeveloped) and transfer six to eight of them into each uterus of 2.5-day pseudopregnant females. Pups should be born 17 days after the transfer.

8. Embryonic stem cells carry the wild-type allele for the *agouti* locus. When using an albino recipient strain, such as the CD1 or NMRI strains, chimeric animals can be identified at birth by the skin and eye pigmentation. When using black recipients, such as the C57BL/6 strain, the identification is possible only from day 5 of life by hair color.

9. Once chimeric animals are identified, adjust the litter size to six to eight pups by removing nonchimeric animals or adding pups from a different litter of the same age.

10. Once the chimeric animals reach sexual maturity, cross them to animals from the recipient strain used in the aggregation. Identify the germline transmitters by the presence of agouti progeny. Identify heterozygous animals by genotyping the agouti progeny and cross them to establish a mouse line carrying the ES modified genome.

Acknowledgments

The author would like to thank Prof. Peter Gruss for his constant support and encouragement during the time these procedures were established in his department. I am grateful to Sharif Mashur for his excellent technical assistance.

References

Bradley, A. (1987). Production and analysis of chimeric mice. *In* "Teratocarcinomas and Embryonic Stem Cells: A Practical Approach" (E. J. Robertson, ed.), pp. 113–151. IRL Press, Oxford, Washington, DC.

Capecchi, M. R. (1989). The new mouse genetics: Altering the genome by gene targeting. *Trends Genet.* **5,** 70–76.

Deng, C., and Capecchi, M. R. (1992). Reexamination of gene targeting frequency as a function of the extent of homology between the targeting vector and the target locus. *Mol. Cell Biol.* **12,** 3365–3371.

Friedrich, G., and Soriano, P. (1991). Promoter traps in embryonic stem cells: A genetic screen to identify and mutate developmental genes in mice. *Genes Dev.* **5,** 1513–1523.

Gossler, A., Doetschman, T., Korn, R., Serfling, E., and Kemler, R. (1986). Transgenesis by means of blastocyst-derived embryonic stem cell lines. *Proc. Natl. Acad. Sci. U.S.A.* **83,** 9065–9069.

Gossler, A., Joyner, A. L., Rossant, J., and Skarnes, W. C. (1989). Mouse embryonic stem cells and reporter constructs to detect developmentally regulated genes. *Science* **244,** 463–465.

Hogan, B., Beddington, R., Constantini, F., and Lacy, E. (1994). "Manipulating the Mouse Embryo" 2nd ed. Cold Spring Harbor Laboratory Press, Cold Spring Harbor, NY.

Mansour, S. L., Thomas, K. R., and Capecchi, M. R. (1988). Disruption of the proto-oncogene *int-2* in mouse embryo-derived stem cells: A general strategy for targeting mutations to non-selectable genes. *Nature* **336,** 348–352.

Mansouri, A., Stoykova, A., Torres, M., and Gruss, P. (1996). Dysgenesis of Cephalic neural crest derivatives in *Pax-7-1*-mutant mice. *Development* **122:** 831–838.

Nagy, A., Rossant, J., Nagy, R., Abramow-Newerly, W., and Roder, J. C. (1993). Derivation of completely cell culture-derived mice from early-passage embryonic stem cells. *Proc. Natl. Acad. Sci. U.S.A.* **90,** 8424–8428.

Riele, H. R., Maandag, E. B., and Berns, A. (1992). Highly efficient gene targeting in embryonic stem cells through homologous recombination with isogenic DNA constructs. *Proc. Natl. Acad. Sci. U.S.A.* **89,** 5128–5132.

Robertson, E. J. (1986). Pluripotent stem cell lines as a route into the mouse germline. *Trends Genet.* **2,** 9–13.

Robertson, E. J. (1987). Embryo-derived stem cell lines. *In* "Teratocarcinomas and Embryonic Stem Cells: A Practical Approach" (E. J. Robertson, ed.), pp. 71–112. IRL Press, Oxford, Washington, DC.

Skarnes, W. C., Auerbach, B. A., and Joyner, A. L. (1992). A gene trap approach in mouse embryonic stem cells: The LacZ reporter is activated by splicing reflects endogenous gene expression, and is mutagenic in mice. *Genes Dev.* **6,** 903–918.

Subramanian, V., Meyer, B. I. and Gruss, P. (1995). Disruption of the murine homeobox gene Cdx1 affects axial skeletal identities by altering the mesodermal expression domains of Hox genes. *Cell* **83:** 641–653.

Torres, M., and Gómez-Pardo, E. G., Dressler, G. R., and Gruss, P. (1995). *Pax-2* controls multiple steps of urogenital development. *Development* **121:** 4057–4065.

Yamada, G., Mansouri, M., Torres, M., Blum, M., Stuart, E. T., Schultz, M., de Robertis, E., and Gruss, P. (1995). Targeted mutation of the mouse *Goosecoid* gene leads to neonatal death and craniofacial defects in mice. *Development* **121:** 2917–2922.

6

Organoculture of Otic Vesicle and Ganglion

Juan J. Garrido, Thomas Schimmang, Juan Represa, and Fernando Giraldez
Instituto de Biología y Genética Molecular (IBGM)
Facultad de Medicina, Universidad de Valladolid
47005-Valladolid, Spain

I. Introduction

The vertebrate inner ear derives from the embryonic ectoderm. The earliest morphologic evidence for the primordium of the ear is the otic placode, a thickened area of ectoderm close to the hindbrain. It can be recognized in apposition to the neural tube as early as from the 3–6-somite embryo, depending on the animal species. The otic placode then invaginates to form the otic vesicle. The latter consists of a cavity lined by a pseudostratified epithelium of high proliferative activity. The otic vesicle is developmentally autonomous and it contains all information concerning the organization of the membranous labyrinth and the different cellular phenotypes (Swanson *et al.*, 1990). An excellent summary of the timing of the early stages of ear development in several vertebrates can be found in Anniko (1983).

Preceding differentiation of the different cell phenotypes that populate the adult ear, the otic vesicle goes through a distinct period of proliferative growth. This process is under the control of growth factors and other molecules that regulate the transition between the proliferative and differentiative states of the otic vesicle. The action of these factors and molecules was first revealed by *in*

Current Topics in Developmental Biology, Vol. 36

0070-2153/98 $25.00

vitro studies that showed that serum and bombesin are able to reinitiate DNA synthesis and cell proliferation of quiescent otic vesicles (Represa *et al.*, 1988). Further studies revealed that other growth factors and retinoic acid (RA) regulated cell proliferation, and they did so through activation of particular intracellular signalling mechanisms and genes (León *et al.*, 1995*b*). The procedures for carrying out these assays and the type of analysis that can be done are the subject of the first part of this chapter.

The cochleovestibular ganglion (CVG) contains the primary afferent neurons that connect the sensory epithelia of the inner ear with the central nervous system. It originates from the otic vesicle and neural crest, and goes through a period of intense cell proliferation until neuroblasts become postmitotic and start to differentiate (D'Amico Martel and Noden, 1983). Cell proliferation, survival, neurite extension, and perhaps many other phenotypic properties of the auditory neurons are under the control of specific molecules, among which the nerve growth factor (NGF) family of neurotrophins appears to play a crucial role. The first indication for these ideas came from *in vitro* culture experiments of CVG that showed that NGF was able to induce cell proliferation of the CVG (Represa & Bernd, 1989). Further studies, including studies with cell dissociates and the analysis of mice carrying null-mutations for the different neurotrophins and their receptors, have outlined the multiple roles played by the different neurotrophins in ear development. The procedures for carrying out experiments with proliferating or differentiating auditory ganglions are discussed, along with the strategies used to analyze the cultured ganglia.

II. Organotypic Culture of Otic Vesicles and the Study of Growth Factors and Cell Proliferation in the Otic Vesicle

We describe here the procedure to culture isolated otic vesicles and perform the basic assays for cell proliferation. Indications on how to prepare the samples to perform other analyses are also outlined. Experiments are described here for the chick embryo, but they can be extended to the mouse embryo without many changes. Organ culture techniques applied to the otic vesicle were first attempted by Orr (1968) and further developed by Van de Water and Ruben (1974), Friedman *et al.* (1977), and Represa *et al.* (1988).

A. Isolation of Otic Vesicles

1. Otic vesicles are isolated from 3-day-old chick embryos corresponding to stage 18 of Hamilton and Hamburger (1951). After removing extraembryonic membranes, the embryo is placed on a Sylgard-coated Petri dish and immobilized with fine pins used for entomology. Using transillumination, the otic vesicle and the branchial arches should be visible and accessible for dissection.

2. Remove the ectoderm overlying the otic vesicle by making two incisions perpendicular to the anteroposterior embryonic axis. One passing immediately ahead of the otic vesicle starts at the first branchial cleft and ends at the dorsum of the embryo. The second starts at the third branchial cleft and goes up to the dorsum behind the otic vesicle. These two incisions are joined by a perpendicular one passing ventrally to the otic vesicle. This makes a small window in the ectoderm that covers the otic vesicle. The ectoderm is then gently lifted up, leaving the otic vesicle exposed to the surface.

3. Remove the otic vesicle by dissecting with fine forceps from the surrounding mesenchyme. Usually, a fine layer of mesenchyme remains surrounding the otic vesicle that for most purposes can be ignored. If further cleaning is required, then a short, 3–5-min incubation in a Ca^{2+}/Mg^{2+}-free medium with 0.05% dispase will help to clean all mesenchyme.

4. Store otic vesicles in standard medium (see later) at room temperature if dissection is fast and the number of vesicles required small (no longer than 0.5 hr and about 12–16 vesicles for an average experienced dissector). If longer periods are required or when vesicles are needed for protein or mRNA measurements, it is better to store them in ice-cold medium.

B. Culture of Explanted Otic Vesicles (Fig. 1)

The standard assay for cell proliferation proceeds by incubating the otic vesicles in serum and growth factor-free medium, which makes them quiescent.

1. Place isolated otic vesicles in the center of 2.5-cm diameter, four-well Petri dishes covered with 250 μl of standard medium. The amount of solution covering the explants appears to be critical to allow the correct gas exchange but avoid collapsing of the otic vesicles.

2. Use M-199 with Hank's salts, without serum and growth factors. Glutamine and, optionally, antibiotics (penicillin and streptomycin) can be supplemented at standard concentrations. The standard culture medium consists of serum-free M-199 medium with Earle's salts (Biowhitaker, Walkersville, MD) supplemented with 20 m*M* glutamine (Biochrom, Berlin, Germany), 25 m*M* HEPES (Sigma, UK) and antibiotics (penicillin, 50 UI/ml and streptomycin, 50 mg/ml; Biochrom). Incubations are carried out at 37°C in a water-saturated atmosphere containing 5% CO_2.

3. Take some Polaroid pictures of otic vesicles in different wells. Further on, this will help when checking the culture.

4. Incubate overnight in the mitogen-free medium. Depending on the starting time, this will give a 16- to 20-hour incubation in the absence of growth factors, which should be enough for growth arrest (the length of the cell cycle at this stage is not longer than 7–8 hr).

5. Observe otic vesicles through the dissecting microscope. They should have an aspect very similar to the one they had when incubation started. Take Polaroid

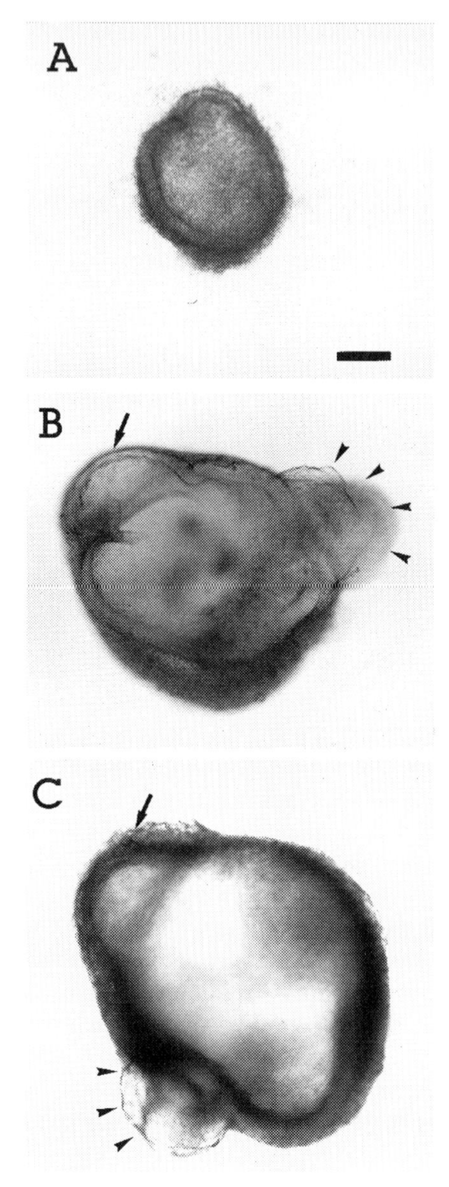

Fig. 1. Reactivation of growth-arrested otic vesicles. Appearance of otic vesicles (OV) obtained from 63-hr (stage 18) chick embryos after 24-hr *in vitro* culture in serum-free medium (M-199) alone (A), or M-199 containing 20% bovine serum (B), or serum plus the growth factor bombesin (C). Note that OV cultured in the presence of M-199 alone do not appear to have grown or differentiated morphologically and resemble OV as they appear in the 63-hr (stage 18) embryo. In contrast, OV cultured in the presence of either serum or bombesin plus serum exhibit a considerable amount of growth and morphologic differentiation, and resemble *in vivo* OV from 87-hr (stage 21–22) chick embryos. Note the enlargement of the endolymphatic appendage (arrow) and the large growing structure in the ventral portion of the OV that will become the cochleovestibular ganglion (arrowheads). Scale bar = 200 μm.

pictures of the otic vesicles that were examined the day before and compare. Check that otic vesicles remain of the same size and with epithelium of uniform thickness.

6. Change the medium to one containing the desired addition and the proliferative effect assayed. We say that we now reactivate growth from a population of cells that are mostly quiescent and that have been isolated from those factors that were present in the embryo (Fig. 1).

7. Incubate for another 24 hr, after which otic vesicles are again examined under the microscope. Once more, a Polaroid picture of the samples gives a good indication of how the experiment has evolved.

8. At this stage, otic vesicles can be processed for different purposes. Most purposes require simply the collection of the otic vesicle with a pipette and treatment as if it were an aliquot of a cell suspension.

C. Analysis of Cultured Otic Vesicles

1. *Evaluation of growth.* A first approximation to evaluate growth can be done by measuring the major and minor diameters of the otic vesicle. For a constant cell density, the cell number of an epithelial layer is proportional to the surface area. Taking the otic vesicle as an ellipsoid, the surface area is proportional to the square of the average radius, and the change in the surface area can be estimated from the ratio of the average radius (León *et al.*, 1995a). This is a convenient approach in those samples that are going to be processed for Western blotting and in which direct measurements of cell number or DNA incorporation cannot be performed.

2. *Determination of DNA synthesis.* This is carried out in the standard manner with ^{3}H-thymidine. Otic vesicles are incubated with labeled thymidine, collected and individually washed in phosphate-buffered saline (PBS), extracted with 10% trichloroacetic acid, and counted in a scintillation counter. This gives a figure for thymidine incorporation for each otic vesicle in a standard and direct way (Represa *et al.*, 1988).

3. *Western blotting.* Western blotting of abundant proteins such as the mitogen-associated proliferating cell nucleus antigen or the Fos oncoprotein, which are highly expressed, allows detection with a single otic vesicle (León *et al.*, 1995a, 1995b).

4. *Northern blotting.* This technique requires collection of many otic vesicles to get sufficient material, and therefore is advisable only if strictly necessary. One hundred micrograms of RNA corresponds to about 50–70 otic vesicles (Represa *et al.*, 1990). RT-PCR should be the method for semiquantitative measurement of gene expression, and it can be performed on 5–10 otic vesicles.

5. *Studies with labeled precursors.* Incubation of otic vesicles with precursors may provide information on the metabolism of particular compounds. Synthesis of glycosyl-phosphatidylinositol (GPI) has been carried out in otic vesicles by

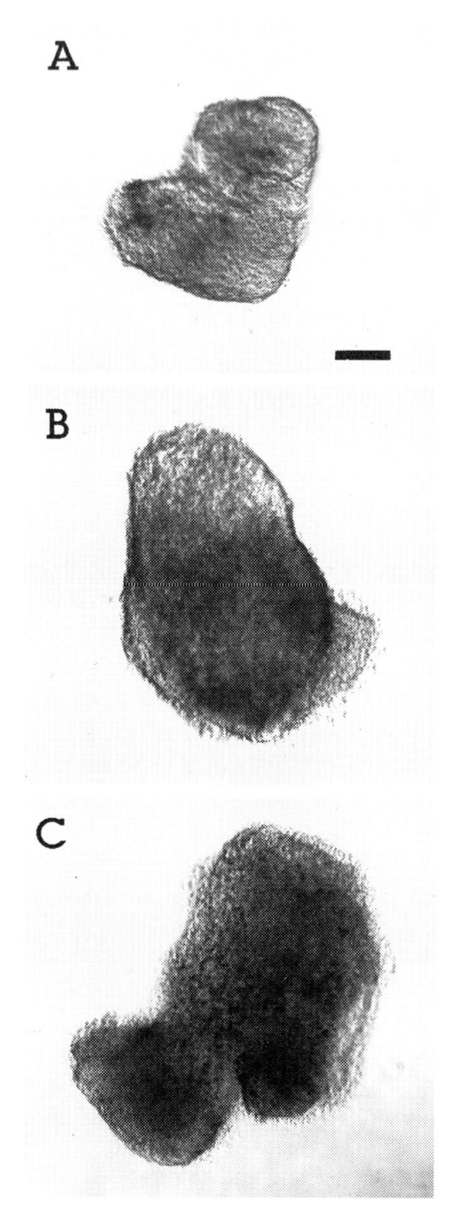

Fig. 2. Growth of cochlear ganglion *in vitro*. Appearance of cochlear ganglion obtained from 72-hr (stage 19–20) chick embryos after 24-hr *in vitro* culture in serum-free medium (M-199) alone (A), M-199 containing 20% bovine serum (B), or M199 plus 50 ng/ml of nerve growth factor (NGF) (C). The size of the ganglion is larger in cultures exposed to serum or NGF. Scale bar = 75 μm.

labeling them with 100 μCi/ml of [^{3}H]glucosamine for 24 hr, in the presence of 1% fetal bovine serum. Hydrolysis of GPI by growth factors can now be performed by rinsing otic vesicles that were washed with PBS and placing them in serum-free medium and stimulating at 37°C with the growth factor (i.e., 10 n*M* insulin-like growth factor I [IGF-1]) for different time periods. Cellular lipids are extracted and labeled GPI purified (see León *et al.*, 1996).

6. *Whole-mount preparations.* Cultured otic vesicles allow whole-mount studies of mRNA and proteins to be carried out with the corresponding probes. They can be performed according to standard procedures (Nieto *et al.*, 1995*b*).

III. Culture of Auditory and Vestibular Ganglia: Whole Ganglia and Dissociated Cells

Culture of explanted CVG can be made for different purposes. We describe here those intended to analyze the proliferative period of the CVG and to study neurite outgrowth and differentiation. Analysis of cell survival and other neuronal features requires dissociation and culture of dispersed neurons, which we also describe briefly.

A. Cultures of Proliferating Cochleovestibular Ganglia (Fig. 2)

1. Ganglia are dissected from stage 20 chick embryos, equivalent to 68- to 72-hr incubation of fertilized eggs. The embryo is placed for the isolation of the otic vesicle and visualized by transillumination. The ectoderm overlying the otic vesicle is removed as described previously, and to the rostral and medial aspect of the otic vesicle the CVG can be identified as a slightly darker oval structure.

2. Separate the ganglion from the otic vesicle by gently tearing apart the otic vesicle with fine forceps while holding the embryo in place. This usually damages the otic vesicle but leaves the CVG, which is held from the neural tube by a short neural connection, intact.

3. Cut this connection between the neural tube and the CVG and isolate the CVG. It is now possible to recognize the two main parts of the ganglion, cochlear and vestibular, that at this stage remain fused. It is important to recognize the geniculate ganglion that sometimes remain adherent to the CVG and needs to be cleaned out by dissection (see Bernd and Represa, 1989).

4. Place isolated ganglia on 35-mm, four-well Petri dishes and cover with 250 μl of medium as described for the otic vesicles.

5. The standard culture medium consists of serum-free M-199 medium with Earle's salts, supplemented with 1% fetal calf serum, (FCS) 20 m*M* glutamine, 15 m*M* $NaCO_3H$, 10 m*M* HEPES, and antibiotics (penicillin, 50 UI/ml and streptomycin, 50 mg/ml). Incubations are carried out at 37°C in a water-saturated atmosphere containing 5% CO_2 (Avila *et al.*, 1993).

Studies on the proliferative phase of the CVG have been done using techniques similar to those used for the otic vesicle (see Fig. 2). Similarly, experiments on ^{3}H-thymidine uptake and precursor metabolite labeling, and protein expression studies, have revealed the regulatory role of neurotrophins (Varela-Nieto *et al.*, 1991; Represa *et al.*, 1991; Avila *et al.*, 1993) and other growth factors such as IGF-1 (León *et al.*, 1995b).

B. Cultures of Whole Cochleovestibular Ganglia for Neurite Outgrowth Assay

After 7 days of development *in ovo*, most neuroblasts in the CVG are postmitotic (D'Amico Martel, 1982). From incubation day 7 (E-7) to hatch is usually the interval selected for survival and neurite outgrowth studies.

1. At these older stages, cochlear and vestibular ganglia can be isolated separately. The head of the embryo is cut off and split by a mediosaggital incision. The brain is removed and the petrous portion of the temporal bone primordia is exposed. The upper wall of the internal auditory meatus and tegment tympani is removed and the vestibular ganglion is excised from the VIII cranial nerve. The whole membranous labyrinth is then isolated, and the cochlear duct opened longitudinally to dissect the cochlear ganglion from the otic epithelium of the basilar papillae.
2. Ganglia are placed on collagen-coated coverslips or collagen matrix gels. Coverslips are prepared by coating sequentially with polyornithine (Sigma; 500 μg/ml in phosphate saline for 2 hr at 20°C) and with the basement membrane protein laminin (Sigma, 10 μg/ml in PBS overnight at 4°C).
3. Matrix collagen gels are prepared from a stock solution of concentrated 3% rat tail collagen mixture in acid (Cellagen, Flow). To prepare 1 ml, this solution is mixed 1:8 with 10× M-199 medium and 100 μl of 0.5 *M* $NaCO_3$, and gently removed (do not stir because the collagen will not polymerize). Keep the mixture and all components on ice while pipetting. This solution can be readily used or frozen at −20°C for use later. The collagen gel mixture is pipetted in the middle of the 35-mm dishes to form a small dome with about 15–20 μl of solution. It requires about 10 min to gellify at room temperature, or less if put in the incubator at 37°C. This is enough time to allow placing the ganglia on the collagen gel matrix and to position them as desired. Typically, one ganglion is dipped in the center of the dome. After 10–20 min the gel matrix sets and the culture solution (see earlier) can be added gently, ensuring that the tip surface of the matrix dome is not submerged but in contact with the surface. This is important for correct gas exchange with the explant to take place.

Experiments on whole ganglion explants allowed the neurite outgrowth ability of neurotrophins and the temporal pattern of their action to be characterized

(Represa and Bernd, 1989; Avila *et al.*, 1993). Limitations to the use of explants of whole ganglia come mainly from the fact that although relatively simple, they are a mixture of at least two different cell types, neurons and glia, which are of different embryologic origins. Neurons derive from the otic placode, whereas glial cells are of neural crest origin. In addition, cells in the ganglion are densely packed and undergo cell–cell interactions. These sometimes may be desirable, but are not suitable for examining the behavior of isolated neurons or glial cells.

C. Culture of Cells Dissociated from Cochleovestibular Ganglia

1. Isolated cochlear or vestibular ganglia are incubated for 20–30 min in Ca^{2+}/Mg^{2+}-free Hank's balanced salt solution (OCa^{2+}/OMg^{2+} HBSS) containing 0.05% trypsin (or, alternatively, 0.5% dispase). Time required and the addition of trypsin may be modified according to the activity of enzymes and the age of explants. Early stages, E-6 to E-9, should not be incubated longer than 15 min, but older stages may require up to 30 min.
2. Trypsin solution is retired and 2 ml of culture medium containing 10% FCS is added to block the remaining trypsin trace. Ganglia are then centrifuged for 2 min at 100*g* and resuspended in standard medium containing 10% FCS, where they are now finally dissociated using a fine flame-polished Pasteur pipette. Dissociation can be controlled under the microscope. If after 10 passes through the pipette there are still any large tissue fragments, allow them to sediment and collect the supernatant. Going on with trituration always produces cell damage.
3. Cell suspension can be partially enriched for neurons by preplating on a poly-D-lysin-coated plastic dish for 30 min and collecting the medium after a gentle shake. In our experience, the number of neurons obtained using this procedure is severely reduced. A good alternative for obtaining neurons to be studied in isolation from possible interactions with glial and non-neural cells is to maintain the original cell suspension but plate at low density, about 500–1000 cells per well.
4. The cell suspension is adjusted with fresh culture medium containing 10% FCS to approximately 1000 cells/ml, and the cells mixed by gently inverting the tube. Cells are dispensed on the coated wells or coverslips and allowed to adhere for about 2 hr.
5. Cell suspension is seeded on acid-washed coverslips or plastic culture dishes treated with 100 μg/ml poly-D-lysine for 2 hr at 37°C. Culture dishes are washed three times with water, and 200 μl of laminin solution in PBS, 20 μg/ml, are added. Laminin is incubated at 37°C for 5 hr. The culture dishes are then washed three times with 1 ml PBS and kept at 37°C with PBS until the cell suspension is ready. Laminin should not be allowed to dry on the coverslips.
6. If a neurotrophic factor is going to be tested, careful and sequential changing of the medium should be done to reduce the serum content after 2 hr in

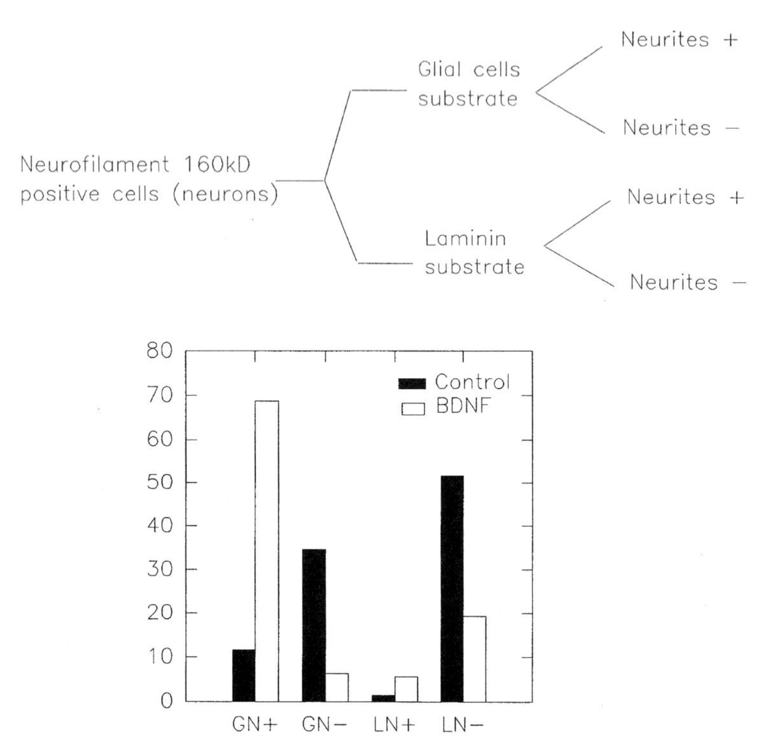

Fig. 3. Analysis of dissociated cell cultures from cochlear ganglia. The percentages of cells that were positive to 160-kDa neurofilament in control (closed bars) and with BDNF (5 ng/ml; open bars) are represented in ordinates. They were classified in four groups: cells growing on glial cells and extending neurites (GN+), cells growing on glial cells but not extending neurites (GN−), cells growing directly on laminin and extending neurites (LN+), and those growing on laminin but showing no neurite extension (LN−).

medium containing 10% FCS. Cells are grown in 500-μl final volume. Desired factors are added by replacing half of the volume of the culture.

7. Culture medium is Dulbecco's modified essential medium and nutrient mixture F-12 (1:1), where glutamate has been replaced by Glutamax-I. Medium is supplemented with glucose (2.8 g/l), transferrin (100 μg/ml), progesterone (60 ng/ml), putrescine (16 μg/ml), sodium selenite (30 n*M*), penicillin (100 U/ml), and streptomycin (100 μg/ml). No insulin is added to the culture medium.

Careful analysis of mixed neuron–glia cell cultures allow information to be obtained on the ability of neurotrophic molecules to support survival and differentiation of neuronal populations, their dependence on cell–cell interactions, and the induction of typical phenotypic properties. Figure 3 provides an example of how the neural response to brain-derived neurotrophic factor (BDNF) in cell

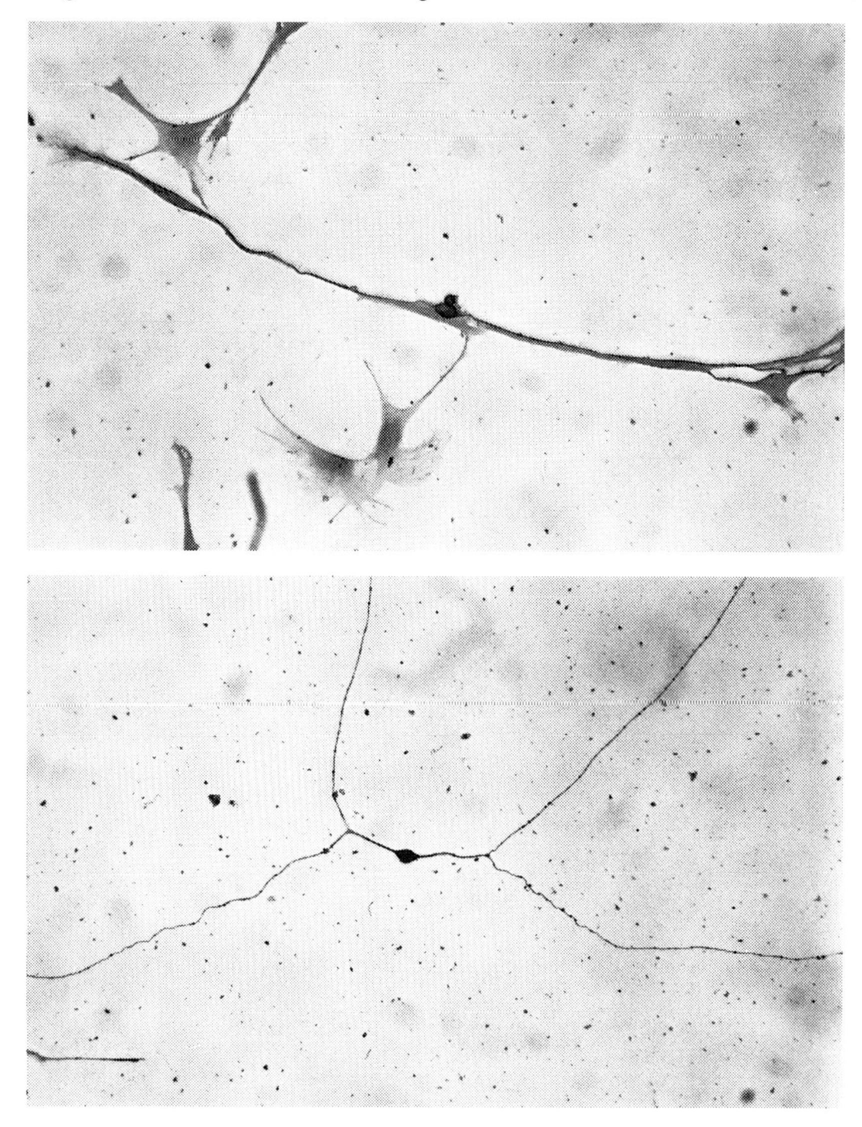

Fig. 4. Infection of CVG neurons by a virus carrying *bdnf*. Two representative photographs of a culture of dissociated CVG neurones from E-34, incubated with 5 ng/ml BDNF (upper photograph), or with a herpes simplex virus type 1 carrying the *bdnf* gene (lower photograph) Original magnification ×500. (Garrido *et al.*, 1996.)

cultures is analyzed. Dissociated cells from cochlear ganglions of E-334 chick embryos were cultured for 48 hr with BDNF. Cells were fixed and stained with an anti-neurofilament (160 kDd) antibody. All neurons in the plate were counted and classified according to the following criteria: 1) their ability to extend neurites longer than at least three cell body diameters, and 2) their contact with glial cells. The number of neurofilament-positive cells may be taken for the neurons that attached to the plate at time zero and gives the original density. That this assumption is most probably correct is indicated by the fact that both control and BDNF-cultured plates show the same number of stained cells. The number of cells that extend neurites should reflect the number of the surviving neurons. It is apparent from Figure 3 that the effect of exogenously added BDNF seems to depend on the interaction between neurons and glial cells. This situation is, however, different when BDNF is provided to the neurons through the viral infection of the cells carrying the *bdnf* vectors, as illustrated in Figure 4. This experiment again shows the ability of BDNF to support survival but, in this case, the population of neurons that is rescued by BDNF is shifted toward that growing directly on the laminin substrate.

The described cultures of cochlear and vestibular neurons are suitable for a number of dynamic studies, such as patch-clamp electrophysiologic analysis (Valverde *et al.*, 1991; Sheppard *et al.*, 1991; Jiménez *et al.*, 1996) and microfluorescence techniques (Jiménez and Nuñez, 1996). The main limitation of this technique is, no doubt, the number of cells available. A single ganglion provides about 15,000 neurons, a figure that is too small for most biochemical and molecular studies. In addition, one always has to bear in mind that dissociated neurons lose their interaction with the normal environment and their target tissues. After a certain period in culture, they might change some of their native properties.

IV. Concluding Remarks

Organ culture techniques applied to the otic vesicle and CVG have allowed progress in the study of the role of growth factors in the regulation of cell proliferation in sensory organs. We discuss here some of these studies.

A. Growth Factors and Cell Proliferation in the Early Developing Inner Ear

Cellular proliferation in the otic vesicle is under the control of several growth factors that are able to reinitiate DNA synthesis and cell proliferation in quiescent otic vesicles (Represa *et al.*, 1988). Mitogenic action is associated with an increase in the turnover of GPI (Varela-Nieto *et al.*, 1991; Represa *et al.*, 1991) and the rapid induction of the c-*fos* gene and protein (Represa *et al.*, 1990; León *et al.*, 1995a).

Insulin-like growth factors are members of a group of structurally homologous peptides encoded by genes that belong to the insulin gene family (Bondy *et al.*, 1990; Liu *et al.*, 1993; Baker *et al.*, 1993). The effects of IGF-1 on cultured otic vesicle have been studied by León *et al.* (1995b). IGF-1 induced DNA synthesis and increased cell number and mitotic rate in a dose-dependent manner with a half-maximal concentration of 10^{-10} *M*. IGF-2 and insulin were also able to induce growth, but with a lower potency. IGF-1 also stimulates growth in the CVG. Binding of IGF-1 to specific receptors occurred with high affinity, and an autoradiographic study of sections from otic vesicles showed the location of labeled IGF-1 in the otic epithelium and the CVG during stages of growth (León *et al.*, 1995b). IGF-1 has many properties that make it an interesting candidate for exerting regulatory control of cell division and growth both in the otic vesicle and the CVG during normal development. Thus far it is the most efficient growth factor for mimicking both serum and normal development during this critical proliferative period. It is well known that media used for long-term cultures of otic explants require high doses of insulin for viability (Van de Water and Ruben, 1974). One possible explanation for this observation is that at such high concentrations, multiple interactions take place between the different receptors and ligands of the insulin family of growth factors. Because IGF-1 is much more effective in inducing growth in the developing ear than any other member of the insulin family and is expressed in the otic vesicle, IGF-1 may be the endogenous insulin-like activity required during the normal development of the ear.

The c-*fos* proto-oncogene is an essential element in the propagation of the mitogenic signal regulating the expression of secondary genes that ultimately leads to cell division (Morgan and Curran, 1991). A relation between c-*fos* and cell proliferation in the otic vesicle was suggested by study of c-*fos* mRNA. Growth factors that stimulate cell proliferation in the otic vesicle induce a rapid and transient increase in c-*fos* expression and Fos protein and show a similar relationship with growth factors and retinoic acid (see later). Fos expression is developmentally regulated, and antisense oligonucleotides targeted against c-*fos* are able to inhibit Fos expression and growth of cultured otic vesicle (León *et al.*, 1995a, 1995b).

Retinoic acid influences the patterning and specification of several cell phenotypes during embryonic development (Conlon, 1995; Maden and Holder, 1992), and it has been suggested as a potential molecule to regulate cell differentiation in ear development (Represa *et al.*, 1990). In the proliferating otic vesicle, RA arrests growth factor-induced cell proliferation and stimulates differentiation of sensory and secretory cell types (Represa *et al.*, 1990). This role for RA has been further substantiated by results showing that the developing ear contains both RA-binding proteins and endogenous RA (Kelley *et al.*, 1993; Ruberte *et al.*, 1993), and that RA leads to formation of supernumerary hair cells "*in vitro*" (Kelley *et al.*, 1993). The c-*fos* proto-oncogene is downregulated by RA in cultured otic vesicles and this occurs in parallel with the inhibition of the rate of cell division. On the other hand, RA-dependent differentiation does not occur if Fos is not allowed to increase by using antisense oligonucleotides (León *et al.*, 1995b). Therefore, c-*fos* increase and downregulation seems to be a necessary sequence for the action of RA. However, c-*fos* regulation is certainly not the only mechanism involved in RA-induced hair cell formation, because c-*fos* downregulation with antisense oligonucleotides could not reproduce RA-induced patches of differentiated otic epithelium (León *et al.*, 1995a).

B. Multiple Roles of Neurotrophins in Ear Development

The NGF family of factors includes NGF, BDNF, and neurotrophins 3, 4, and 5 (NT3, NT4, and NT5), which are structurally homologous polypeptides sharing about 60% of amino acid identity and membrane receptors, as well as biologic effects (Hallbök *et al.*, 1991; Chao, 1992). These neurotrophins display distinct biologic effects on the developing CVG of mouse and avian embryos. First, they are able to stimulate cell division of isolated CVG during early proliferative periods of development (Varela-Nieto *et al.*, 1990; Represa *et al.*, 1991; Represa et al., 1993). This effect is coupled to the hydrolysis of a membrane GPI and the generation of a biologically active inositol-phosphoglycan (IPG), which acts as a powerful mitogen for the CVG (Represa *et al.* 1991). Second, neurotrophins are able to induce neurite outgrowth in both ganglion explants and isolated neurons. The effect is stage specific, and maximal during innervation and synaptogenesis (Avila *et al.*, 1993; Vazquez *et al.*, 1994). Finally, neurotrophins support survival of isolated neurons in culture (Avila *et al.*, 1993; Vazquez *et al.*, 1994). Analysis of dissociated neurons from CVG reveals that, along with maintaining survival, neurotrophins are able to rescue many phenotypic properties of the neurons. For instance, neurofilaments and microfilament-associated proteins are induced differentially by neurotrophins (unpublished observations), and expression of calcium channels is maintained in the presence of NT3 (Jiménez *et al.*, 1996), as are glutamate receptors (Jiménez and Nuñez, 1996).

The importance of neurotrophins for normal development of the ear has been strengthen by the analysis of mice carrying null-mutations for neurotrophic factors (BDNF or NT3) or their high-affinity receptors, the *trk* proto-oncogene family (Snyder, 1994; Barbacid, 1994). Loss of function of these genes results in a massive cell death and reduction in the number of cochlear and vestibular neurons. Careful analysis throughout development revealed that vestibular neurons are able to establish their peripheral connections in the absence of *trk*B expression, but fail to maintain this target innervation, suggesting that neurotrophins are indispensable for securing synaptic contacts (Ernfors *et al.*, 1995; Schimmang *et al.*, 1995, Minichiello *et al.*, 1995).

References

Anniko, M. (1983). Embryonic development of vestibular sense organs and their innervation. *In* "Development of Auditory and Vestibular Systems" Vol. 2. (R. Romand, ed.), pp. 375–423 Elsevier, Amsterdam.

Avila, M., Varela-Nieto, I., Romero, G., Mato, J. M., Giraldez, F., Van De Water, T. R., and Represa, J. (1993). Brain-derived neurotrophic factor and neurotrophin-3 support the survival and neuritogenesis response of developing cochleovestibular ganglion neurones. *Dev. Biol.* **159,** 266–275.

Baker, J., Liu, J., Robertson, E. J., and Efstratiadis, A. (1993). Role of insulin-like growth factors in embryonic and postnatal growth. *Cell* **75,** 73–82.

Barbacid, M. (1994). The *Trk* family of receptors. *J. Neurobiol.* **25,** 1386–1403.

Bernd, P., and Represa, J. (1989). Characterization and localization of nerve growth factor receptors in the embryonic otic vesicle and cochleovestibular ganglion. *Dev. Biol.* **134,** 11–20.

Bondy, C. A., Warner, H., Roberts, C., and LeRoith, D. (1990). Cellular pattern of insulin-like growth factor-I and type-I IGF receptor gene expression in early organogenesis: Comparison with IGF-II gene expression. *Mol. Endocrinol.* **4,** 1386–1398.

Chao, M. V. (1992). Neurotrophin receptors: A window into neuronal differentiation. *Neuron* **9,** 583–593.

Conlon, R. A. (1995). Retinoic acid and pattern formation in vertebrates. *Trends Genet.* **11,** 314–319.

D'Amico Martel, A. (1982). Temporal patterns of neurogenesis in avian cranial sensory and autonomic ganglia. *Am. J. Anat.* **163,** 351–372.

D'Amico Martel, A., and Noden, D. M. (1983). Contributions of placodal and neural crest cells to avian peripheral ganglia. *Am. J. Anat.* **366,** 445–468.

Ernfors, P., Van De Water, T. R., Loring, J., and Jaenisch, R. (1995). Complementary roles of BDNF and NT-3 in vestibular and auditory development. *Neuron* **14,** 1153–1164.

Friedman, I., Hodges, G. M., and Riddle, P. N. (1977). Organ culture of the mammalian and avian embryo otocyst. *Ann. Otol. Rhinol. Laryngol.* **86,** 371–381.

Garrido, J. J., Alonso, M. T., Lim, F., Represa, J., Giraldez, F., and Schimmang, T. (1996). Using herpes simplex virus type 1 (HSV-1) mediated gene transfer to study neurotrophins in cochlear neurons. *Int. J. Dev. Biol.* (suppl. 1), 149–150.

Hallbök, F., Ibañez, C. F., and Person, H. (1991). Evolutionary studies of the nerve growth factor reveal a novel member abundantly expressed in *Xenopus* ovary. *Neuron* **6,** 845–858.

Hamburger, V., and Hamilton, H. L. (1951). A series of normal stages in the development of the chick embryo. *J. Morphol.* **88,** 49–92.

Jiménez, C., and Nuñez, L. (1996). Glutamate receptors in the developing cochlear ganglion. *Int. J. Dev. Biol.* (suppl. 1), 159–160.

Jiménez, C., Represa, J., Giraldez, F., and García-Díaz. (1996). Calcium currents in dissociated cochlear neurons from the chick embryo and their modification by neurotrophin-3. *Neuroscience* **77,** 673–682.

Kelley, M. W., Xiao-Mei, X., Wagner, M. A., Warchol, M. E., and Corwin, J. T. (1993). The developing organ of Corti contains retinoid acid and forms supernumerary hair cells in response to exogenous retinoid acid in culture. *Development* **119,** 1041–1053.

Liu, J., Baker, J., Perkins, A. S., Robertson, E. J., and Efstratiadis, A. (1993). Mice carrying null mutations of the genes encoding insulin-like growth factor I and type 1 IGF receptor. *Cell* **75,** 59–72.

León, Y., Sánchez, J. A., Miner, C., McNaughton, L. M., Represa, J., and Giraldez, F. (1995a). Developmental regulation of Fos protein during the proliferative growth of the otic vesicle. *Dev. Biol.* **167,** 75–86.

León, Y., Vazquez, E., Sanz, C., Vega, J. A., Mato, J. M., Giraldez, F., Represa, J., and Varela-Nieto, I. (1995b). Insulin-like growth factor-I regulates cell proliferation in the developing inner ear, activating glycosyl-phosphatidylinositol hydrolysis and Fos expression. *Endocrinology* **136,** 3494–3503.

Maden, M., and Holder, N. (1991). The involvement of retinoic acid in the development of the vertebrate nervous system. *Development Supplement* **2,** 87–94.

Minichiello, L., Piehl, F., Vazquez, E., Schimmang, T., Hökfelt, T., Represa, J., and Klein, R. (1995). Differential effects of combined *trk* receptor mutations on dorsal root ganglion and inner ear sensory neurones. *Development* **121,** 4067–4075.

Morgan, J. I., and Curran, T. (1991). Stimulus–transcription coupling in the nervous system: Involvement of the inducible proto-oncogenes *fos* and *jun*. *Ann. Rev. Neurosci.* **14,** 421–451.

Nieto, M. A., Patel, K., and Wilkinson, D. G. (1996). *In situ* hybridisation analysis of chick embryos in whole-mount and tissue sections. *Methods Cell Biol.* **51,** 219–235.

Orr, M. F. (1986). Histogenesis of sensory epithelium in reaggregates of dissociated embryonic chick otocysts. *Dev. Biol.* **17,** 39–54.

Represa, J., Avila, M. A., Miner, C., Giraldez, F., Romero, G., Clemente, R., Mato, J. M., and Varela-Nieto, I. (1991). Glycosyl-phosphatidylinositol/inositol phosphoglycan: A signalling system for the low affinity nerve growth factor receptor. *Proc. Natl. Acad. Sci. U.S.A.* **88,** 8016–8019.

Represa, J., Avila, M. A., Romero, G., Mato, J. M., Giraldez, F., and Varela-Nieto, I. (1993). Neurotrophic factors regulate cell proliferation during inner ear development through glycosyl-phosphatidylinositol hydrolysis. *Dev. Biol.* **159,** 257–265.

Represa, J., and Bernd, P. (1989). Nerve growth factor and serum differentially regulate development of the otic vesicle and cochleovestibular ganglion in vitro. *Dev. Biol.* **134,** 21–29.

Represa, J., Miner, C., Barbosa, E., and Giraldez, F. (1988). Bombesin and other growth factors activate cell proliferation in chick embryo otic vesicles in culture. *Development* **102,** 87–91.

Represa, J., Sánchez, A., Miner, C., Lewis, J., and Giraldez, F. (1990). Retinoic acid modulation of the early development of the inner ear is associated with the control of c-*fos* expression. *Development* **110,** 1081–1090.

Ruberte, E., Friederich, V., Morris-Kay, and Chambon, P. (1993). Differential distribution of CRABP I and CRABP II transcripts during mouse embryogenesis. *Development* **115,** 973–987.

Schimmang, T., Minichiello, L., Vazquez, E., San Jose, I., Giraldez, F., Klein, R., and Represa, J. (1995). Developing inner ear sensory neurons require *trk*B and *trk*C receptors for innervation of their peripheral targets. *Development* **121,** 3381–3391.

Sheppard, D. N., Valverde, M. A., Represa, J., and Giraldez, F. (1991). Transient outward current in cochlear neurones of the chick embryo. *Neuroscience* **50,** 631–639.

Snyder, W. D. (1994). Functions of the neurotrophins during nervous system development: What the knockouts are teaching us. *Cell* **77,** 627–638.

Swanson, G. J., Howard, M., and Lewis, J. (1990). Epithelial autonomy in the development of the inner ear of a bird embryo. *Dev. Biol.* **137,** 243–257.

Valverde, M. A., Sheppard, D. N., Represa, J., and Giraldez, F. (1991). Development of Na and K currents in the cochlear ganglion of the chick embryo. *Neuroscience* **50**, 621–630.

Van de Water, T. R., and Ruben, R. J. (1974). Growth of the inner ear in organ culture. *Ann. Otol. Rhinol. Laryngol.* **83,** 1–16.

Varela-Nieto, I., Alvarez, L., and Mato, J. M. (1993). Intracellular mediators of peptide hormone action. *In* "Handbook of Endocrine Research Techniques." (F. de Pablo F. and C. G. Scanes, eds.), Academic Press, New York.

Varela-Nieto, I., Represa, J., Avila, M. A., Miner, C., Mato, J. M., and Giraldez, F. (1991). Inositol phospho-oligosaccharide stimulates cell proliferation in the early developing inner ear. *Dev. Biol.* **143,** 432–435.

Vazquez, E., Van de Water, T. R., Del Valle, M., Vega, J. A., Staeker, H., Giraldez, F., and Represa, J. (1994). Pattern of *trk*B protein-like immunoreactivity and in vitro effects of brain-derived neurotrophic factor on developing cochlear and vestibular neurons. *Anat. Embryol.* **189,** 157–167.

7

Organoculture of the Chick Embryonic Neuroretina

Enrique J. de la Rosa, Begoña Díaz, and Flora de Pablo
Department of Cell and Developmental Biology
Centro de Investigaciones Biológicas
Consejo Superior de Investigaciones Científicas
E-28006 Madrid, Spain

I. Introduction

Primary cell cultures are an essential tool to test in a controlled environment the effects of growth factors and other extracellular signals on cellular processes, such as proliferation, differentiation, and survival. Culture systems allow the study of multiple aspects of cellular physiology, from extracellular signals (factor dose–response curves, binding to receptors), through the transduction cascade (activation of receptors and intracellular substrates, translocations of mediators), to the targets (enzymatic modulation, gene transactivation), and finally the resulting cellular response. Dissociated cell cultures represent the maximal achievement in reducing the endogenous signals reaching the target cells, thus permitting easier identification of primary effects of added molecules. However, dissociated cell culture conditions may be far from the physiologic situation, especially in the case of complex tissues such as the central nervous system, because the very important cell-to-cell and matrix-to-cell interactions are destroyed. In contrast,

Current Topics in Developmental Biology, Vol. 36

0070-2153/98 $25.00

many of these anatomic and functional interactions are well preserved in organotypic cultures (e.g., whole-organ explants, tissue slices), providing a system closer to the physiologic situation. We are studying retinal neurogenesis by exploiting the advantages of both dissociated cell cultures (De la Rosa *et al.*, 1994; Hernández-Sánchez *et al.*, 1994) and organotypic cultures (Hernández-Sánchez *et al.*, 1995). A thoughtful evaluation of the possible use of organotypic cultures should be done in the approach to each particular biologic problem.

The chick neuroretina is a portion of the central nervous system well characterized in terms of neuroanatomic patterns and stages of development (Kahn, 1974; Rager, 1980; Spence and Robson, 1989; Prada *et al.*, 1991; Belecky-Adams *et al.*, 1996). Much larger than the rodent retina and accessible to *in vivo* manipulation *in ovo*, the chick embryo retina is suitable for a comprehensive approach that includes biochemical, molecular, and cellular studies, as well as for establishing correlation between *in vitro* and *in vivo* studies. Between days 4 and 7 of embryonic development (E4–E7), the chick retina consists largely of a proliferating neuroepithelium, and differentiation proceeds in a central-to-peripheral gradient. A number of factors controlling those processes have been characterized because neurogenesis also proceeds *in vitro* (reviewed in Cepko *et al.*, 1996).

This chapter focuses on the chick whole-retina culture system and on the assays to evaluate some of the cellular processes occurring in the culture. The protocols can be adapted either to retinas of other vertebrates or to other embryonic organs.

II. Whole Neuroretina Culture

A previously existing protocol to culture avian neuroretina (Halfter and Deiss, 1986) has been adapted to serum-free conditions, an important requisite when testing the biologic effects of growth factors present in the serum. Serum-free conditions ideally reduce the complexity of the cellular response, facilitating analysis of the primary effects of the added growth factors.

A. Dissection of the Neuroretina

Work should be performed under the aseptic working conditions usual for cell and tissue cultures. All reagents must be sterile. The dissection tools are cleaned with 70% (v/v) ethanol. The use of a horizontal-flow sterile hood is recommended when the cultures are going to be kept longer than 1 day. Shorter culture time requires only a clean area (wipe the binoculars and the surrounding area with 70% [v/v] ethanol as well).

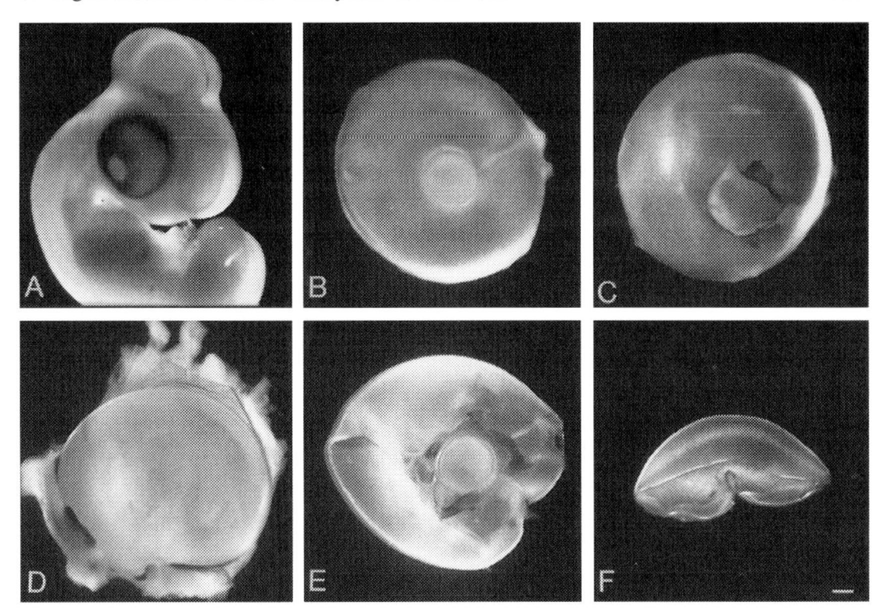

Fig. 1. Stepwise dissection of the chick neuroretina. (A) E5 chick embryo out of the shell. (B) Eye removed from the embryo; anterior view with the lens in the center. (C) Starting dissection; posterior view with the optic nerve head in the center. The pigmented epithelium is first detached around the optic nerve head. (D) Proceeding with dissection; posterior view. The pigmented epithelium is removed from the posterior to the anterior pole. (E) Lens and vitreous humor are then removed from the front. (F) Dissected neuroretina. Scale bars = 1 mm in (A), and 0.3 mm in (B–F).

1. Wipe the egg shell with 70% (v/v) ethanol.
2. Cut the egg shell at the air chamber pole with curved scissors and remove it to expose the embryo.
3. Take the embryo out by the neck, using curved forceps, and place it into a 35-mm Petri dish filled with phosphate-buffered saline (PBS) prewarmed to 37°C (Fig. 1A). Cut the head and discard the body.
4. From now on, work under magnifying binoculars. Remove the eyes from the embryo head by gently tearing apart the surrounding connective tissue with two dissecting forceps (Dumont No. 5) and place the eyes in prewarmed PBS (Fig. 1B). If an eye is punctured and starts collapsing during the dissection procedure, it should be discarded.

5. The neuroretina is then microdissected free of its pigmented epithelium and surrounding connective tissue. Start by tearing and detaching the pigmented epithelium around the optic nerve head (Fig. 1C) and from there peel off the retina like a fruit, as if the pigmented epithelium were the skin (Fig. 1D). The lens and vitreous body are then removed from the front side (Fig. 1E). The neuroretina can now be placed in the culture medium (Fig. 1F). To avoid tissue disruption, a P1000 micropipette should be used for the transfer, with a blue tip cut adjusted to the retina size.

For optimal retina preservation, try to complete the dissection in less than 5 min per retina.

Alternatively, to study the influence of the pigmented epithelium on the neuroretina, this layer can be left in its position in close contact with the neuroretina. In this case, a very careful removal of the connective tissue surrounding the eye should be performed.

B. Culture Media

The standard medium used in this organoculture consists of F-12–Dulbecco's modified Eagle medium supplemented with 5 μg/ml insulin, 100 μg/ml transferrin, 16 μg/ml putrescine, 6 ng/ml progesterone, 5.2 ng/ml sodium selenite, and 50 μg/ml gentamicin (all can be obtained from Sigma or Gibco), except if any of them is the tested factor, in which case it is, obviously, excluded. Prepare sterile stocks 100× for insulin in HC1 10 m*M*, 100× for transferrin in H_2O, 10,000× for putrescine and selenite in H_2O, and 10,000× for progesterone in ethanol.

With this medium, the early embryonic retina maintains growth and proliferation longer even than 48 hr (see later, and Fig. 2A). However, we prefer short-term studies (Fig. 2B) to avoid medium conditioning by endogenous synthesis of growth factors. In cultures longer than 24 hr, this can be partially overcome by using large volumes of medium (2 ml per E5 retina) and by changing medium daily. In addition, the longer the culture time, the more distant it becomes from physiologic conditions. A careful evaluation should be done to determine the shortest culture time to obtain significant results in evaluating the cellular response to the applied treatment.

If desired, when working with factors not present in serum, basal F-12–Dulbecco's modified Eagle medium can be supplemented with 10% fetal calf serum (FCS) and 2% chicken serum. Alternatively, 10% egg white, a more physiologic source of growth factors for the avian embryo, can be added to the basal medium.

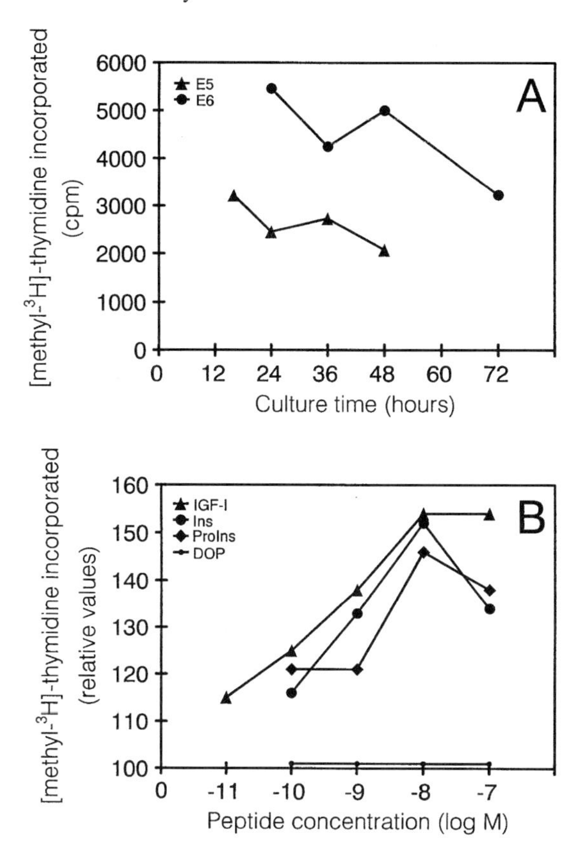

Fig. 2. DNA biosynthesis in cultured neuroretinas. (A) Time course of [methyl-^{3}H]-thymidine incorporation in cultured E5 and E6 retinas. [methyl-^{3}H]-thymidine was added 2 hr before the collection of the tissue at the indicated time points (data reproduced from A. López-Carranza, 1995, Doctoral Thesis, Universidad Complutense de Madrid, Spain). (B) Dose–response of E6 retinas to insulin (Ins), proinsulin (ProIns), insulin-like growth factor-I (IGF-I), and desoctapeptide insulin (DOP) added to the culture medium (in this case without basal insulin). At the end of a 16-hr culture period, the retinas were pulsed for 2 hr with [methyl-^{3}H]-thymidine (data reproduced from Hernández-Sánchez *et al.*, 1995, with permission).

III. Evaluation of Cellular Processes in the Whole Neuroretina

Once the whole neuroretina is in culture, the cellular response to culture conditions can be evaluated either in the whole retina or in dissociated single cells. Evaluating the neuroretina as a whole gives an average view of how the retina is responding to the culture treatments. This procedure is faster than the evaluation at individual cell level (see IV) and, therefore, ideal for preliminary approaches or studies with a large number of experimental points.

A. Proliferation Assay

In highly proliferating tissues, such as the embryonic retina, cell proliferation directly correlates with DNA synthesis. Therefore, proliferation can be evaluated by measuring the incorporation of [methyl-^{3}H]-thymidine as follows.

1. Add to the medium 0.15–0.5 μCi/ml of [methyl-^{3}H]-thymidine (70 Ci/mmol; Amersham) at the desired time. If the culture medium contains FCS, a higher amount of [methyl-^{3}H]-thymidine should be added to overcome dilution by the unlabeled thymidine present in FCS.
2. Collect the neuroretina after the desired labeling interval using a P1000 micropipette with a blue tip cut adjusted to the retina size. Two-hour pulses allow for a better comparison of DNA synthesis rates than continuous labeling (Fig. 2). Wash the retina twice with ice-cold PBS, and precipitate it in 0.5 ml of ice-cold 5% (w/v) trichloroacetic acid for at least 30 min at 4°C.
3. Slightly homogenize the tissue in an Eppendorf tube with a fitting homogenizing microdevice and spin it down (20,000 *g*, 5 min, 4°C).
4. Wash the precipitated tissue three times with 1 ml of cold 5% (w/v) trichloroacetic acid and twice with 1 ml of cold ethanol by spinning down in an Eppendorf tube.
5. Dissolve the tissue pellet in 1% (w/v) sodium dodecyl sulfate (SDS), 0.1 N NaOH (100 μl per E6 retina), by heating at 60°C until dissolution is completed.
6. Cool down to room temperature, add scintillation liquid (check the required amount depending on the water compatibility of your product), and determine the incorporated [methyl-^{3}H]-thymidine in a β-scintillation counter.

B. Differentiation Assay

Average cell differentiation in the whole cultured retina can be evaluated by the level of expression of neuronal and glial markers in Western blots. We usually use the axonal protein G4 (Rathjen *et al.*, 1987; De la Rosa *et al.*, 1990; Hernández-Sánchez *et al.*, 1995), but the same assay is valid using any monoclonal antibody (mAb) or antiserum detecting differentiation antigens by immunoblot (Pimentel *et al.*, 1996).

1. Collect the neuroretina after the desired culture interval. Because this assay measures total amount of the marker, to obtain a significant difference between various treatments may require longer culture times than those previously described to evaluate proliferation rates using radioactive pulses. An alternative

protocol to measure the expression rates of the marker could be the combination of radioactive metabolic labeling and immunoprecipitation.

2. Solubilize the tissue by boiling in SDS-polyacrylamide gel electrophoresis (PAGE) sample buffer at 25 mg of retina wet weight per milliliter. For early embryonic chicken retina, or when using rodent retina, it may be necessary to pool several of them.

3. Fractionate the sample by SDS-PAGE and transfer the proteins to nitrocellulose or Immobilon-P (PVDF, Millipore) membranes, using the conditions adequate for your marker. Blots should be stained to confirm equal protein load in every lane. Soak the blot for 5 min in Ponceau S (Sigma; 0.1% [w/v] in 5% [v/v] acetic acid). Take a photograph if required, and destain in PBS.

4. Block the membrane by incubation in 8% (wt/vol) powdered skim milk in PBS for a minimum of 2 hr at room temperature.

5. Incubate the blot sequentially with mAb G4 (mouse ascites, 1/3000 dilution) for 60 min at room temperature, and peroxidase goat anti-mouse IgG + IgM (Jackson Laboratories, 1/8000 dilution) for 45 min at room temperature. Four washing steps of 5 min between incubations, as well as the dilutions, are done with PBS containing 0.1% (w/v) Tween 20 and 2 mg/ml bovine serum albumin (BSA). The previous conditions are indicative and should be optimized for each different marker. Blots can be stained sequentially for several markers.

6. For maximal sensitivity, develop the blot with the ECL system (Amersham) or any other chemiluminescence kit, according to the manufacturer's instructions, and expose it to autoradiography film, such as Kodak X-Omat or BioMax.

7. The specific bands can be quantified in a densitometer. Accurate quantification requires prefogging of the film.

C. Detection of Apoptosis

If a treatment induces massive apoptosis, this can be also evaluated in the whole neuroretina by the characteristic pattern of the DNA electrophoretic ladder (Morales *et al.*, 1997).

1. Collect the neuroretina after the desired culture interval. Homogenize the tissue (one E6 retina) in 400 μl of lysis buffer (10 m*M* Tris-HCl, pH 7.5, 1 m*M* [EDTA], 0.2% Triton X-100) for 15 min at 4°C.
2. Isolate the fragmented DNA from the intact nuclei by centrifugation at 20,000*g*, 15 min, 4°C.
3. Incubate the supernatant containing the fragmented DNA at 55°C for 6 hr with 400 μl of 10 m*M* Tris-HC1, pH 7.5, 300 m*M* NaCl, 80 m*M* EDTA, 2% SDS, and 200 mg/ml proteinase K.
4. Purify the DNA by standard phenol–chloroform extraction, precipitation in ethanol, washing, and drying.

5. Dissolve the pellet in 20 μl of TE buffer (10 m*M* Tris-HCl, pH 7.5, 1 m*M* EDTA), and electrophorese the DNA in a 1.5% (w/v) agarose gel, containing 0.1 mg/ml ethidium bromide. The nucleosome-size DNA fragments can be visualized and photographed under ultraviolet light.

IV. Evaluation of Cellular Processes in Dissociated Retinal Cells

While culturing the retina as a whole, it is feasible to dissociate it afterward to evaluate multiple markers in a single cell. This approach allows for in-depth analysis of the progression of neurogenesis. For instance, cells labeled with both [methyl-^{3}H]-thymidine and a neuronal or glial marker are those that have left the cell cycle and have differentiated under the culture treatments. Apoptotic cell death can be observed and quantified in specific cell subpopulations.

A. Retina Dissociation Procedure

Before any detection, the whole cultured neuroretina has to be dissociated to single cells. Additional considerations about tissue dissociation can be found in Chapter 12. Here, we refer a protocol valid for E5–E6 chick retina.

1. Collect the neuroretina after the desired culture interval. Rinse twice in cold PBS.
2. Optimal preservation of most of the markers is achieved by immediate fixation in 4% (w/v) paraformaldehyde in 0.1 *M* phosphate buffer, pH 7.1 for 1 hr.
3. After fixation, wash the retina twice with 1 ml PBS containing 3 mg/ml BSA by spinning down for 5 sec in an Eppendorf tube.
4. Incubate the retina at 37°C in 1 ml of PBS containing 1.5 mg/ml of BSA and 2.5 mg/ml of trypsin (freshly prepared; TRL from Worthington Biochemical Corporation, NJ) for 30 min, passing it three to five times every 10 min through a siliconized Pasteur pipette.
5. Stop the reaction by adding either a twofold excess of soybean trypsin inhibitor (Sigma), or 30 mg/ml BSA.
6. Aliquots of the cell suspension, tipically 50,000 cells in 100 μl, are deposited on a slide using a cytospin at 700 rpm for 7 min.

Some markers may require an alternative fixation, such as methanol or acetone. Dissociation can also be performed in fresh tissue (as previously, but reducing the trypsin concentration to 0.5 mg/ml), and the cells fixed afterward in the appropriate fixative or even kept alive for surface labeling.

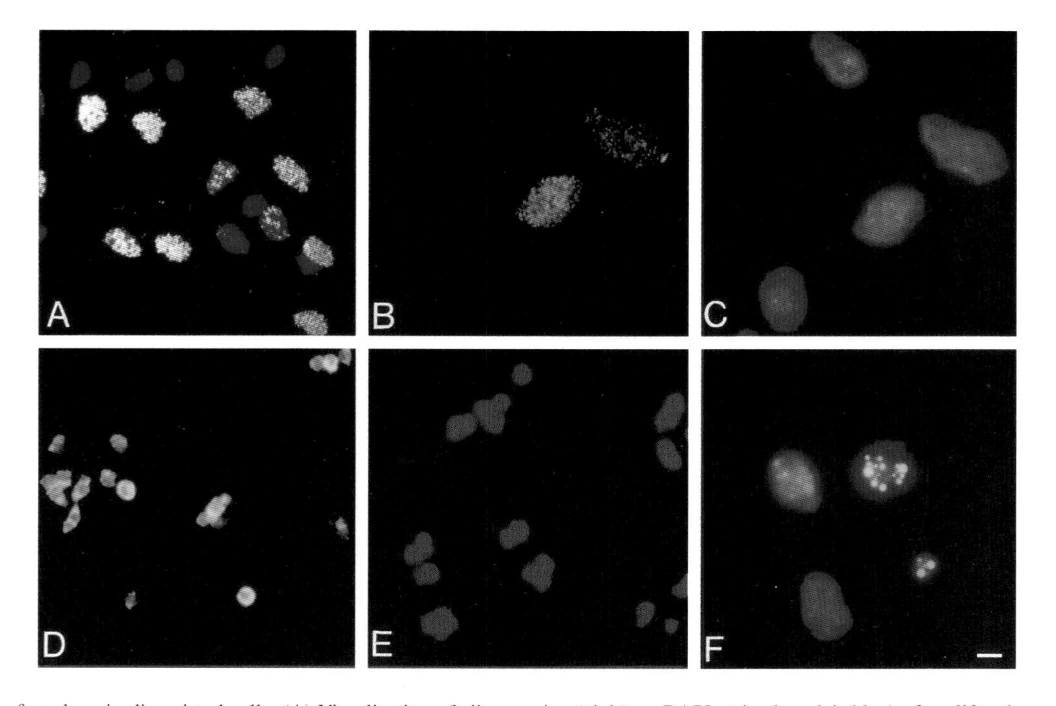

Fig. 3. Visualization of markers in dissociated cells. (A) Visualization of silver grains (pink) on DAPI-stained nuclei (blue) of proliferating cells by combining dark-field and fluorescence. (B) Cells in cycle can be evaluated by immunostaining for PCNA, using Cy2 as fluorochrome. (C) The same field as in B counterstained with DAPI; note that there are two more cells negative for PCNA. (D) Expression of the cytosolic precursor marker PM1 by immunofluorescence with Cy2; note the different levels of PM1 staining. (E) Expression of β-tubulin by immunofluorescence with Texas red; note the intense staining in all cells that cover the nuclei. (F) Apoptosis can be evaluated by the presence of pyknotic nuclei after DAPI staining, as in the two cells on the right. Scale bars = 3 μm in B, C, and F, 6 μm in A, and 12 μm in D and E.

B. Proliferation Assay

Proliferating cells can be visualized after [methyl-^{3}H]-thymidine incorporation and autoradiography (Fig. 3A). Alternatively, Br-d-uridine labeling followed by immunostaining can be used. For radioactive labeling, proceed as follows:

1. Culture the retina in the presence of 2 μCi/ml of [methyl-^{3}H]-thymidine (70 Ci/mmol; Amersham). Note that a higher activity or longer labeling time is required than in the case of whole retina processing (see III.A).
2. After the desired culture and labeling interval, process the cells as described in IV.A.
3. Cells are then dehydrated by washing through ethanol of increasing concentration (15 sec in each 50% [v/v], 70%, 85%, 96%, and twice in absolute ethanol).
4. Perform the dipping step in a darkroom. Cells on the slides are covered by nuclear emulsion NTB2 (Kodak) melted at 41°C and diluted 1:2 with H_2O containing 2% (v/v) glycerol.
5. Dipped slides are allowed to dry for 2 hr in the dark before being kept in a black box with desiccant for 7 days at 4°C.
6. The emulsion is developed in D-19 developer (Kodak) for 3 min at 14°C, fixed in 25% (w/v) sodium thiosulfate for 5 min, and exhaustively washed for 15 min in running water.
7. The cells can be observed using phase-contrast microscopy and the silver grains in bright or dark field, once mounted by any standard procedure. However, counterstaining of the cells allows a faster counting. Particularly convenient is 4′, 6-diamidino-2-phenylindole (DAPI) staining of the nuclei (Fig. 3A; see IV.D), which allows quicker quantification and simultaneous evaluation of apoptosis (Fig. 3F).

C. Detection of Cellular Markers

Many different markers can be detected by immunocytochemistry, including proliferating cell markers, such as proliferating cell nuclear antigen (Fig. 3B; Boehringer-Mannheim), neural precursor markers, such as PM1 (Fig. 3D; Hérnandez-Sánchez *et al.*, 1994), or tubulin (Fig. 3E; Sigma).

1. Collect the neuroretina after the desired cultured interval and process the cells as described in IV.A.

2. The defection of cytoplasmic antigens requires further permeation after fixation. Incubate the cells in 0.05% (wt/vol) Triton X-100, 2 mg/ml BSA, and 100 m*M* glycine in PBS (permeation-washing buffer) for 30 min at room temperature.

3. Block the preparations by incubation with 15% (vol/vol) normal goat serum in permeation-washing buffer for 15 min at room temperature.

4. Incubate the cells sequentially in the primary antibody (test dilutions around 1/1000 for mouse ascites and 1/5000 for rabbit antiserum, optimizing for each marker) for 45 min at room temperature; then in biotinylated goat anti-mouse IgG + IgM or biotinylated goat anti-rabbit (Amersham, 1/200 dilution) for 30 min at room temperature; and finally in Cyanine 2-streptavidin or Cyanine 3-streptavidin (Amersham, 1/200 dilution) for 30 min at room temperature. Four washing steps between incubations, as well as the dilutions, are done with the permeation-washing buffer.
5. Mount the slides for epifluorescence microscopy. The very low fading and bleaching of cyanines allows for mounting in simple glycerol:PBS (9:1).
6. Again, DAPI staining of the nuclei (see IV.D) allows easier quantification (Fig. 3B,C).

Immunostaining of surface molecules can be performed in intact cells as described previously (see also comment in IV.A), but the incubation times should be reduced by a half and the washing buffer replaced by PBS. The cells are fixed at the end by incubation with 4% (wt/vol) paraformaldehyde in 0.1 *M* phosphate buffer, pH 7.1 for 30 min at room temperature.

D. Measuring Apoptosis

As a first approach, apoptosis can be easily evaluated by the presence of pyknotic nuclei (Fig. 3F).

1. Collect the neuroretina after the desired culture interval and process the cells as described in IV.A.
2. Stain the cells by mounting the slides with 4 μg/ml DAPI (Sigma) and 1 mg/ml o-phenylenediamine (Sigma) in glycerol:PBS (9:1).

We advise in addition the use of more sensitive protocols, such as TUNEL labeling, detailed in Chapter 15.

V. Concluding Remarks

The protocols described here apply to E5–E6 chick neuroretinas. Culturing or processing younger or older retinas requires adequate scaling of the methods. In addition, these protocols could be applicable to retinas from other vertebrates after optimization. Comparable systems, culture of the whole chick embryo and culture of the otic vesicle, are presented in Chapters 2 and 6.

The degree of physiologic integrity provided by organocultures lies between that provided by dissociated cell culture and *in ovo* manipulation. The protocols described here for evaluation of proliferation, differentiation, and apoptosis can be directly used or easily adapted to retinas treated *in ovo* by intravitreal injection or systemic delivery of factors, antibodies, antisense oligonucleotide molecules, or any other compound. Detailed and well controlled studies in culture can be physiologically validated in their essential aspects by parallel *in ovo* manipulations.

Future technical developments pursue the *in toto* analysis of cellular processes, because the developmental gradients of the retina are not taken into consideration by the protocols described here. Protocols based on those presented here, but including confocal imaging and image processing and quantification, will add the very important positional information to the evaluation of response to growth factors.

Acknowledgments

The participation of Ana López-Carranza, who provided Figure 2A, and Catalina Hernández-Sánchez in the establishment of the protocols described here, is gratefully acknowledged. We also thank José M. García-Pichel for critical reading of the manuscript. Research in our laboratory is supported by grants from DGICYT (PM94-152 and PM96-0003), FIS (94/151), and CAM (AE 376/95).

References

Belecky-Adams, T., Cook, B., and Adler, R. (1996). Correlations between terminal mitosis and differentiation fate of retinal precursor cells *in vivo* and *in vitro*: Analysis with the "window-labeling" technique. *Dev. Biol.* **178,** 304–315.

Cepko, C. L., Austin, C. P., Yang, X., Alexiades, M., and Ezzeddine, D. (1996). Cell fate determination in the vertebrate retina. *Proc. Natl. Acad. U.S.A.* **93,** 589–595.

De la Rosa, E. J., Arribas, A., Frade, J. M., and Rodríguez-Tébar, A. (1994). Role of neurotrophins in the control of neural development: Neurotrophin-3 promotes both neuron differentiation and survival of cultured chick retinal cells. *Neuroscience* **58,** 347–352.

De la Rosa, E. J., Kayyem, J. F., Roman, J. M., Stierhof, Y. D., Dreyer, W. J., and Schwarz, U. (1990). Topologically restricted appearance in the developing chick retinotectal system of Bravo, a neural surface protein: Experimental modulation of environmental cues. *J. Cell Biol.* **111,** 3087–3096.

Halfter, W., and Deiss, S. (1986). Axonal pathfinding in organ-cultured embryonic avian retinae. *Dev. Biol.* **114,** 296–310.

Hernández-Sánchez, C., Frade, J. M., and de la Rosa, E. J. (1994). Heterogeneity among neuroepithelial cells in the chick retina revealed by immunostaining with monoclonal antibody PM1. *Eur. J. Neurosci.* **6,** 105–114.

Hernández-Sánchez, C., López-Carranza, A., Alarcón, C., de la Rosa, E. J., and de Pablo, F. (1995). Autocrine/paracrine role of insulin-related growth factors in neurogenesis: Local expression and effects on cell proliferation and differentiation in retina. *Proc. Natl. Acad. Sci. U.S.A.* **92,** 9834–9838.

Kahn, A. J. (1974). An autoradiographic analysis of the time of appearance of neurons in the developing chick retina. *Dev. Biol.* **38,** 30–40.

Morales, A. V., Serna, J., Alarcón, C., de la Rosa, E. J., and de Pablo, F. (1997). Role of prepancreatic (pro)insulin and the insulin receptor in prevention of embryonic apoptosis. *Endrocrinology* (in press).

Pimentel, B., de la Rosa, E. J., and de Pablo, F. (1996). Insulin acts as an embryonic growth factor for *Drosophila* neural cells. *Biochem. Biophys. Res. Commun.* **226,** 855–861.

Prada, C., Puga, J., Pérez-Méndez, L., López, R., and Ramírez, G. (1991). Spatial and temporal patterns of neurogenesis in the chick retina. *Eur. J. Neurosci.* **3,** 559–569.

Rager, G. H. (1980). Development of the retinotectal projection in the chicken. *Adv. Anat. Embryol. Cell Biol.* **63**.

Rathjen, F. G., Wolff, J. M., Frank, R., Bonhoeffer, F., and Rutishauser, U. (1987). Membrane glycoproteins involved in neurite fasciculation. *J. Cell Biol.* **104,** 343–353.

Spence, S. G., and Robson, J. A. (1989). An autoradiographic analysis of neurogenesis in the chick retina in vitro and in vivo. *Neuroscience* **32,** 801–812.

8

Embryonic Explant and Slice Preparations for Studies of Cell Migration and Axon Guidance

Catherine E. Krull and Paul M. Kulesa
Division of Biology
California Institute of Technology
Pasadena, California 91125

I. Introduction

One of the fundamental aims in developmental biology is to understand the cellular and molecular mechanisms that underlie the correct patterning of various cell types and tissues in the embryo. In some regions of the developing nervous system, neural precursors or neuronal processes travel extensively along stereotypical pathways to their target regions. Substantial effort has been devoted to examining how these migrating cells or cell processes navigate to their proper positions in the embryo. One approach has been to evaluate the effects of various molecules localized to the migratory environment on the movement or process outgrowth of isolated cells in culture. The advantage of this approach is that one can observe cell behavior in a simplified *in vitro* environment. This approach,

Current Topics in Developmental Biology, Vol. 36

0070-2153/98 $25.00

however, does not allow examination of cell behavior in the normal environmental context, where a wide array of factors and molecular influences is present.

Since the early 1990s, remarkable advances have enabled these early developmental processes to be followed in more complete tissue preparations. Various explant systems have provided a rich arena in which to examine early events, including inductive interactions, cell migration, and neural connectivity (Barber *et al.*, 1993; Dickinson *et al.*, 1995; Krull *et al.*, 1995). In addition, advances in microscopy have allowed for high-resolution images, capturing migrating cells or growing axons over time in these types of tissue preparations (Hotary *et al.*, 1996). We have developed procedures for culturing explanted embryonic chick and mouse tissue (Barber *et al.*, 1993; Krull *et al.*, 1995; Krull *et al.*, submitted; Kulesa and Fraser, submitted; see also Hotary *et al.*, 1996). These explants have served as an advantageous bioassay with which to examine the mechanisms that influence the migration of neural precursors in the peripheral nervous system, the neural crest, and the development of rhombomeric units in the hindbrain.

Here, we describe procedures for preparing explants and transverse slices from the chicken embryo. In addition, we describe methods for culturing explants of embryonic mouse tissue. Each culture system possesses distinct advantages and should be adaptable to many studies of developmental processes that occur over a wide spatiotemporal range. These preparations maintain several features common to the intact embryo, including tissue structural integrity and the correct distribution of various molecular constituents for at least 2–3 days.

II. Preparing and Culturing Chick Embryo Explants

Materials and equipment necessary include the following:

- Set of dissecting tools
  - Scissors
  - Needle holder with electrolytically sharpened tungsten needle (0.010 mm)
  - Two pairs of fine forceps (No. 5)
- India ink (Pelikan Fount)
- 70% and 100% ethanol
- 18- and 25-gauge needles; 1- and 3-ml syringes
- Hamilton syringe
- Pulled glass micropipettes (capillary tubing: Kimble; 0.8–1.10 × 100 mm)
- Sterile glass Petri dishes
- One each, angled and straight probes (Fine Science; No. 10032-13, No. 10030-13)
- Millicell inserts (Millipore; Millicell-CM, PICM 030 50)
- Six-well culture plates (Falcon)
- Culture medium (see later)
- Sterile Howard Ringer's solution
- Beaker for alcohol–tool cleaning

Albumen waste container
DiI (Molecular Probes; Cell Tracker CM-DiI, C7000)
Circular glass coverslips (25-mm diameter, Fisher)
Parafilm
Pipettemen
Sterile plastic transfer pipettes
Scotch tape (Magic)
Filter paper
Human fibronectin (Collaborative Biomedical)
Sterile Pasteur pipettes

A. Embryo Preparation

1. Incubate eggs at 38°C in a humidified rocking incubator to the desired stage of the development. Typically, embryos at stage 11–12 of development (Hamburger and Hamilton, 1951) are used to make explants for trunk neural crest migration assays. Embryos at stage 9–10 are used to examine hindbrain neural crest development. Rinse incubated chick eggs with 70% ethanol. Remove 3 ml of albumen with a 18-gauge needle attached to a 3-ml syringe from the blunt end of each egg. Apply Scotch tape to the egg surface. ("Candling" eggs before the ethanol rinse allows one to determine the embryo position before egg opening.)

2. Dilute India ink 1:10 in Howard Ringer's solution. Bend 25-gauge needle to a 90° angle, draw up the India ink solution in a 1-ml syringe, and attach the needle to the syringe. Carefully cut a small circular window in the taped area of the egg with a small scissors to view the embryo. Insert the needle of the ink-filled syringe at the interface between the yolk and blastoderm, sliding the needle below the embryo. Carefully expel a small quantity of the ink solution to help visualize the embryo. This procedure is useful for young embryos but not necessary for older embryos. Alternatively, the application of a weak solution of neutral red can help to make the older embryo easily visible. Add a couple drops of Ringer's to the embryo and seal the eggs with Scotch tape until ready to use.

B. Trunk and Hindbrain Explants

1. Typically, for studies of neural crest migration, the developing neural tube is prelabeled with DiI. Add 10 μl of 100% ethanol to one ampule of Cell Tracker CM-DiI. Add 90 μl of 10% sucrose–distilled water to the ampule. Backfill micropipettes with this DiI solution using a Hamilton syringe and attach to a pressure injection apparatus (house air or a pico-spritzer) with a needle holder and micromanipulator (Narashige). Carefully break the tip of the micropipette to allow DiI to flow to the tip. Cut through the Scotch tape to reopen the egg window and position the micropipette over the neural tube. Tear a hole in the vitelline mem-

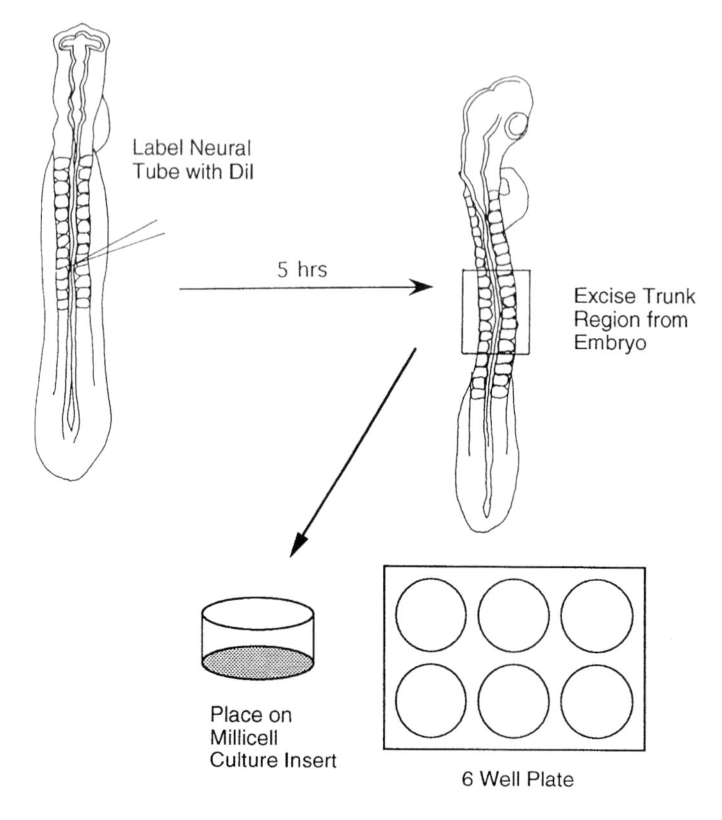

Fig. 1. Preparation of chicken embryonic trunk explants. To label premigratory neural crest, DiI is injected into the lumen of the neural tube of a stage 11 embryo. After 5 hr, a region of the trunk is excised from the embryo. This region contains bilateral pairs of somites, the neural tube and associated neural crest, ectoderm, and endoderm. The tissue is carefully positioned on a Millicell culture insert, which is then placed in a well of a six-well tissue culture plate. Culture medium underlies the Millicell membrane and does not overlie the explant.

brane over the region to be labeled with a tungsten needle. Carefully penetrate with the micropipette to the lumen of the neural tube and expel a small amount of dye. For labeling of trunk neural crest, DiI is injected at the level of the fourth to fifth most newly formed somite (Fig. 1). For labeling of hindbrain neural crest, DiI is injected at the level of rhombomere 1–2, or as desired. Withdraw the micropipette, add a couple drops of Ringer's solution to the embryo, and reseal with tape. Place eggs in a humidified 38°C egg incubator for 3–5 hr.

2. Reopen taped window over egg with scissors. Using filter paper, make several O-shaped rings. The outside diameter of the rings should be longer than the major axis of the embryo. The inner diameter should provide ample space between the filter paper and the embryo. Place the O-ring on top of the blasto-

derm with a pair of forceps so that the embryo is positioned at the center of the O-ring. Cut through the blastoderm with scissors around the perimeter of the O-ring. With forceps, carefully lift the O-ring with embryo attached to a waiting dish filled with Ringer's solution. Gently move the embryo through the Ringer's solution to remove India ink and any remaining yolk. Alternatively, the embryo may be cleared of ink and yolk by gently expelling Ringer's across the ventral surface with a pipetteman. Often, moving the embryo will disattach the embryo from the O-ring. Gently shake embryo from O-ring if necessary or use forceps to peel the embryo from the O-ring. Separate the overlying vitelline membrane from the embryo. Carefully transfer embryo using forceps or a plastic transfer pipette to a dish with fresh Ringer's solution.

3. Position embryo in the dish so that the region of interest is exposed. For trunk explants (Fig. 1), make two transverse cuts through the embryo: the first cut is at the level of the 4th most newly formed somite, the second is at the 10th most newly formed somite. Two longitudinal cuts are then made, extending from the transverse cuts, just lateral to the somites. These cuts free the trunk region from the embryo proper.

For hindbrain explants (Fig. 2), two transverse cuts are made through the embryo: one at the level of rhombomere 1 and the other at the first or second somite. Two cuts are then made lateral to the hindbrain area that remove this region from the embryo. Using forceps and the tungsten needle, carefully remove ventral tissues from the hindbrain region. This allows better visual access to the

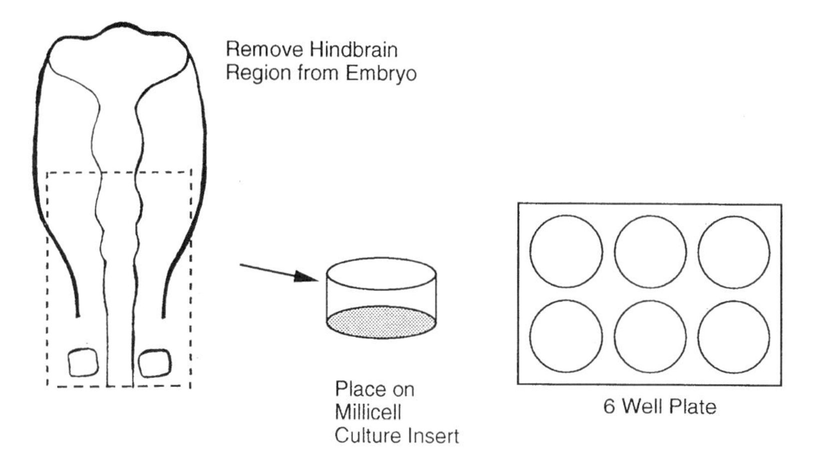

Fig. 2. Preparation of hindbrain explants. After removing the embryo from its embryonic membranes and labeling the neural tube with DiI, if desired, excise the hindbrain area from the chick or mouse embryo. The rostral limit of the explant extends to the first rhombomere, whereas the caudal limit lies posterior to the first somite. Trim the ventral surface of the explant to remove unnecessary tissues. Position the explant carefully on the Millicell insert and place it in a six-well tissue culture plate. Apply culture medium below the Millicell membrane; do not submerge the explant in medium.

labeled cells in the hindbrain and promotes tissue health. Using a sterile Pasteur pipette, the trunk or hindbrain regions are placed into culture medium on ice until explantation.

4. Add 1 ml of culture medium to each well of a six-well plate. Place a Millicell insert in each well. Gently jiggle the insert to coat the membrane with medium and remove any air bubbles. Using a sterile Pasteur pipette, place an explant with a small amount of medium onto the Millicell membrane. Under the dissecting microscope, orient the explant, unfolding any areas as needed with the angled probe. Place trunk explants with the ventral surface adjacent to the membrane and the dorsal surface up. Hindbrain explants do well with either dorsal or ventral surfaces overlying the membrane. Remove excess medium around the explants with a sterile Pasteur pipette. This helps to spread out the explants. Place three to four explants in each well.

C. Whole Embryo Explants

Whole embryos (Fig. 3) can be explanted into culture using techniques similar to the trunk–hindbrain explant procedures described previously. This technique is successful with embryos as young as stage 8 of development. Whole embryo explants develop and maintain excellent tissue integrity for approximately 2 days in culture.

1. Coat the Millicell insert membrane with 200 μl of a fibronectin solution (20 μg/ml fibronectin in phosphate buffer). Place the Millicell insert in a covered dish, avoiding drying of the fibronectin, until ready to use. Prepare eggs and label regions of interest with dye, if required, as described in II.A and II.B, no. 1. Remove embryos from egg with O-rings, as described in II.B, no. 2. Place the whole embryos with surrounding blastoderm attached into tissue culture medium on ice.

2. Aspirate any excess fibronectin from the Millicell membrane with a sterile Pasteur pipette. Place each Millicell insert into a well of a six-well plate and add 1 ml of culture medium below each insert. Cut off the bottom one third of a plastic transfer pipette and use this pipette to transfer a whole embryo onto the Millicell membrane. To assist with orienting and positioning the embryo, add a few drops of Ringer's to the insert membrane. Gently spread the embryo and surrounding blastoderm with forceps or probes on the membrane. Excess solution should be removed using a pipetteman. Place the tip of the pipetteman at the rostral and caudal ends of the explanted tissue so that as the solution drains, the embryo maintains a straight-line posture along the rostrocaudal axis. This is an important step that naturally spreads out the explanted tissue without flattening the embryo and attempts to mimic the tension of the blastoderm, which is normally stretched over the yolk sac of the intact embryo. Each whole embryo explant covers approximately two thirds (~2.8 cm^2) of the total membrane area

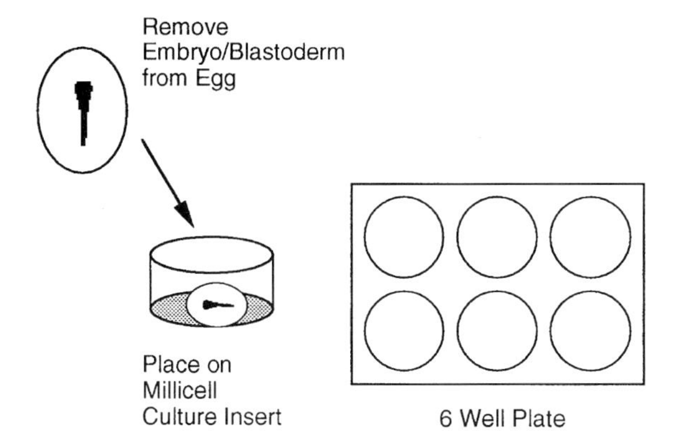

Fig. 3. Whole chicken embryo explants. Remove the whole embryo, including a substantial portion of the adjacent blastoderm, from the egg and rinse well. Carefully place the embryo explant on a Millicell culture insert, adding extra fluid if needed, and gently position the embryo on the membrane. Place Millicell insert in a well of a six-well plate and add culture medium to the area below the membrane.

of the Millicell insert. Therefore, a single whole embryo explant can be placed in each well.

D. Culture Conditions

The culture medium that supports best the overall integrity and health of the chicken explants is a defined medium (see Brewer *et al.*, 1993). The recipe is as follows: Neurobasal Medium (Gibco, No. 21103-015), 98 ml, B27 supplement (Gibco, No. 17504), 2 ml, and 0.5 m*M* L-glutamine. Grow explants in a 5% CO_2 incubator at 38°C. Change culture medium every 1–2 days.

III. Chick Embryo Slice Preparation

Materials and equipment necessary include the following:

Vibratome (TPI; series 1000) and overhead microscope
Plastic ice cubes (or refrigerated vibratome unit)
Sterile phosphate buffer
Double-edge razor blades (Ted Pella; No. 121-6)
Metal vibratome mounting blocks
Metal molds
Low–melting-temperature agarose (Sigma; A-5030)
Hanks' balanced salt solution with phenol red (HBSS)

Scalpel blade (Fine Science, No. 15) and holder
One each, straight and angled probes (Fine Science; No. 10030-13, No. 10032-13)
Set of dissecting tools
Glass coverslips (18 mm^2)
Krazy glue
5% agar (Prepare in advance: Dissolve 5% agar in water, pour into glass Petri dishes and autoclave to sterilize. Allow to harden.)
Cotton gauze pads
Culture medium (see later)
Other materials listed in II

A. Embryo Preparation

1. Prepare vibratome bath by rinsing with distilled water and drying. Apply 70% ethanol to the bath with cotton gauze pads and allow to dry. Place vibratome if possible within a sterile hood. Alternatively, place vibratome on bench near the sterile hood and maintain as clean conditions as possible. Position a dissecting microscope over the vibratome cutting surfaces to improve viewing of the slicing procedure.

Prepare 7.5% agarose: prewarm 30 ml of HBSS (with phenol red) in glass beaker in a microwave oven. Slowly add 2.2g of low–melting-temperature agarose and stir continually with metal spatula. Warm but do not boil in microwave and stir until agarose has dissolved. Agarose will eventually harden at room temperature; remelt in microwave as needed.

2. Prepare eggs–embryos as described in II. Label premigratory neural crest by injecting DiI into the lumen of the neural tube, as described previously. Place embryos in sterile phosphate buffer on ice.

3. Excise region of interest from embryo, as described in II.B.

B. Cutting Transverse Slices (Fig. 4)

1. Embed an excised region of embryo in the 7.5% agarose as follows: pour a layer of melted agarose into a metal mold and allow slightly to harden. While hardening occurs, transfer embryo to a sheet of parafilm and wick off with a Kimwipe as much liquid as possible. Place embryo in agarose block and add additional melted agarose carefully over embryo. View embryo placement under the dissecting microscope and gently position with hooked probe if needed. Place block on ice to harden for 2–5 min. Embed each region of embryo just before vibratome sectioning; do not embed several at the same time.

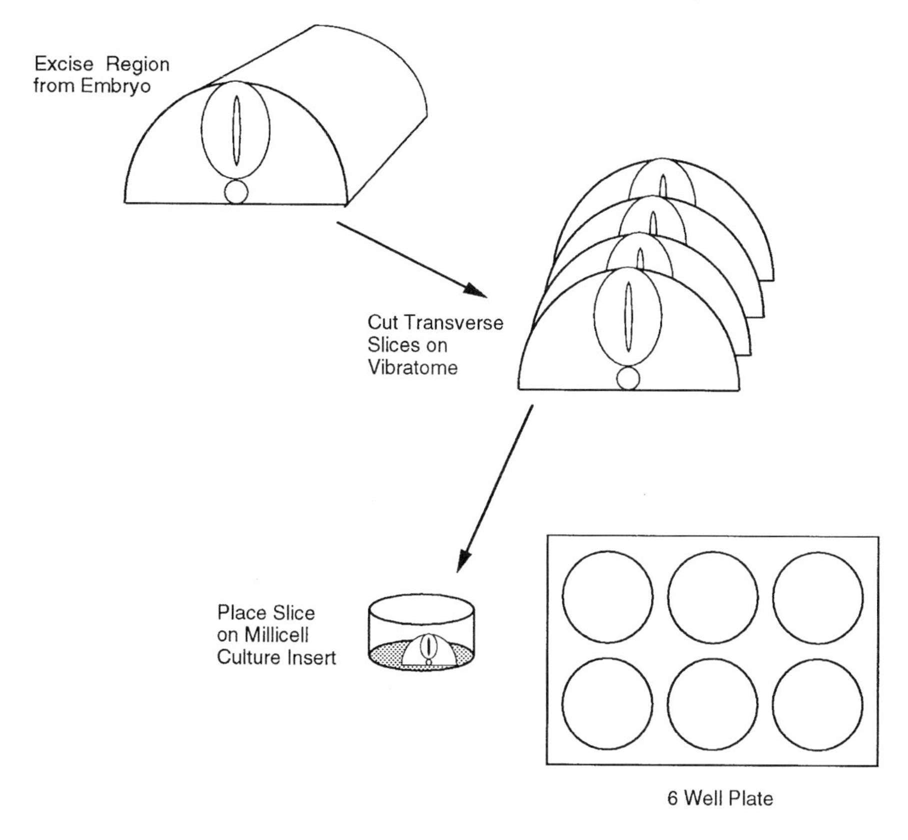

Fig. 4. Embryonic chicken slice preparation. Regions of interest are excised from the embryo and embedded in agarose, in preparation for vibratome sectioning. Transverse slices are cut at 100–150 μm and carefully transferred to a Millicell culture insert, which is placed in a six-well tissue culture plate. Culture medium is added below and not above the Millicell membrane.

2. Fill each well of a six-well plate with 1 ml of sterile phosphate buffer. Place a Millicell in each well with sterile forceps. Jiggle them to remove air bubbles and make the insert membrane transparent.

3. Attach a glass coverslip to vibratome mounting block with 1 drop of Krazy glue, allowing it to dry for a few minutes. Place block–coverslip in 70% alcohol for a few minutes and allow to air-dry under sterile hood. Cut a block from the sterilized hardened 5% agar and attach it with Krazy glue to the glass coverslip. This agar block will serve as a support for the agarose-embedded embryo. Remove embryo in agarose block from metal mold and, under a dissecting microscope, cut a block containing the embryo from the agarose. Orient the block so that transverse sections will be cut from the embryo on the vibratome and attach it with Krazy glue in front of the agar block on the vibratome mounting block.

4. Fill vibratome bath with cold sterile phosphate buffer and mount one half of the razor blade in vibrantome blade holder. (If vibratome unit is not refrigerated, add plastic frozen cubes to maintain a cold bath temperature.) Place mounted embryo into chuck and tighten. Cut 100- to 150-μm slices from the embryo, viewing sectioning as it proceeds through an overhead microscope. Cut slices at low speed and medium-high-amplitude settings. Leave cut slice on the razor blade and do not allow it to float in the bath. Remove the surrounding agarose from the embryo slice by gently teasing it away with the two probes. Carefully move the slice onto a scalpel blade with one of the probes.

5. Transfer the slice to the Millicell membrane by pushing it gently from the blade with a probe. A small amount of buffer will also surround the slice, and this helps in the transfer. Orient the slice on the membrane with a probe, unraveling any folded areas. It is critically important to manipulate the slice as little as possible to maintain the best tissue integrity. Any excess liquid around the slice should be aspirated off using a sterile glass Pasteur pipette. Place approximately three to four slices per culture insert. Remove the phosphate buffer below the Millicell membrane with a P1000 pipetteman and replace with 1 ml of culture medium.

This technique is also adaptable for studies of developmental events in older embryos. For studies of axon patterning, regions of the slice can be labeled with dye after placement on the Millicell insert. Typically, DiI is microinjected into ventral regions of the developing neural tube to label motor neurons (refer to II.B, no. 1).

C. Culture Conditions

The culture medium that supports best the health and integrity of the chicken embryonic slices is as described in II.D (see Brewer *et al*., 1993). Grow slices in a 5% CO_2 incubator at 38°C. Change the culture medium every day. The slices will maintain their structural integrity for 1–2 days in culture.

IV. Preparing and Culturing Mouse Embryo Explants

Materials and equipment necessary include the following:

Set of dissecting tools
One each, angled and straight probes (Fine Science; No. 10032-13, No. 10030-13)
Culture medium (see later)
Human fibronectin (Collaborative Biomedical)
Insect pins
35-mm Sylgard-coated plastic tissue culture plate

Sterile phosphate buffer or Gey's buffer, refrigerated
Scalpel blades (Fine Science, No. 15) and holder
Other materials listed in II

A. Embryo Preparation

1. For studies of hindbrain neural crest migration, embryos at approximately 8–8.5 days of development are required. In a sterile hood, euthanize pregnant mouse and remove uterus. In sterile cold phosphate buffer, dissect out embryos and remove all membranes under a dissecting microscope. Rinse embryos in fresh phosphate buffer. Store embryos in culture medium on ice.

B. Hindbrain Explants

1. Under a sterile hood, add 200 μl of fibronectin solution (20 μg/ml fibronectin in sterile phosphate buffer) to the Millicell culture insert membrane; tip insert if needed to coat the membrane. Place inserts in covered glass sterile dish to avoid drying of fibronectin.

2. Cut a "trench" in the center of a Sylgard-coated dish using a scalpel; trench serves to stabilize the mouse embryo during DiI labeling. (Sylgard dish can be reused; rinse with water and spray dish with 70% ethanol, allowing it to dry before use.) Fill dish with cold phosphate buffer.

3. Place mouse embryo into Sylgard dish, using a probe to position its dorsal surface up in the trench. Further stabilize embryo with insect pins if needed. Backfill micropipettes with DiI and attach to pressure injection apparatus (house air or pico-spritzer). Break slightly the tip of the micropipette with a fine forceps to allow DiI to flow. Under a dissecting microscope, position micropipette over the dorsal surface of the hindbrain neural tube. Using the micromanipulator, carefully insert micropipette into the lumen of the neural tube and expel a small amount of DiI, thereby labeling most neural tube cells and premigratory neural crest. Remove micropipette and insect pins and place labeled embryo in culture medium on ice.

4. After all embryos are labeled, place a single embryo in a phosphate buffer-filled glass dish. Using a tungsten needle, make two transverse cuts: the first is made rostral to the hindbrain and the second just caudal to the first or second somite. These cuts should extend from the dorsal to ventral surfaces, removing the hindbrain region and surrounding tissues from the embryo. Using forceps to hold the tissue in place, carefully cut away the ventral tissues from the hindbrain explant. Cuts should be distinct and should not consist of any tearing or teasing away of the tissue.

5. Fill each well of six-well plate with 1 ml of culture medium. Aspirate excess fibronectin solution from the Millicell inserts and place inserts in six-well

plate. Using a sterile Pasteur pipette, place each hindbrain explant onto Millicell membrane with a small amount of culture medium. Explants can be oriented using a probe with either ventral or dorsal surfaces facing the membrane. Typically, three or four explants are placed on each insert.

C. Culture Conditions

Culture the mouse explants in the following medium: Dulbecco's modified Eagle medium (DMEM)–F12, (Gibco; No. 11321-015, 500 ml), 45 ml, PenStrep (Sigma; No. P-0781), 500 μl, and fetal bovine serum (FBS; Gibco; No. 16000-036), 5 ml. The PenStrep and FBS can be premixed, aliquoted, and frozen at −20°C. Place cultures in a 5% CO_2 incubator at 38°C. The mouse explants maintain their structural integrity for at least 3 days in culture.

V. Evaluation of Explant and Slice Features

Explant and slice preparations should be examined to determine whether characteristics common to the intact embryo are retained in these preparations, suggesting that they can satisfactorily mimic the *in vivo* environment. A critical aspect of these types of preparations is that they retain an overall tissue integrity and health. For example, the neural tube and somites in the trunk explants develop in association with each other but are easily visualized as distinct structures during their development in culture. Another important feature to assess is whether molecular constituents of the explant and slice preparations possess their typical distributions. It is also necessary to evaluate whether normal developmental events occur according to their usual spatial and temporal scales in these preparations. For example, do epithelial somites develop into compartmentalized units consisting of dermomyotome and sclerotome, and do neural crest cells migrate along their typical pathways?

We have used various molecular markers to evaluate the distributions of some cell surface and extracellular matrix molecules in the explant–slice preparations. Normal developmental events, including cell migration and axon outgrowth, can be examined using time-lapse videomicroscopy with a confocal or inverted microscope (Kulesa and Fraser, 1997). In addition, we have examined static sections of explants grown for different times in culture to determine whether, for example, somite compartmentalization proceeds in a manner comparable to that observed *in vivo*. Procedures for fixing and labeling explants–slices with antibodies or other labeling agents follow. In addition, procedures for sectioning fixed explants are included.

A. Fixation and Staining

1. Remove culture medium from each well with a Pasteur pipette or pipetteman. Add 1 ml of phosphate-buffered saline (PBS) to the well below the culture insert, in place of the medium, two times for 15 min each. Remove PBS and replace with fixative agents. Typically, we fix in 4% paraformaldehyde for 1 hr to overnight.

2. Remove the paraformaldehyde and rinse each well three times with PBS for 15–30 min. This rinse can be done both above and below the Millicell membrane. Care must be taken not to dislodge the explant or slice with the first application of PBS to the top of the membrane; gently move the PBS with a probe toward the tissue. If the explant or slice becomes dislodged from the membrane, staining can still proceed and the results will be readily visualized.

3. Some antibodies require a blocking step to eliminate nonspecific antibody staining. Typical blocking agents include normal goat serum, bovine serum albumen, or nonfat dry milk. Add blocking solution to the explants–slices both above and below the Millicell membrane and allow to incubate for 30 min to 1 hr. Remove blocking solution and add primary antibody (~1.5 ml total) both above and below the Millicell membrane. Cover the six-well plate with foil and incubate 1–2 days at 4°C.

4. Remove the primary antibody with a Pasteur pipette and rinse explants–slices four times, 15 min each, with 1 ml of PBS. Add secondary antibody (~1.5 ml total) both above and below the Millicell membrane. Rhodamine- and fluorescein-tagged secondary antibodies work well in this system; we have not evaluated other tagged secondary antibodies. Cover the plate with foil and incubate for 4 hr to overnight at 4°C.

5. Rinse explants–slices with 1 ml of PBS three times, 15 min each. View antibody labeling using an inverted microscope and record results photographically.

B. Sectioning

1. Remove culture medium from well with a Pasteur pipette. Add 1 ml of PBS to the well, in place of the culture medium, two times for 15 min each. Remove PBS and replace with fixative agent. Typically, we fix in 4% paraformaldehyde for 1 hr to overnight. Remove fixative and rinse with PBS three times, 15 min each. All rinsing and fixation should occur below the membrane and not above it.

2. Under the dissecting microscope, gently dislodge each explant to be sectioned from the Millicell membrane using an angled probe, being careful not to damage the explant. The explant is now ready to be removed from the culture insert, and this can be accomplished in one of two ways: 1) Using a scalpel blade, cut a rectangle through the Millicell membrane, around the explant. With fine

forceps, carefully grasp the membrane rectangle and lift the explant out of the culture dish. Place the membrane with explant into a vial containing PBS and gently shake the membrane to move the explant into the PBS. 2) Add a small drop of PBS to the dislodged explant. Use a Pasteur pipette to remove the floating explant into PBS.

3. Prepare the explants for standard cryosectioning by immersing them in 5% sucrose–phosphate buffer followed by 15% sucrose–phosphate buffer until explants sink. Place explants in a melted solution of 15% sucrose–7.5% gelatin at 37°C for 4 hr. Place an explant into plastic mold with fresh gelatin and allow to harden. Freeze molds in liquid nitrogen. Carefully trim blocks before attaching to cryostat chuck for sectioning. Allow blocks to equilibrate in a cryostat set at −25°C and section blocks at 10–30 μm.

VI. Explant and Slice Experimental Perturbations

One of the strengths of these explant and slice preparations, in contrast to the *in vivo* environment, is a high accessibility to blocking reagents. Typically, a blocking reagent is added during a preincubation step, before explanation of the tissue onto the Millicell membrane, and during the entire culture period as an additive to the culture medium. Function-blocking antibodies, plant lectins, various enzymes, and fusion proteins have been used successfully in this paradigm to disrupt certain aspects of cell migration. General procedures are described in the following, and these should be adaptable to a particular reagent.

1. Before explantation on the Millicell inserts, place the chick tissue in 500 μl–1 ml of solution (sterile filtered) containing the blocking reagent and incubate for 4 hr at 38°C in the 5% CO_2 incubator. The blocking reagent can be diluted in the Neurobasal medium for this preincubation step.

2. Explant the embryonic tissue onto the Millicell inserts, as described previously, and add the same concentration of blocking reagent to the culture medium as was added to the preincubation step. Some proteins may possess a short half-life and should be reapplied to the medium at the appropriate intervals.

3. For mouse tissue, add the blocking reagent to DMEM–F12 in the appropriate concentration and incubate the tissue at 38°C for 4 hr in a 5% CO_2 incubator. Position the mouse embryonic tissue on the Millicell membrane, adding the same concentration of blocking reagent to the culture medium as was added to the preincubation step.

VII. Conclusions

This chapter describes methods for preparing explants and slice preparations of the embryonic chick and mouse. Of critical importance, the explants and slices retain their tissue architecture for 1–3 days in culture and possess molecular characteristics typical of intact embryos. Furthermore, these types of preparations allow one to disrupt the embryonic environment through the addition of various blocking reagents and observe the subsequent effects of these manipulations via static images and using time-lapse videomicroscopy. These types of paradigms should continue to offer a rich cellular and molecular arena in which to study early developmental events, including cell migration and process outgrowth.

References

Barber, R. P., Phelps, P. E., and Vaughn, J. E. (1993). Preganglionic autonomic motor neurons display normal translocation patterns in slice cultures of embryonic rat spinal cord. *J. Neurosci.* **13,** 4898–4907.

Brewer, G. J., Torricelli, J. R., Evege, E. K., and Price, P. I. (1993). Optimized survival of hippocampal neurons in B27 supplemented Neurobasal, a new serum-free medium combination. *J. Neurosci. Res.* **35,** 567–576.

Dickinson, M. E., Selleck, M. A., McMahon, A. P., and Bronner-Fraser, M. (1995). Dorsalization of the neural tube by the non-neural ectoderm. *Development* **121,** 2099–2106.

Hamburger, V., and Hamilton, H. L. (1951). A series of normal stages in the development of the chick embryo. *J. Morphol.* **88,** 49–92.

Hotary, K. B., Landmesser, L. T., and Tosney, K. W. (1996). Embryo slices. *Methods in Avian Embryology* **51,** 109–124.

Krull, C. E., Collazo, A., Fraser, S. E., and Bronner-Fraser, M. (1995). Segmental migration of trunk neural crest: Time-lapse analysis reveals a role for PNA-binding molecules. *Development* **121,** 3733–3743.

Krull, C. E., Lansford, R., Gale, N. W., Marcelle, C., Collazo, A., Yancopoulos, G. D., Fraser, S. E., and Bronner-Fraser, M. (1997). Interactions of EPH-related receptors and ligands confer rostrocaudal pattern to trunk neural crest migration. (Submitted).

Kulesa, P. M., and Fraser, S. E. (1997). Cell dynamics revealed by time-lapse videomicroscopy of chick hindbrain explant cultures. (Submitted).

Serbedzija, G. N., Bronner-Fraser, M., and Fraser, S. E. (1992). Vital dye analysis of cranial neural crest cell migration in the mouse embryo. *Development* **116,** 297–307.

9

Culture of Avian Sympathetic Neurons

Alexander v. Holst and Hermann Rohrer
Max-Planck-Institut für Hirnforschung
Abteilung Neurochemie
60528 Frankfurt/Main
Germany

I. Introduction

Sympathetic neuron cultures have been widely used to study many different aspects of neuronal development. Aspects of neuron survival (Barde, 1989), in particular the action of the classical neurotrophin, nerve growth factor (NGF; Levi-Montalcini, 1987), the developmental control of neurotrophin responsiveness (Ernsberger *et al.*, 1989a; Rodriguez-Tébar and Rohrer, 1991; Birren *et al.*, 1993; v. Holst *et al.*, 1995), neurotrophin signal transduction mechanisms (Heumann, 1994), neurite outgrowth (Ernsberger and Rohrer, 1994), and the control of neurotransmitter phenotype (Patterson and Chun, 1974; Rao and Landis, 1993), have been major issues studied with cultured sympathetic neurons. An exceptional property of immature sympathetic neurons is their ability to proliferate *in vivo*

Current Topics in Developmental Biology, Vol. 36

0070-2153/98 $25.00

(Rothman *et al.*, 1978) and *in vitro* (Rohrer and Thoenem, 1987). This allows the analysis of sympathetic neuronal proliferation (Zackenfels *et al.*, 1995) and is advantageous for the expression of exogenous genes using retroviral vectors (Reissmann *et al.*, 1996). Two major sources of sympathetic neurons are commonly used, the neonatal superior cervical ganglia (SCG) of the rat and mouse and avian lumbosacral chain ganglia. The advantage of the latter ganglia is that ganglia containing mainly proliferating, immature neurons and ganglia containing differentiated, postmitotic neurons are easily accessible, and large numbers of cells can be obtained. Recent data indicate that sympathetic neurons of SCG and trunk sympathetic ganglia belong to different cell lineages (Durbec *et al.*, 1996) and may display different properties (Durbec *et al.*, 1996; Borasio *et al.*, 1993; Nobes *et al.*, 1996; Kobayashi *et al.*, 1994; v. Holst *et al.*, 1995). Thus, both types of sympathetic neurons should be analyzed in each species. For this reason, a procedure to dissect avian superior cervical ganglia is included.

Conditions for culturing avian sympathetic neurons have been described previously in the primary literature (Edgar *et al.*, 1981; Rohrer and Thoenen, 1987; Ernsberger *et al.*, 1989a). The purpose of this chapter is to provide a summary of techniques that have proven to be successful over many years in our laboratory in the study of sympathetic neuron development, and that can be easily established by newcomers in the field. A particular property of immature sympathetic neurons is the possibility of obtaining exogenous gene expression by transfection (Heller *et al.*, 1995; v. Holst *et al.*, 1995). Because this method is much less time consuming and difficult than the single-cell injections widely used for primary neuron cultures (e.g., Allsopp *et al.*, 1993), details of this procedure are also provided.

II. Dissection of Ganglia

A. Embryonic Lumbosacral Sympathetic Ganglia

Fertilized chick eggs, incubated in a commercial incubator at 38.5°C and approximately 70% humidity for the appropriate number of days, are put in an upright position and opened by cutting a hole (2 cm diameter) at the blunt end, using medium-sized pointed scissors. The embryo is held at the neck with curved forceps and transferred to a Petri dish (10 cm diameter). After cutting off the head, the embryo is put on its back, the body opened and the internal organs removed.

These operations are carried out under a stereomicroscope with fiberoptics using fine straight forceps (Dumont No. 5). To increase the contrast, the dish is set on a black sheet of plastic and the fiberoptics are arranged so that the light hits the embryo somewhat tangentially. Most important, to remove blood and tissue fragments, the embryo is flushed repeatedly with sterile phosphate-buffered saline (PBS) during the dissection procedure.

Although ganglia are removed only from the lumbosacral region, the internal organs of the thoracic region (lung, esophagus) are also usually removed. Kidneys and adrenals are dissected with particular care to avoid removal of the sympathetic ganglia together with these tissues. This is particularly a problem at the earlier stages of development (embryonic days E6.5–E9). In the lumbosacral region, the sympathetic ganglia are easily observed as a chain of ganglia that lies on both sides of the vertebral column (Fig. 1). Each ganglion is connected to the ventral roots by two nerves, the central and peripheral rami communicantes. To remove the ganglia, both nerves must be cut. The central, preganglionic nerves are cut by moving closed forceps between vertebral column and the sympathetic chain. The peripheral, postganglionic connections of sympathetic ganglia to the ventral root are ripped individually. After these manipulations, a chain of about 10 ganglia can be removed (Figs. 1 and 2). The chains are collected in sterile PBS supplemented with glucose (1 mg glucose/ml PBS). With some experience, 10 chains can be collected in about 30 minutes. The earliest stage at which we are able to dissect sympathetic chains from the lumbosacral region is E6.5 (stage 29 according to Hamburger and Hamilton, 1951). At E8–E10, dissection is easiest; at later stages connective tissues are more solid, which makes dissections more time consuming.

Sympathetic ganglia can also be dissected at thoracal levels. However, the ganglia have to be dissected individually, and are tightly associated with dorsal root ganglia. Thus, contamination with sensory dorsal root ganglion neurons is difficult to avoid.

B. Embryonic Superior Cervical Ganglia

Embryos are obtained as described previously and transferred to a Petri dish (10-cm diameter). After cutting the neck close to the trunk, the head of the embryo is placed on its dorsal side such that the lower beak faces the experimenter. This is trivial at more advanced stages of development (E8–E10) because the head is rather flat and balanced. If earlier stages are used (e.g., E6), the almost round bulges of the tectal hemispheres have to be removed to balance the head of the embryo on its dorsal side. The lower beak and the tongue are carefully removed by gentle pulling with a fine forceps. The SCG is slightly elongated in shape and lies near the lower beak junction. It is situated close to and slightly below the common carotid artery at the level where the artery separates into two branches, the inner and outer carotid. The arteries are very useful and important landmarks for orientation, and care should be taken not to destroy them. The arteries are freed from the connective tissue on top of them. The SCG then becomes accessible from the lateral side after carefully sliding along the common carotid artery

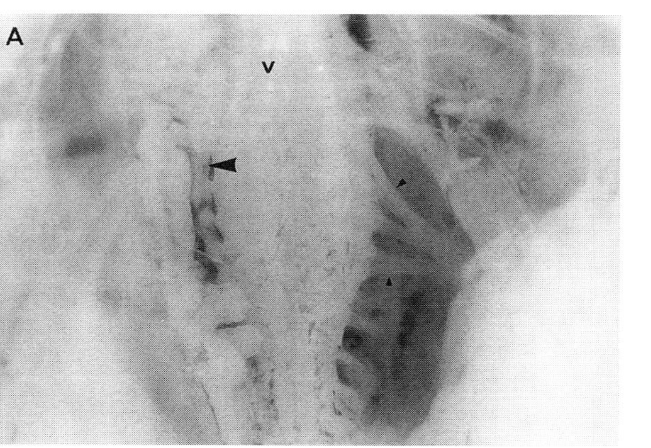
A
v

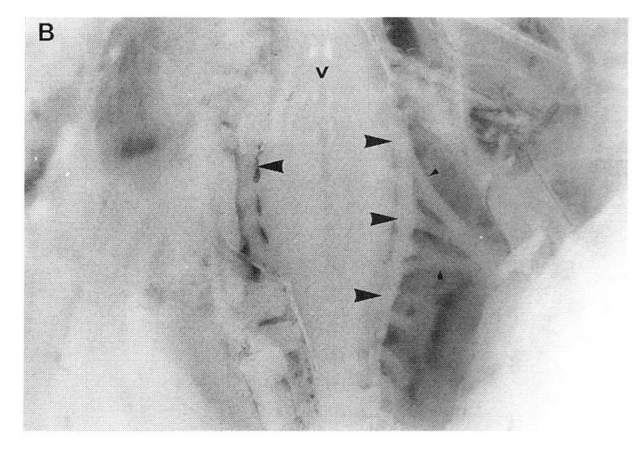
B
v

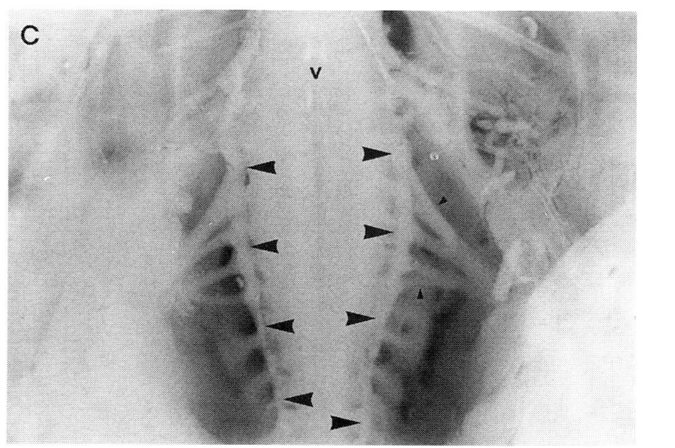
C
v

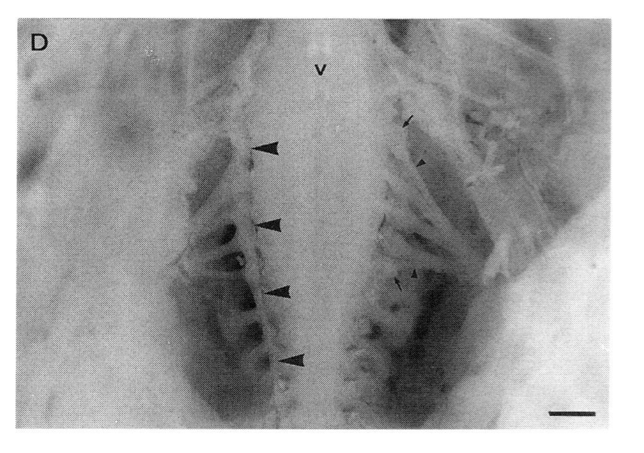
D
v

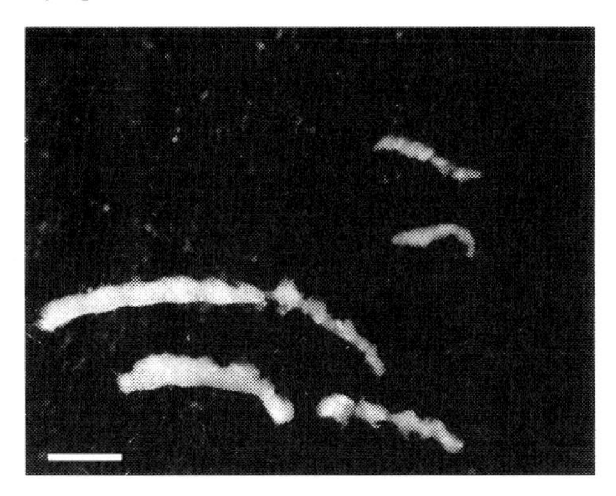

Fig. 2. Lumbosacral sympathetic chains from E12 and E6.5 after removal from equivalent embryos. Two examples each of E12 and E6.5 are shown. The ganglia are cleaned of connective tissue. Note the large size difference between the two developmental stages (both are removed over the same length of segments in the lumbosacral region). The lower E12 chain was ripped into two pieces during removal from the embryo. The scale bar is 1 mm.

with a fine forceps. The ganglion lies below the artery and is mechanically removed and placed in sterile PBS–glucose. We have somewhat limited experience with SCG cultures because we only very recently started preparing and cultivating these neurons. We find E8–E9 the easiest age to obtain the SCG, and so far E6 has been the earliest stage cultivated. One potential problem is the proximity of cranial sensory ganglia (e.g., petrosal ganglion), which may lead to contamination with sensory neurons if they are mistakenly prepared as or with the SCG. This is avoided by starting with single-cell cultures of individual ganglia. SCG neurons are distinguished from sensory neurons by morphology, which may be difficult for novices. Definitive identification of SCG neurons is possible by using double immunocytochemistry (see V) with antibodies against the enzyme tyrosine hydroxylase (a marker for sympathetic neurons) and a

←

Fig. 1. Sequence of events during the preparation of sympathetic chain ganglia (A–D). The lumbosacral region of an E12 embryo is shown. Rostral is to the top and caudal to the bottom in every figure, with the vertebral column (v) running vertically through the entire length of the visual field. The paravertebral sympathetic chains are marked with large arrowheads. Many of the ventral roots can be seen, and some are indicated by small arrowheads. (A) The situation after the removal of the internal organs is shown. A part of the sympathetic chain on the left side (large arrowhead) can already be seen. (B) The sympathetic chain on the right side was freed from surrounding tissue, and is marked with multiple large arrowheads, whereas the situation on the opposite side is unchanged. (C) Both sympathetic chains are visible after the other sympathetic chain was also freed. (D) The sympathetic chain on the right side is removed and the sensory spinal ganglia can be seen below the ventral roots (two examples are marked with arrows), whereas the sympathetic chain on the opposite side is still *in situ*. The scale bar is 1 mm.

neuronal marker like neurofilament or Q211 (v. Holst, unpublished observations).

Materials necessary for these procedures include the following:

1. Fertilized chicken eggs
2. Egg incubator
3. Sterilized instruments (scissors; curved forceps; fine straight forceps, Dumont No. 5).
4. Dissecting microscope with fiberoptics
5. Sterile PBS, pH 7.4
6. Sterile PBS supplemented with 0.1% glucose (PBS–glucose, filter sterilized, not autoclaved)
7. Monoclonal anti-tyrosine hydroxylase antibody (Boehringer Mannheim)
8. Monoclonal anti-neurofilament 160 antibody (Boehringer Mannheim)
9. Monoclonal antibody Q211 (Henke-Fahle, 1983)

III. Dissociation of Ganglia

A. Lumbosacral Sympathetic Ganglia

Sympathetic chains are collected in PBS–glucose. The chains are carefully examined for contaminating connective tissue and eventually cleaned. The ganglia are then transferred to 15 ml conical Falcon tubes using a siliconized, fire-polished Pasteur pipette and allowed to settle. PBS–glucose is removed, and the chain ganglia are washed once in 1 ml PBS before 1 ml of freshly diluted trypsin in PBS (1:10) is added. (We keep frozen aliquots of 100 μl trypsin [1890 U/ml] in PBS and add 900 μl PBS immediately before use.) The ganglion chains are gently suspended in the trypsin and incubated at 37°C in a water bath. (The chains tend to stick to the Pasteur pipette during transfer. Sticking chains are easily recovered by slowly pipetting in the trypsin solution before starting the incubation.)

The incubation period is

22 minutes for E6.5 sympathetic ganglia
25 minutes for E7 sympathetic ganglia
27 minutes for E9 sympathetic ganglia
30 minutes for E11–E13 sympathetic ganglia

The trypsinization is terminated by the addition of 100 μl of trypsin inhibitor. During trypsinization, some cells may be lysed, causing the extrusion of DNA that occasionally leads to clumping of sympathetic chains, slow settling, and difficulties in the removal of trypsin. To avoid this, 20 μl of DNase solution is routinely added together with the trypsin inhibitor, and the volume is adjusted to 2 ml with complete medium (F14 with 10% horse serum [HS] and 5% fetal calf serum [FCS]). After settling of the ganglia, the supernatant is removed and 1 ml of complete medium is added. This step is repeated once. The ganglia are then dissociated to single cells by triturating with a siliconized, fire-polished Pasteur pipette. The ganglia are dissociated after 5–10 slow pipetting strokes. The pipette tip should always be near the bottom of the tube to avoid foaming. The cell number is determined using a hemocytometer (counting chamber) and cells are then either used directly for culture or are enriched for neuronal cells (see IV). An average yield of $8.3 \pm 2.4 \times 10^4$ cells per embryo (two sympathetic chains) should be obtained at E7.

B. Superior Cervical Ganglia

The procedure is identical to the one described previously, with slight variations of the trypsinization times:

20 minutes for E6 SCG
30 minutes for E8–E12 SCG

Materials necessary for these procedures include the following:

1. Sterile trypsin (Worthington, cat. No. 3707) stock solution ([1890 U/ml], 0.1 ml aliquots) in PBS, pH 7.4, stored at −20°C.
2. Sterile PBS, pH 7.4
3. Sterile soybean trypsin inhibitor (Worthington 3571, 0.1 ml aliquots) in PBS, pH 7.4, stored at −20°C. The concentration is adjusted to block completely the amount of trypsin used for the digestion of the ganglia.
4. DNaseI (Sigma) 2 mg/ml in PBS (sterile 20 μl aliquots), stored at −20°C.
5. Counting chamber
6. 15 ml conical tubes (Falcon)
7. Siliconized Pasteur pipette. The sharp and uneven opening of the pipette is smoothed by fire polishing. This also results in a narrowing of the opening. The final diameter of the opening is about 500 μm.

Pasteur pipettes are siliconized by a short incubation in Silicone solution (in isopropanol; Serva) in a glass beaker. Care should be taken to avoid air bubbles, especially in the tips of the Pasteur pipettes. The solution is decanted and the Pasteur pipettes are inverted onto a piece of paper to remove excess fluid. They are then incubated in an oven at 150°C for 1 hour, cooled down to room temperature (RT), bathed again in Silicone solution, and finally incubated at 150°C overnight. They can be stored at RT, but must be autoclaved before use if storage is under nonsterile conditions.

IV. Selection of Neurons or Non-neuronal Cells

At E7, lumbosacral sympathetic chain ganglia contain only a very small proportion of non-neuronal cells (approximately 2% of O4-positive glial cells and 3% fibroblast-like cells) (Rohrer and Thoenen, 1987; Zackenfels *et al.*, 1995). With increasing age of the embryos, the proportion of glial cells increases because of glial cell proliferation and naturally occurring neuronal cell death. At E16, non-neuronal cells represent 64 ± 8% of the total ganglion cell population (Rohrer, 1985).

To enrich for specific cell populations, we have used three different procedures. The same procedures can very likely also be used for SCG. We have, however, not yet tested them in our laboratory. In the literature, a differential centrifugation protocol for the removal of non-neuronal cells from E8–E12 SCG on a Percoll density gradient has been described (Zurn and Mudry, 1986).

A. Preplating

The easiest way to enrich for the neuronal cell population relies on the differential adhesivity of neurons and non-neuronal cells to tissue culture plastic surfaces. Non-neuronal cells rapidly attach to uncoated tissue culture dishes, whereas neurons do not settle on this substrate (Barde *et al.*, 1980).

Ganglion dissociates are plated on uncoated tissue culture dishes for 2–3 hours at 37°C in a humidified incubator in serum-containing medium (F14 10% HS) with a cell density not exceeding 8×10^5 cells/10 ml/100 mm culture dish. In contrast to the usual procedure, the ganglia are dissociated in F14 medium with 10% HS. When most of the non-neuronal cells have attached (they can be recognized by their small size, phase-dark appearance, and uneven cell surface, compared with the smooth cell surface of phase-bright, larger-diameter neurons), the dishes are gently flushed several times with medium to harvest the neuronal population. This procedure works well with cells from older sympathetic ganglia (E10–E12). A disadvantage of the procedure is that the cells have to be diluted for the preplating (otherwise the neurons stick to the attached non-neuronal cells), and even further during the flushing step. We try to avoid concentrating the cells by centrifugation because this results in the formation of small aggregates that cannot be disassembled into single cells. The preplating procedure does not work with cells from young embryos (E7) because not only non-neuronal cells but also a large proportion of immature sympathetic neurons attach to tissue culture plastic.

Materials necessary for this procedure include the following:

1. Falcon tissue culture dishes (type 3003)
2. F14 medium 10% HS

B. Selection of Neurons or Glial Cells by Panning

This procedure takes advantage of cell surface antigens selectively expressed by neurons (Q211 antigen; Henke-Fahle, 1983; Rösner *et al.*, 1985) and glial cells (O4 antigen; Sommer and Schachner, 1981) that are recognized by mouse monoclonal antibodies (Rohrer *et al.*, 1985; Rohrer and Thoenen, 1987). Very pure populations of neurons or glial cells are obtained. The method is not quantitative, however; the cell yield is relatively low. A further problem arises from the fact that the selected cell population is very firmly attached to the panning dish and needs to be harvested by trypsinization. The following protocol has been used to select cell populations for the analysis of mRNA expression (Zackenfels *et al.*, 1995) and is based on the procedure used by Barres and colleagues for the purification of glial cell types from dissociated rat optic nerves (Barres *et al.*, 1992).

Tissue culture dishes (100 mm, Falcon 1029) are coated overnight at 4°C with polyclonal anti-mouse IgM antibody (50 μg/10 ml of 50 m*M* Tris-HCl, pH 9.5). The dishes are washed three times with 8 ml PBS and then incubated for 1 hour at RT with the Q211 antibody (ascites 1:1000) or O4 antibody (concentrated hybridoma supernatant 1:400), diluted in HEPES-buffered F14 medium supplemented with 1 mg/ml bovine serum albumin (BSA). The dishes are washed with PBS (3× with 8 ml). The PBS is removed, and cell suspensions are plated in HEPES-buffered F14 medium–0.1% BSA. In contrast to the usual procedure, the ganglia are dissociated in HEPES-buffered F14 medium–0.1% BSA. The cells are allowed to settle for 45–60 minutes. The dishes are gently swirled every 15 minutes. At the end of the panning procedure, the supernatant with nonattached cells is removed. For sequential panning, the supernatant is transferred to a panning dish coated with the second antibody (we usually first select for O4-positive cells and then for Q211-positive cells). Cells that are only loosely attached are removed by thoroughly flushing with PBS (8× 8 ml). The cells bound to the antibody-coated dishes are harvested by trypsinization for 15 minutes at 37°C in 5 ml trypsin–ethylenediaminetetraacetic acid (EDTA). Trypsinization is stopped by the addition of 5 ml complete medium (F14–10% HS–5% FCS), and the cells are collected by centrifugation (10 minutes, 150g). The purity of the selected cell populations has to be analyzed by plating small aliquots and staining for the Q211 and O4 antigens (see V). Starting from a mixed cell population containing 95% neurons, 2% glial cells, and 3% fibroblast-like cells at E7 (Rohrer and Thoenen, 1987), we obtained a pure neuronal population (99.8 ± 0.2%) and a strongly enriched glial cell population (96 ± 1.7%; Zackenfels *et al.*, 1995).

Materials necessary for this procedure include the following:

1. Falcon Petri dishes (type 1029)
2. Sterile PBS, pH 7.4

3. Anti-mouse IgM, 50 μg/ml in 50 m*M* Tris-HCl, pH 9.5 (Dianova 115 065 075)
4. HEPES-buffered F14 medium–0.1% BSA
5. Q211 or O4 antibody (Henke-Fahle, 1983; Sommer and Schachner, 1981; respectively) diluted in HEPES-buffered F14 medium–0.1% BSA. The optimal concentration of the antibody used for panning has to be determined empirically.
6. Trypsin–EDTA (Gibco BRL)
7. F14 medium–10% HS–5% FCS
8. Tabletop clinical centrifuge

C. Negative Selection of Cell Populations by Complement-Mediated Cytotoxic Kill

Q211-positive neuronal cells or O4-positive glial cells can be selectively eliminated from ganglion dissociates by incubating the cells with antibodies in the presence of complement (Rohrer, 1985; Rohrer *et al.*, 1986; Zackenfels *et al.*, 1995). The combined treatment with both antibodies results in the selection of the non-neuronal, fibroblast-like cell population of sympathetic ganglia. This population cannot be enriched by positive selection because of the lack of appropriate cell surface markers. This procedure is very selective and eliminates virtually completely the antigen-expressing cells (Zackenfels *et al.*, 1995). The high efficiency of the procedure may be due to the fact that both antibodies belong to the IgM subclass, and that the antigens are glycolipids expressed in high numbers on the cell surface. An additional advantage is that the selected, antigen-negative cell population is immediately available for culturing (in contrast to the panned cells).

After trypsinization, soybean trypsin inhibitor and DNaseI are added (see III). The ganglia are washed in F14 medium supplemented with 1 mg/ml BSA before dissociation in the same medium to single cells. The cell suspension is incubated with appropriate dilutions of Q211 and O4 antibody and 200 μl complement (total volume 1 ml in a 15 ml Falcon tube with open lid) for 35 minutes at 37°C in a humidified 5% CO_2 incubator. The cells are harvested by centrifugation (150g, 10 minutes), the supernatant is discarded, and the pellet is dissociated in about 700 μl of F14–0.1% BSA. Intact cells are separated from cellular debris by centrifugation through a BSA cushion: the cell suspension is carefully layered on top of 3 ml F14 medium supplemented with 3% BSA (Ernsberger and Rohrer, 1988). After centrifugation (150g, 25 minutes), the supernatant is carefully removed. The pellet containing the fibroblast-like cells is resuspended in complete medium and may be used for culture. A small aliquot is plated on dishes coated with poly-DL-ornithine (PORN) and analyzed for efficiency of cytotoxic kill by staining the cells for the Q211 (neuronal) or O4 (glial) antigens (see V).

Materials necessary for this procedure include the following:

1. F14 medium with 1 mg/ml BSA (0.1%, w/v)
2. F14 medium with 30 mg/ml BSA (3%, w/v)
3. Guinea pig complement (see VIII). The amount of complement required for efficient kill has to be determined empirically.
4. O4 and Q211 monoclonal antibodies (Sommer and Schachner, 1981; Henke-Fahle, 1983; respectively). The antibody dilutions required for efficient kill have to be optimized. At least 10-fold lower antibody concentrations are used for killing compared to cell-surface staining (see V).
5. Humidified 5% CO_2 incubator
6. PORN-coated tissue culture dishes (see VI)

V. Identification of Neuronal and Glial Cell Populations

The cellular composition of sympathetic ganglion dissociates and selected cell populations can be determined using markers for neurons and glial cells in short-term cultures. Cells are plated in F14–10% HS–5% FCS on tissue culture dishes coated with PORN (see VI). On this substrate, cells are firmly attached after 2–3 hours and can be stained with monoclonal antibodies that recognize cell type-specific cell surface antigens. We routinely use the Q211 antibody (Henke-Fahle, 1983) as marker for neurons and the O4 antibody (Sommer and Schachner, 1981) as marker for glial cells. The antigens recognized by these antibodies are glycolipids (polysialogangliosides for the Q211 antibody, Rösner *et al.*, 1985; sulfatide for the O4 antibody, Bansal *et al.*, 1989) that are not affected by the trypsin treatment. Instead of the Q211 antibody, tetanus toxin can be used (Rohrer *et al.*, 1985; Rohrer and Thoenen, 1987). It is expected that the commercially available marker Neurotag, a fluorescein-labeled tetanus toxoid, gives similar results, but we have no experience with this marker.

Live, unfixed cultures are washed with Krebs–Ringer's–HEPES-buffer supplemented with 0.1% BSA (KRH/A) and then incubated for 20 minutes with monoclonal antibodies diluted in KRH/A. After repeated washing with KRH/A, the cultures are fixed with 4% paraformaldehyde in PBS, pH 7.4 (20 minutes). The cultures are washed twice with PBS supplemented with BSA (PBS/A), incubated with biotinylated anti-mouse antibody (30 minutes), washed again in PBS/A, followed by fluorescein isothiocyanate (FITC)-labeled Streptavidin dissolved in PBS/A (20 minutes). After final washing in PBS, the cells are mounted in PBS/glycerol, covered with coverslips, and viewed under a microscope equipped with the appropriate fluorescence optics. For staining of cultured cells, we find 35 mm dishes containing four wells (10 mm diameter; Greiner) very useful. Each well can be used for a different staining protocol using a small volume (80 μl). A slight disadvantage of this procedure is that the wall of the 35 mm dish has to be broken after mounting of the cells to be able to view the cells under a conventional, noninverted fluorescence microscope. To remove the wall of the dish, we use a small pair of tongs. An alternative possibility for staining cultured cells are coverslips that contain 8 or 16 compartments (Chamberslides; Costar). In our experience, we get higher background with these, most likely because of less uniform washing compared with the four-well dishes.

Materials necessary for this procedure include the following:

1. PORN-coated four-well dishes (see VI)
2. Complete culture medium (F14–10% HS–5% FCS)
3. KRH/A (see VIII)
4. Monoclonal antibodies Q211 (Ascites 1:100, Henke-Fahle, 1983) and O4 (1:20, Sommer and Schachner, 1981)
5. PBS/A (see VIII)
6. 4% paraformaldehyde in PBS, pH 7.4 (see VIII)
7. Biotinylated anti-mouse antibody (1:100; Amersham)
8. FITC-labeled Streptavidin (1:100; Amersham)
9. PBS–glycerol (1:1; v/v)
10. Coverslips (10 mm diameter)

VI. Cell Culture

A. Preparing Tissue Culture Dishes

For optimal culture conditions, immature sympathetic neurons have to be seeded on coated tissue cultures dishes. The combination of PORN (Sigma) and laminin (Gibco BRL) as substratum has successfully been established (Edgar *et al.*, 1984). The tissue culture dishes are first coated with PORN (0.5 mg/ml in borate buffer, sterile filtered) over night at RT. They are air dried after twice washing with sterile H_2O and can be stored up to 1 month at 4°C. Before ganglia preparation, these dishes are coated with laminin (10 μg/ml in PBS) at RT. We routinely use 35 mm dishes with four wells (Greiner) and coat only the wells because this allows the use of small volumes (75–100 μl/well). The following slightly varying procedures are also performed in our laboratory: first, incubation at 37°C in a humidified 5% CO_2 incubator if the time until plating of the neurons is below 3 hours (with a minimum of 90 minutes); second, if for experimental reasons other dishes and therefore larger volumes of laminin have to be applied (e.g., 900 μl for standard 35 mm tissue culture dishes; 350 μl/well in 24-well plates), the concentration of laminin is lowered to 5 μg/ml. To compensate for the lowered concentration, the dishes are incubated overnight at RT in a humidified chamber to prevent drying. Before plating of the cells, the dishes are washed twice with sterile PBS (usually during trypsinization of the ganglia). It is important to avoid drying the laminin-coated wells because this results in loss of the biological activity of laminin.

Sympathetic neurons are then plated at the appropriate density and in the appropriate medium, depending on the experimental design.

Materials necessary for these procedures include the following:

1. PORN (Sigma)
2. 0.15 *M* borate buffer, pH 8.3 (see VIII)
3. Sterile H_2O
4. Sterile PBS, pH 7.4
5. Sterile laminin (EHS-laminin, Gibco BRL)
6. Tissue culture dishes (e.g., 35 mm dishes with four wells [Greiner], conventional 35 mm dishes [Nunc], 24-well plates [Costar], 96-well plates [Nunc])
7. Medium

B. Serum-Supplemented Media

Under standard conditions, sympathetic neurons are cultivated in Ham's F14 medium containing 10% HS (heat inactivated) and 5% FCS (heat inactivated) at 37°C in a humidified 5% CO_2 incubator (Ernsberger *et al.*, 1989a). Sympathetic neurons are critically dependent on the neurotrophin NGF for survival *in vitro* and *in vivo* (Levi-Montalcini and Booker, 1960; Rohrer *et al.*, 1988; Crowley *et al.*, 1994). Thus, sympathetic neuron cultures have to be supplemented with NGF. We usually use saturating concentrations of mouse NGF (10 ng/ml), purified from male mouse salivary glands according to Suda *et al.* (1978). However, recombinant NGF obtained from various commercial sources can also be used.

An exception are the immature sympathetic neurons from E6.5 lumbosacral chain ganglia. These cells depend on laminin as substratum, but survive independent of NGF for 4 days *in vitro* (Ernsberger *et al.*, 1989a). They do not become NGF dependent during this time and die from day 5 onward, which is correlated with a lack of high-affinity NGF receptors (Rodriguez-Tébar and Rohrer, 1991). If all-*trans* retinoic acid (RA), an oxidized form of vitamin A, is added to these immature sympathetic neurons, they respond with an increased expression of trkA (v. Holst *et al.*, 1995), the signal-transducing NGF receptor (Kaplan *et al.*, 1991), which leads to NGF-dependent survival (Rodriguez-Tébar and Rohrer, 1991; v. Holst *et al.*, 1995). The single addition of 5 n*M* RA at the beginning of the culture period is sufficient to allow for long-term cultures (10 or more days) in the presence of NGF.

High potassium concentrations (35–40 m*M*) also support the survival of sympathetic neurons, even in the absence of NGF (Ernsberger *et al.*, 1989a). Such high potassium concentrations lead to depolarization of the membrane potential and calcium influx (DiCicco-Bloom and Black, 1989). However, the mechanism of survival is still unknown.

Serum batches are known to vary not only from different sources but from lot to lot. We have found it useful to test a large number of serum batches (especially FCS) for their effects on neuronal morphology, survival (cell number), and proliferation (incorporation of tritiated thymidine) of E7 lumbosacral sympathetic neurons. We then order larger volumes of the most suitable batch and store them at −20°C.

The volume of medium needed depends on the kind of cell culture dishes used, and is 1.5 ml in 35-mm dishes. If the cultures are to be maintained for longer times, the medium has to be exchanged every 3–4 days. Usually two thirds of the medium volume is removed by careful pipetting. The cultures are replenished with prewarmed (37°C) medium and resupplemented with NGF (10 ng/ml). The number of cells plated depends mainly on the experimental purpose, and has to be determined empirically, as does the best-suited tissue culture means. The following cell numbers, dishes, and assays should give an indication for beginners, but are in no way to be seen as rigid values.

$2–3 \times 10^4$ cells/35-mm dish: Survival assay/immunocytochemistry
2.5×10^4 cells/well (24-well plate): RNA extraction
$3–5 \times 10^4$ cells/well (35 mm dish): Transfection
$>1 \times 10^5$ cells/35 mm dish (no wells): Biochemical assays

Materials necessary for these procedures include the following:

1. Ham's F14 medium (Gibco BRL)
2. HS (heat inactivated; Boehringer Mannheim; Gibco BRL)
3. FCS (heat inactivated; Boehringer Mannheim)
4. Complete medium (F14–10% HS–5% FCS)
5. PORN- and laminin-coated tissue culture dishes
6. NGF diluted in F14–10% HS–5% FCS. We prepare a 2 μg/ml stock solution from frozen aliquots of NGF (1 mg/ml in PBS/A), store it at 4°C, and use it for 7 days.
7. Humidified 5% CO_2 incubator
8. Waterbath

Optional:

9. RA (Sigma) is dissolved in dimethyl sulfoxide (DMSO, Sigma) at a 50-m*M* concentration, stored at −20°C, and serially diluted in medium before use. Note that RA is a lipophilic teratogen and should be handled very carefully with gloves only. In addition, RA is sensitive to light and oxidation (i.e., its activity is lost). We therefore dissolve and aliquot RA under safelight conditions in the dark. Even after storage for 4

years in DMSO at $-70°C$ we did not notice any loss of activity in the induction of NGF-dependent survival in E6.5 lumbosacral sympathetic neuron cultures (v. Holst, unpublished observation).

10. 3 *M* KCl stock solution. A dilution of 1:100 into F14 culture medium results in a final potassium concentration of 35 m*M*.

C. Serum-Free Medium

Lumbosacral sympathetic neurons have the advantage that they can be cultivated under serum-free (i.e., defined) conditions. This may be important under certain circumstances in which serum supplementation occludes possible effects of growth factor addition. For example, approximately 40% of E7 sympathetic neurons proliferate in the presence of serum, a proportion that cannot be increased by the addition of various growth factors (Ernsberger *et al.*, 1989b). However, under serum-free conditions, it was possible to show that the cells responded to some of the growth factors with an increase in proliferation (Zackenfels *et al.*, 1995). Serum components may also exert negative effects by blocking or sequestering reagents under investigation.

For serum-free cultures, Ham's F14 medium is supplemented with defined concentrations of transferrin, sodium selenite, progesterone, putrescine, insulin, and BSA, all of which can be stored at $-20°C$ (see VIII). This serum-free medium is prepared according to Bottenstein and Sato (1979), but please note that we now use different (lower) concentrations than in the original N2 formula (Bottenstein and Sato, 1979) and than previously published (Zackenfels *et al.*, 1995). The medium has to be freshly made with basal F14 medium (see VIII) not more than 2 weeks old. We usually do not use it longer than 3 days. In contrast to the usual procedure, the ganglia have to be dissociated in serum-free F14 medium.

Also under serum-free conditions, sympathetic neurons require NGF for survival and can be cultivated for 10 days or longer (v. Holst, unpublished observation).

Materials necessary for this procedure include the following:

1. Ham's F14 medium (Gibco BRL)
2. Transferrin (Sigma), 0.5 mg/ml stock solution in F14 medium
3. Sodium selenite (Sigma), 5.2 mg/10 ml stock solution in H_2O
4. Progesterone (Sigma), 6.3 mg/10 ml stock solution in ethanol (abs.)
5. Putrescine (Sigma), 1.6 mg/ml stock solution in F14 medium
6. BSA (Sigma), 10% (w/v) stock solution in F14 medium
7. Insulin (Sigma), 5 mg/ml stock solution in 0.01 HCl
8. PORN- and laminin-coated tissue culture dishes

9. NGF diluted in serum-free medium. We prepare a 2 μg/ml stock solution from frozen aliquots of NGF (1 mg/ml in PBS/A), store it at 4°C, and use it for 7 days.
10. Humidified 5% CO_2 incubator

VII. Expression of Exogenous Genes by Transfection

Immature sympathetic neuron cultures (E6.5–E7) are suitable for transient transfection of the neurons using the calcium phosphate precipitation method (Chen and Okayama, 1987) according to the modifications introduced by Gabellini *et al.* (1992). The method takes advantage of the fact that DNA is complexed to calcium phosphate during the formation of the precipitate. The precipitate covers the cells in the tissue culture dish and is eventually taken up. This is more likely to happen during cell division, and thus probably works quite well with immature sympathetic neurons because about 40% of the neurons divide during the second day *in vitro* (Ernsberger *et al.*, 1989b). The following points of the procedure are important and should be considered. The formation and size of the precipitate critically depends on the pH of the BES-buffered saline (2× BBS). We tested a series of different 2× BBS that differed only in their pH in 0.02 steps between 6.88 and 7.06, and found that a pH of 6.94 resulted in the highest transfection efficiency. Thus, it is important to determine empirically the optimum pH for this method. In addition, serum in the medium reduces the transfection efficiency and must be lowered to 0.5%. Moreover, a high density of neurons is needed to obtain a reasonable number of transfection events. In principle, any DNA of interest can be introduced, and we have successfully used the method for the expression of antisense constructs (Heller *et al.*, 1995) as well as for overexpression purposes (v. Holst *et al.*, 1995).

Sympathetic neurons (E6.5–E7) are prepared as described previously and plated on PORN-and laminin-coated dishes at a density of $3–5 \times 10^4$ cells in 100 μl/well (in 35 mm dishes containing four wells). We use minimal essential medium (MEM) throughout the culture period instead of F14, because the latter gives rise to huge precipitates during transfection. The ganglia are dissociated therefore in MEM–10% HS–5% FCS. After 2 hours, when the cells are attached, the dishes are flooded with prewarmed MEM–10% HS–5% FCS. After cultivation overnight, the medium is changed to 1.8 ml MEM–0.5% serum (MEM–10% HS–5% FCS diluted 30-fold with MEM), and 200 μl of the transfection mixture (2× BBS [100 μl], 1 *M* $CaCl_2$ [25 μl] and 10 μg DNA in 75 μl H_2O preincubated for 10 minutes at RT) is added dropwise to each 35-mm dish, preferentially over the wells. The dishes are placed for 6–12 hours into a humidified CO_2 incubator with 3% CO_2, which assists the actual transfection. After the incubation period, the medium is changed back to MEM–10% HS–5% FCS and supplemented with 35 m*M* potassium chloride for 1 day to increase neuron survival after the harsh transfection conditions. The transfection efficiency can be determined by cotransfection with a *lacZ* encoding expression plasmid that is added in 1/10 of the amount of the DNA of interest (Heller *et al.*, 1995). The expression of β-galactosidase (β-gal, the gene product of *lacZ*) can be detected by conventional immunocytochemistry (the cells are fixed in 4% paraformaldehyde, permeabilized with PBT1, (PBS/A with 0.1% Triton), and incubated with a monoclonal anti-β-gal antibody. After detection with an anti-mouse FITC-conjugated antibody, the cultures are mounted in PBS–glycerol and viewed under a microscope using epifluorescence ultraviolet illumination). Approximately 1–10% of the cells are transfected (immunoreactive for β-gal) when analyzed 2 days after transfection.

The cells can routinely be cultivated for 4 days, and for longer time periods if 5 n*M* RA is added before the transfection and the cultures are supplemented with NGF (10 ng/ml). The method is not suited for quantitative monitoring of the survival of live neurons under a phase-contrast microscope because the dark calcium phosphate precipitate covers most of the cells and cannot be washed away. Because of the relatively low absolute numbers of transfected cells and the many cells not surviving the transfection procedure, biochemical analyses also are not possible. Therefore, the analysis of the effects caused by introducing certain expression plasmids is mainly restricted to questions that can be addressed in single cells by immuncytochemical approaches (e.g., Heller *et al.*, 1995; v. Holst *et al.*, 1995).

Materials necessary for this procedure include the following:

1. MEM (Gibco BRL)
2. HS (heat inactivated; Boehringer Mannheim; Gibco BRL)
3. FCS (heat inactivated; Boehringer Mannheim)
4. MEM–10% HS–5% FCS
5. MEM–0.5% serum (MEM–10% HS–5% FCS diluted 30-fold with MEM)
6. 2× BBS (see VIII)
7. 1 *M* $CaCl_2$
8. DNA in H_2O (10 μg expression plasmid)
9. 3 *M* KCl stock solution. A dilution of 1:100 into MEM culture medium results in a final potassium concentration of 35 m*M*.
10. NGF diluted in MEM–10% HS–5% FCS. We prepare a 2 μg/ml stock solution from frozen aliquots of NGF (1 mg/ml in PBS/A), store it at 4°C, and use it for 7 days.
11. PORN- and laminin-coated tissue culture dishes
12. Humidified 5% CO_2 incubator
13. Humidified 3% CO_2 incubator

14. 4% paraformaldehyde in PBS, pH 7.4 (see VIII)
15. PBT1 (PBS/A with 0.1% Triton; see VIII)
16. Monoclonal anti-β-gal antibody (1:1000, Promega)
17. Anti mouse FITC-conjugated antibody (1:100, Dianova)
18. PBS–glycerol (1:1; v/v)

VIII. Solutions and Media

A. Solutions

2× BBS: 50 m*M* *N,N*-bis(2-hydroxyethyl)-2-aminoethansulfonic acid (BES, Sigma), 280 m*M* NaCl, 1.5 m*M* Na_2HPO_4, pH 6.94

Borate buffer: 150 m*M* boric acid (Merck) is dissolved and the pH adjusted to 8.3 with 30% NaOH.

Complement: Complement is isolated from guinea pig blood obtained by puncture of the heart. After coagulation for 2 hours at 4°C, the clot is separated into smaller pieces before centrifugation (2000 rpm, 10–15 minutes 4°C) to maximize the yield of serum. The serum (supernatant) is subjected to a second round of centrifugation as previously. The clear serum is taken and diluted 1:3 with F14 medium. To reduce the nonspecific cytotoxicity of the complement, the diluted serum is incubated with 80 mg noble agar (Difco) per 3 ml for 1 hour on ice with mixing by inversion every 10 minutes. After centrifugation (3000 rpm, 4°C, 15 minutes), the clear serum is filter sterilized and stored in small aliquots at −70°C. Complement is not stable and can be thawed only once.

DNaseI: DNaseI (Sigma) is dissolved in PBS (2 mg/ml), sterile filtered, and stored in 20 μl aliquots at −20°C.

Krebs–Ringer's–HEPES buffer (KRH): 125 m*M* NaCl, 4.8 m*M* KCl, 1.3 m*M* $CaCl_2 \cdot 2H_2O$, 1.2 m*M* $MgSO_4 \cdot 7H_2O$, 1.2 m*M* KH_2PO_4, 5.6 m*M* D-glucose, 25 m*M* *N*-2-hydroxy-ethylpiperazin-*N'*-2-ethansulfonic acid (HEPES), pH 7.3

KRH/A: KRH, 0.1% (w/v) BSA (Sigma)

Laminin (Gibco BRL, sterile): Laminin is diluted in sterile PBS under a tissue culture hood and added to PORN-coated dishes for 3 hours at RT at a final concentration of 10 μg/ml.

4% Paraformaldehyde: 45 ml H_2O is heated to 60°C before the addition of 2 g paraformaldehyde. After stirring, the solution is clarified with one or two drops of NaOH (1 *N*). Undissolved particles are filtered through a paper filter. When the solution has cooled to RT, 5 ml of 10× PBS are added, the pH is adjusted to 7.3 if necessary, and the volume adjusted to 50 ml with water. We store the fixative in brown bottles at 4°C and use it no longer than 2 weeks.

PBS: 137 m*M* NaCl, 3.0 m*M* KCl, 6.5 m*M* $Na_2HPO_4 \cdot 2H_2O$, 1.5 m*M* KH_2PO_4, pH 7.4. PBS can be prepared as a 10× stock and stored at RT. The pH is adjusted to 7.4 after dilution to 1× PBS, if necessary. For tissue culture purposes, PBS is autoclaved or sterile filtered.

PBS/A: PBS, 0.1% (w/v) BSA (Sigma)

PBT1: PBS/A, 0.1% (w/v) Triton X-100 (Roth)

PORN (Sigma): Poly-DL-ornithine (0.5 mg/ml) is dissolved in borate buffer, sterile filtered, and used for coating tissue culture dishes overnight at RT. The sterile solution can be frozen and stored at −20°C.

Soybean trypsin inhibitor (SBTI, Worthington 3571): SBTI is dissolved in PBS, pH 7.4, sterile filtered, and stored in 100 μl aliquots at −20°C. The concentration is adjusted to block completely the amount of trypsin used for the digestion of the ganglia.

Trypsin (Worthington 3707): Trypsin is dissolved in PBS pH 7.4 (1890 U/ml), sterile filtered, and stored in 100 μl aliquots at −20°C.

B. Media

Basal F14 medium (Ham's F14, Gibco BRL): F14 medium is purchased as a powder consisting of modified F12 medium (i.e., 2× amino acids, 2× pyruvate, 85 μ*M* ascorbic acid, 11 m*M* glucose, 5 m*M* KCl, 0.85 m*M* $MgCl_2$, 0.15 m*M* $MgSO_4$, 2 m*M* $CaCl_2$, and 0.5 μ*M* $ZnSO_4$), and dissolved completely in 900 ml millipore water. After addition of the antibiotics penicillin (60.6 mg/l) and streptomycin (133 mg/l), the pH is adjusted with NaOH (1 *N*) to 7.3 followed by the addition of bicarbonate according to the supplier's indication (i.e., 1.96 g $NaHCO_3$/l). It is important first to adjust the pH to 7.3 and then add the bicarbonate. The volume is adjusted to 1 liter with water, and the medium is gassed with CO_2 until it acquires a salmon-like color before sterile filtration. It can then be stored up to 4 weeks at 4°C.

HEPES-buffered F14 medium: F14 is prepared as described earlier to the point where the pH is adjusted to 7.3. Then, however, instead of bicarbonate, HEPES is added to give a final concentration of 20 m*M* (5.2 g/l). Before the volume is brought to 1 liter with water, the pH is again adjusted to 7.3 using HCl (1 *N*). The medium is sterile filtered and can be stored up to 4 weeks at 4°C.

Complete F14 medium: 10% (v/v) heat-inactivated HS and 5% (v/v) heat-inactivated FCS are added to basal F14 medium. The complete medium is sterile filtered, stored at 4°C, and used within 1–2 weeks. Heat inactivation is achieved by incubation of the serum-containing bottles at 56°C for 30 minutes before smaller aliquots are stored at −20°C.

Serum-free F14 medium: Insulin (5 μg/ml), transferrin (0.5 μg/ml), putrescine (10 μ*M*), sodium selenite (30 n*M*), progesterone (20 n*M*), and 0.1–1.0% (w/v) BSA (all components from Sigma) are added from stock solutions to basal F14 medium to obtain the indicated final concentrations. After sterile filtration and storage at 4°C, the medium can be used for a limit of 3 days.

Minimal essential medium (Gibco BRL): MEM is purchased as fluid medium and supplemented with 1% (v/v) of a 100× antibiotics solution (penicillin 5000

U/ml and streptomycin 5 mg/ml) and sterile glutamine (2 m*M* final concentration). If serum components are necessary, the procedure is the same as for the complete F14 medium.

Acknowledgments

The authors wish to thank everyone who contributed to the development of the methods over the years. We also thank N. Zimmermann, D. Junghans, M. Geissen, C. Schneider, H. Patzke, and C. Thum for helpful suggestions, and C. Morgans for critical reading of the manuscript. The work in our laboratory has been supported by grants from the Deutsche Forschungsgemeinschaft (SFB 269), Fonds der chemischen Industrie and Deutsche Krebshilfe.

References

Allsopp, T. E., Robinson, M., Wyatt, S., and Davies, A. M. (1993). Ectopic *trkA* expression mediates a NGF survival response in NGF-independent sensory neurons but not in parasympathetic neurons. *J. Cell Biol.* **123,** 1555–1566.

Bansal, R., Warrington, A. E., Gard, A. L., Ranscht B., and Pfeiffer, S. E. (1989). Multiple and novel specificities of monoclonal antibodies of O1, O4, and R-mAb used in the analysis of oligodendrocyte development. *J. Neurosci. Res.* **24,** 548–557.

Barde, Y.-A. (1989). Trophic factors and neuronal survival. *Neuron* **2,** 1525–1534.

Barde, Y.-A., Edgar, D., and Thoenen, H. (1980). Sensory neurons in culture: Changing requirements for survival factors during embryonic development. *Proc. Natl. Acad. Sci. U.S.A.* **77,** 1199–1203.

Barres, B. A., Hart, I. K., Coles, H. S. R., Burne, J. F., Voyvodic, J. T., Richardson, W. D., and Raff, M. C. (1992). Cell death and control of cell survival in the oligodendrocyte lineage. *Cell* **70,** 31–46.

Birren, S. J., Lo, L., and Anderson, D. J. (1993). Sympathetic neuroblasts undergo a developmental switch in trophic dependence. *Development* **119,** 597–610.

Borasio, G. D., Markus, A., Wittinghofer, A., Barde, Y.-A., and Heumann, R. (1993). Involvement of ras p21 in neurotrophin-induced response of sensory, but not sympathetic neurons. *J. Cell Biol.* **121,** 665–672.

Bottenstein, J. E., and Sato, G. H. (1979). Growth of a rat neuroblastoma cell line in serum-free supplemented medium. *Proc. Natl. Acad. Sci. U.S.A.* **76,** 514–517.

Chen, C., and Okayama, H. (1987). High-efficiency transformation of mammalian cells by plasmid DNA. *Mol. Cell. Biol.* **7,** 2745–2752.

Crowley, C., Spencer, S. D., Nishimura, M. C., Chen, K. S., Pitts-Meek, S., Armanini, M. P., Ling, L. H., McMahon, S. B., Shelton, D. L., Levinson, A. D., and Phillips, H. S. (1994). Mice lacking nerve growth factor display perinatal loss of sensory and sympathetic neurons yet develop basal forebrain cholinergic neurons. *Cell* **76,** 1001–1011.

DiCicco-Bloom, E., and Black, I. B. (1989). Depolarization and insulin-like growth factor-I (IGF-I) differentially regulate the mitotic cycle in cultured rat sympathetic neuroblasts. *Brain Res.* **491,** 403–406.

Durbec, P. L., Larsson-Blomberg, L. B., Schuchardt, A., Costantini, F., and Pachnis, V. (1996). Common origin and developmental dependence on c-ret of subsets of enteric and sympathetic neuroblasts. *Development* **122,** 349–358.

Edgar, D., Barde, Y.-A., and Thoenen, H. (1981). Subpopulations of cultured chick sympathetic neurones differ in their requirements for survival factors. *Nature* **289,** 294–295.

Edgar, D., Timpl, R., and Thoenen, H. (1984). The heparin-binding domain of laminin is responsible for its effects on neurite outgrowth and neuronal survival. *EMBO J.* **3,** 1463–1468.

Ernsberger, U., Edgar, D., and Rohrer, H. (1989a). The survival of early chick sympathetic neurons in vitro is dependent on a suitable substrate but independent of NGF. *Dev. Biol.* **135,** 250–262.

Ernsberger, U., and Rohrer, H. (1988). Neuronal precursor cells in chick dorsal root ganglia: Differentiation and survival in vitro. *Dev. Biol.* **126,** 420–432.

Ernsberger, U., and Rohrer, H. (1994). Neurotrophins and neurite outgrowth in the peripheral nervous system. *Semin. Dev. Biol.* **5,** 403–410.

Ernsberger, U., Sendtner, M., and Rohrer, H. (1989b). Proliferation and differentiation of embryonic chick sympathetic neurons: Effects of ciliary neurotrophic factor. *Neuron* **2,** 1275–1284.

Gabellini, N., Minozzi, M.-C., Leon, A., and Roso, R. D. (1992). Nerve growth factor transcriptional control of c-fos promoter transfected in cultured spinal sensory neurons. *J. Cell Biol.* **118,** 131–138.

Hamburger, V., and Hamilton, H. L. (1951). A series of normal stages in the development of the chick embryo. *J. Exp. Zool.* **88,** 49–92.

Heller, S., Finn, T. P., Huber, J., Nishi, R., Geissen, M., Püschel, A. W., and Rohrer, H. (1995). Analysis of function and expression of the chick GPA receptor (GPARa) suggests multiple roles in neuronal development. *Development* **121,** 2681–2693.

Henke-Fahle, S. (1983). Monoclonal antibodies recognize gangliosides in the chick brain. *Neurosci. Lett.* **14,** 160.

Heumann, R. (1994). Neurotrophin signalling. *Curr. Opin. Neurobiol.* **4,** 668–679.

Kaplan, D. R., Hempstead, B. L., Martin-Zanca, D., Chao, M. V., and Parada, L. F. (1991). The *trk* proto-oncogene product: A signal transducing receptor for nerve growth factor. *Science* **252,** 554–558.

Kobayashi, M., Kurihara K., and Matsuoka I. (1994). Retinoic acid induces BDNF responsiveness of sympathetic neurons by alteration of Trk neurotrophin receptor expression. *FEBS Lett.* **356,** 60–65.

Levi-Montalcini, R. (1987). The nerve growth factor: Thirty-five years later. *EMBO J.* **6,** 1145–1154.

Levi-Montalcini, R., and Booker, B. (1960). Destruction of the sympathetic ganglia in mammals by an antiserum to a nerve-growth protein. *Proc. Natl. Acad. Sci. U.S.A.* **46,** 384–391.

Nobes, C. D., Reppas, J. B., Markus, A., and Tolkovsky, A. M. (1996). Active p21Ras is sufficient for rescue of NGF-dependent rat sympathetic neurons. *Neuroscience* **70,** 1067–1079.

Patterson, P. H., and Chun, L. L. Y. (1974). The influence of non-neuronal cells on catecholamine and acetylcholine synthesis and accumulation in cultures of dissociated sympathetic neurons. *Proc. Natl. Acad. Sci. U.S.A.* **71,** 3607–3610.

Rao, M. S., and Landis, S. C. (1993). Cell interactions that determine sympathetic neuron transmitter phenotype and the neurokines that mediate them. *J. Neurobiol.* **24,** 215–232.

Reissmann, E., Ernsberger, U., Francis-West, P. H., Rueger, D., Brickell, P. M., and Rohrer, H. (1996). Involvement of bone morphogenetic proteins-4 and -7 in the specification of the adrenergic phenotype in developing sympathetic neurons. *Development* **122,** 2079–2088.

Rodriguez-Tébar, A., and Rohrer, H. (1991). Retinoic acid induced NGF-dependent survival response and high-affinity NGF receptors in immature chick sympathetic neurons. *Development* **112,** 813–820.

Rohrer, H. (1985). Nonneuronal cells from chick sympathetic and dorsal root sensory ganglia express catecholamine uptake and receptors for nerve growth factor during development. *Dev. Biol.* **111,** 95–107.

Rohrer, H., Acheson, A. L., Thibault, J., and Thoenen, H. (1986). Developmental potential of quail dorsal root ganglion cells analyzed in vitro and in vivo. *J. Neurosci.* **6,** 2616–2624.

Rohrer, H., Henke-Fahle, S., El-Sharkawy, T., Lux, H. D., and Thoenen, H. (1985). Progenitor cells from embryonic chick dorsal root ganglia differentiate in vitro to neurons: Biochemical and electrophysiological evidence. *EMBO J.* **4,** 1709–1714.

Rohrer, H., Hofer, M., Hellweg, R., Korsching, S., Stehle, A. D., Saadat, S., and Thoenen, H. (1988). Antibodies against nerve growth factor interfere in vivo with the development of avian sensory and sympathetic neurons. *Development* **103,** 545–552.

Rohrer, H., and Thoenen, H. (1987). Relationship between differentiation and terminal mitosis: Chick sensory and ciliary neurons differentiate after terminal mitosis of precursor cells whereas sympathetic neurons continue to divide after differentiation. *J. Neurosci.* **7,** 3739–3748.

Rösner, H., Al-Aqtum, M., Henke-Fahle, S. (1985). Developmental expression of GD3 and polysialogangliosides in embryonic chicken nervous tissue reacting with monoclonal antiganglioside antibodies. *Dev. Brain Res.* **18,** 85–95.

Rothman, T. P., Gershon, M. D., and Holtzer, H. (1978). The relationship of cell division to the acquisition of adrenergic characteristics by developing sympathetic ganglion cell precursors. *Dev. Biol.* **65,** 321–341.

Sommer, I., and Schachner, M. (1981). Monoclonal antibodies (O1 to O4) to oligodendrocyte cell surfaces: An immunocytological study in the central nervous system. *Dev. Biol.* **83,** 311–327.

Suda, K., Barde, Y.-A., and Thoenen, H. (1978). Nerve growth factor in mouse and rat serum: Correlation between bioassay and radioimmunoassay determinations. *Proc. Natl. Acad. Sci. U.S.A.* **75,** 4042–4046.

v. Holst, A., Rodriguez-Tébar, A., Michaille, J.-J., Dhouailly, D., Bäckström, A., Ebendal, T., and Rohrer, H. (1995). Retinoic acid-mediated increase in trkA expression is sufficient to elicit NGF-dependent survival of sympathetic neurons. *Mol. Cell. Neurosci.* **6,** 185–198.

Zackenfels, K., Oppenheim, R. W., and Rohrer, H. (1995). Evidence for an important role of IGF-I and IGF-II for the early development of chick sympathetic neurons. *Neuron* **14,** 731–741.

Zurn, A. D., and Mudry, F. (1986). Conditions increasing the adrenergic properties of dissociated chick superior cervical ganglion neurons grown in long-term culture. *Dev. Biol.* **117,** 365–379.

10

Analysis of Gene Expression in Cultured Primary Neurons

Ming-Ji Fann
Institute of Neuroscience
National Yang-Ming University
Taipei, Taiwan 11221
Republic of China

Paul H. Patterson
Biology Division
California Institute of Technology
Pasadena, California 91125

I. Introduction

The nervous system is the most complex organ in the body. There are more than 10^{12} neurons in humans and thousands of phenotypes. In addition, neurons make precise patterns of connections and these networks control an enormous diversity of functions. The outflow of these networks is regulated, in turn, through neuronal activity and local environmental factor influences on neuronal gene expression (Patterson and Nawa, 1993; McConnell, 1995; Thoenen, 1995). Primary cultures of particular populations of dissociated neurons have been used to study this complexity in a more defined situation. Through systematically manipulating culture conditions, such systems have provided valuable information concerning cell proliferation, differentiation, survival, process growth, transmitter secretion, and neuronal death (Furshpan *et al.*, 1987; Yamamori *et al.*, 1989; Ray *et al.*, 1993; Gage *et al.*, 1995; Koh *et al.*, 1995). The use of primary neurons is important in such experiments because they presumably reflect the response of normal cells more accurately than do cell lines. A major difficulty with the use of primary cells is, however, the relatively small amounts of material. Moreover,

Current Topics in Developmental Biology, Vol. 36

0070-2153/98 $25.00

dissection is time consuming and tedious, and considerable care is needed to master the techniques.

Fortunately, recent progress in molecular biologic techniques has enhanced the utility of primary cultures. Amplification of signals by polymerase chain reaction (PCR) after reverse transcription (RT) of RNA to cDNA provides a way to obtain more information than previous methods, and allows the use of small numbers of cultured cells (Belyavsky *et al.*, 1989). Furthermore, the RT-PCR may be used to simultaneously determine changes in the expression of multiple neuronal genes (Fann and Patterson, 1993; Greenlund *et al.*, 1995b). Thus, this approach circumvents some of the limitations of primary culture and greatly enhances the ability to study neuronal gene expression after experimental manipulation.

Because sympathetic ganglia are relatively easy to dissect and various culture methods have been developed to maintain homogeneous neuronal populations *in vitro* (Hawrot and Patterson, 1979; Wolinsky *et al.*, 1985), these cells have been used to study neuronal gene expression in various experimental paradigms (Kessler *et al.*, 1993; Fann and Patterson, 1994a; Greenlund *et al.*, 1995a; Mahanthappa and Patterson, 1996). Particularly useful is the finding that sympathetic cells can be maintained in small numbers in serum-free medium. We review here the use of the RT-PCR assay in analyzing the effects of cytokines on neuronal gene expression.

II. Culture of Small Numbers of Sympathetic Neurons in Chemically Defined Medium

Neurons are prepared and cultured as described previously (Hawrot and Patterson, 1979; Wolinsky *et al.*, 1985; Mahanthappa and Patterson, 1996). Superior cervical ganglia (SCG) are removed from neonatal Sprague-Dawley rats, using a dissection microscope with 10× magnification, and placed in Hank's balanced salt solution without Ca^{2+} and Mg^{2+} (Gibco/BRL, Grand Island, NY). The position of SCG and detailed instructions for their dissection are described by Mahanthappa and Patterson (1996). After SCG are removed from the body, blood vessels and other adherent tissues attached to the ganglia are removed and the sheaths that surround the ganglia are peeled away with fine forceps in Hank's solution in a clean environment (a sterile hood is preferred), using a dissection microscope with 30× magnification.

Enzymatic dissociation of ganglia usually results in good cell separation in less than an hour with high yields. It is important to prerinse the pipettes used in the dissociation process with 1 mg/ml bovine serum albumin (BSA) to minimize loss of neurons due to adherence to glass. Dissociation begins with transferring roughly 60 ganglia (from 30 pups) to a 15-ml tube containing 4 ml of 0.1% trypsin in Hank's balanced salt solution without Ca^{2+} and Mg^{2+}. Digestion proceeds for 5 minutes at 37°C without shaking. Ganglia are then triturated gently with a BSA-coated, fire-polished, and cotton-plugged (in the back, to minimize contamination) Pasteur pipette for 20 strokes (avoiding any bubble

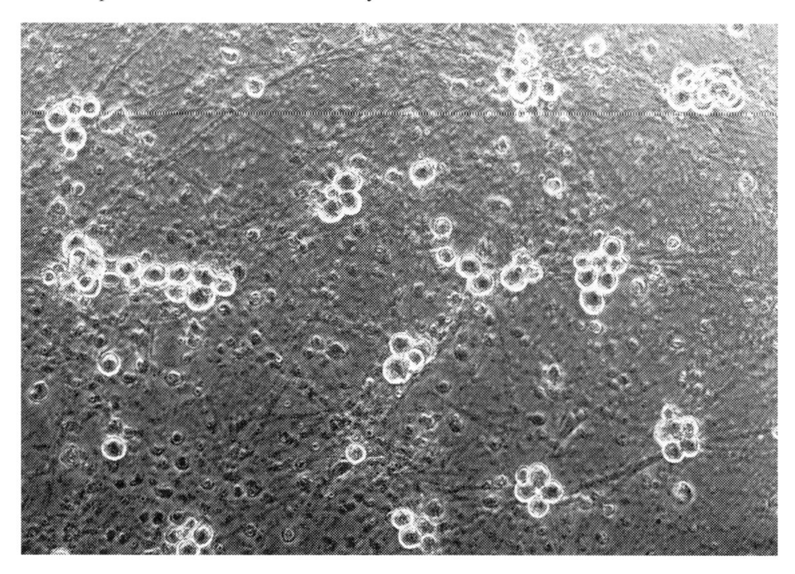

Fig. 1. Representative photomicroscopy of sympathetic neurons cultured in serum-free medium in 96-well plates for 6 days. Scale bar = 50 μm.

formation). The tube is then left undisturbed for 1 minute to let undigested tissue settle to the bottom. Alternatively, the tube can be subjected to centrifugation for less than 10 seconds at 100*g*. The supernatant is transferred to a new tube containing 2 ml of 2 mg/ml trypsin inhibitor (Sigma, St. Louis, MO) and maintained in 4°C until further processing. The remaining chunks of undigested ganglia are treated again with trypsin and trituration, following the same procedures, until they are totally dissociated. All cells are then combined, centrifuged, resuspended in 12 ml culture medium, and filtered into a 50-ml tube through a cell strainer with 70-μm pore size (Falcon, Oxnard, CA) to obtain single-cell suspension. Two hundred microliters of cells is plated into each well of collagen-coated, 96-well plates, and incubated with 5% CO_2. After 7 days, each well contains approximately 3000 cells (Wolinsky *et al.*, 1985), and more than 95% of surviving cells are neurons (Fig. 1).

The culture medium we favor is termed L15-CO_2 (for preparation, see Hawrot and Patterson, 1979) supplemented with 100 μg/ml transferrin, 5 μg/ml insulin, 16 μg/ml putrescine, 30 n*M* sodium selenite, 4 μg/ml aphidicolin (all from Sigma), and 100 ng/ml nerve growth factor (Boehringer Mannheim, Indianapolis, IN). Aphidicolin is a relatively nontoxic antimitotic agent in this serum-free condition and is used to eliminate non-neuronal cells (Wallace and Johnson, 1989). Because aphidicolin can be present in the medium during the entire culture period, preplating the dissociated cells in uncoated tissue culture plates is not necessary to eliminate fibroblasts, Schwann cells, and the like. Another advantage of this protocol is that during the dissociation process and subsequent culture period, neurons are never in contact with serum, which has powerful effects on gene expression (Wolinsky and Patterson, 1985). For ease and reproducibility, neurons are seeded in 96-well plates, at a density of one ganglion per well. These plates are precoated with collagen by adding 40 μl of 500 μg/ml type I collagen (Sigma) and air dried. Type I collagen can also be prepared from rat tail tendons (Mahanthappa and Patterson, 1996). In renewing the medium during long-term culture, half of the medium is changed every 36 hours. This method minimizes the disturbance of the neurons.

III. Preparation of RNA and cDNA

Because the PCR is notorious for contamination, it is wise to take some precautions from the start of the RNA extraction. Plugged Pipetman tips and pipettes should be used, and it is better to have a set of Pipetmen dedicated for this RT-PCR assay.

Total RNA from cultured neurons is prepared by disruption of the cells in 4 *M* guanidinium thiocyanate and extraction with acidic phenol and chloroform (Chomczynski and Sacchi, 1987; Fann and Patterson, 1993). Culture medium is first carefully removed from the wells by suction. One hundred sixty microliters of lysis buffer (4 *M* guanidinium thiocyanate, 25 m*M* sodium citrate, pH 7.0, 100 m*M* 2-mercaptoethanol, 0.5% sodium lauroyl sarcosinate) is added to each well. The lysate is pipetted up and down a few times and transferred to a 1.7-ml siliconized Eppendorf tube and vortexed vigorously to shear DNA. Ten microliters of 1 *M* sodium acetate (pH 4.0), 200 μl water-saturated phenol, and 40 μl chloroform are added sequentially, with vigorous vortexing after each addition. The samples are kept at 4°C for 20 minutes and centrifuged at 12,000*g* for 15 minutes. The water phase is transferred to a fresh siliconized tube containing 10 μl of 3 mg/ml glycogen (Boehringer Mannheim). Glycogen aids the precipitation of RNA and visualization of pellets after centrifugation. RNA is precipitated by addition of 200 μl isopropanol and stored at −20°C overnight.

To produce cDNA, total RNA is centrifuged at 12,000*g* for 15 minutes, washed once with 70% ethanol, and dried in a Speedvac (Savant, Farmingdale, NY). The RNA is directly dissolved in 10 μl of 13 m*M* methylmercury hydroxide for 10 minutes and an additional 2 μl of 75 m*M* 2-mercaptoethanol is added for 5 minutes. These steps disrupt RNA secondary structure. Because methylmercury hydroxide is a toxic chemical, an alternative is to dissolve the RNA in 10 μl of diethylpyrocarbonate-treated H_2O, denature the RNA in 75°C water bath for 15 minutes, and after brief centrifugation at 4°C, 2 μl of 100 m*M* dithiothreitol is added to each sample. Samples are maintained at 4°C to avoid RNA renaturation. Reverse transcription is peformed by adding 8 μl of reaction mixture to each sample such that the final volume is 20 μl and contains 50 m*M* Tris-HCl, pH 8.3, 75 m*M* KCl, 3 m*M* $MgCl_2$, 0.5 m*M* deoxynucleotide triphosphate (dNTP; Boehringer Mannheim), 0.5 μg of oligo dT (Pharmacia, Piscataway, NJ), 100 U RTase (Superscript II, Gibco/BRL) and 20 U RNasin (Promega, Madison, WI). Samples are reacted for 1 hour at 42°C, and kept at −20°C after reaction to stop enzyme action and maintain cDNA integrity.

IV. Polymerase Chain Reaction

Amplification of cDNA is performed in a thermal cycler (MJ Research, Watertown, MA) with each tube containing a final volume of 20 μl, consisting of 1 μl of cDNA, 1× PCR buffer (50 m*M* KCl, 1.5 m*M* $MgCl_2$, 100 m*M* Tris-HCl, pH 9.0, and 0.1% Triton X-100), 0.5 U *Taq* DNA polymerase (Promega), 0.25 m*M* dNTP, and one set of primers (final concentration, 200 n*M*). Reagents are prepared and kept at 4°C before reaction. The standard PCR conditions used to amplify DNA fragments less than 500 bp in length are: 94°C for 45 seconds to denature templates, annealing temperature (temperatures for each cDNA varies according to primers selected) for 75 seconds, and 72°C for 30 seconds to extend fragments. These conditions require modification when different thermal cyclers are used. As an example, if the Perkin-Elmer 9600 model (Norwalk, CT) is used, duration for denaturation and annealing steps can be shortened by 75%. The numbers of cycles used to amplify each cDNA are chosen to allow the PCR to proceed in a linear range, but with sufficient yield to be visualized by ethidium bromide staining after gel electrophoresis. The appropriate cycling profiles (numbers of cycles and annealing temperatures) are determined empirically. Eight microliters of each PCR sample is analyzed on a 2% agarose gel (Gibco/BRL), and the products visualized with ethidium bromide staining and ultraviolet illumination.

V. Selection of Primers

Oligonucleotides are selected as primers according to the criteria described by Lowe *et al.* (1990). Primers designed by this method have good hybridization avidity to target sequences and form primer–dimer products at low frequency. It is critical to select primers that recognize cDNA-derived templates but not genomic DNA templates. This property is important, because minute amounts of genomic DNA may still exist in the acidic-phenol-extracted RNA samples, and this will interfere the interpretation of PCR results. Elimination of genomic template priming can be achieved by choosing sequences that either cover exon–intron junctions or have large introns in between them. Computer software is

Table I Sequences of the Primers, Predicted Sizes of PCR Fragments, and Amplification Cycles Used in the PCR

Gene	Sequence	Predicted size (bp)	Annealing temp.	Amplification cycles	Reference
β-actin					
Sense	5′-TCATGAAGTGTGACGTTGACATCCGT	285	58	18	Nudel *et al.*, 1983
Antisense	5′-CCTAGAAGCATTTGCGGTGCACGATG				
CCK					
Sense	5′-GACTCCGCATCCGAAGAT	366	50	35	Deschenes *et al.*, 1985
Antisense	5′-CTACGATGGGTATTCGTA				
ChAT					
Sense	5′-GCTTACTACAGGCTTTAC	338	50	30	Hahn *et al.*, 1992
Antisense	5′-GACAAACCGGTTGCTCAT				
ENK					
Sense	5′-ATCAACTTCCTGGCATGC	429	50	28	Rosen *et al.*, 1984
Antisense	5′-GCTCGTGCTGTCTTCATC				
NPY					
Sense	5′-GCTAGGTAACAAACGAATGGGG	288	58	22	Larhammar *et al.*, 1987
Antisense	5′-CACATGGAAGGGTCTTCAAGC				
SOM					
Sense	5′-CCAGACTCCGTCAGTTTCTGC	238	58	28	Montminy *et al.*, 1984
Antisense	5′-AGTTCTTGCAGCCAGCTTTGC				
SP					
Sense	5′-ATGAAAATCCTCGTGGCG	$\alpha = 321$, $\beta = 375$	58	27	Carter and Krause, 1990
Antisense	5′-GTAGTTCTGCATTGCGCT	$\gamma = 331$, $\delta = 276$			
TH					
Sense	5′-CGGGCTATGTAAACAGAATGGG	418	58	24	Grima *et al.*, 1985
Antisense	5′-GATGGAGACTTTGGGAAAGGC				
VIP					
Sense	5′-AGTGTGCTGTTCTCACAGTCG	216	58	28	Nishizawa *et al.*, 1985
Antisense	5′-GCTGGTGAAAACTCCATCAGC				

Modified from Fann and Patterson (1993).

available to select primers that fulfill these requirements. For those genes without available genomic DNA sequences, several distinct oligonucleotides should be synthesized and tested in PCR pilot experiments using cDNA from tissues or cells where genes of interest are expressed. Those primers that discriminate cDNA from genomic DNA are selected for further study. Another point to be considered is the size of PCR fragments generated. It is preferable to design primers that generate fragments less than 500 bp; this results in higher amplification efficiency in each cycle and avoids problems with long products, which can be difficult to amplify.

For illustration, we use nine sets of oligonucleotide primers to detect mRNAs for β-*actin*, the neurotransmitter synthetic enzymes choline acetyltransferase (*ChAT*), and tyrosine hydroxylase (*TH*), and the neuropeptides cholecystokinin (*CCK*), enkephalin (*ENK*), neuropeptide Y (*NPY*), somatostatin (*SOM*), substance P (*SP*), and vasoactive intestinal peptide (*VIP*). For *ENK* and *SP*, which are expected to yield more than one transcript because of alternative splicing (Krause *et al.*, 1987; Darmon *et al.*, 1988; Garrett *et al.*, 1989), the primers selected recognize regions common to each transcript and generate a single PCR product for *ENK*, and may generate four different PCR products for *SP*. The primer sequences, annealing temperatures, amplification cycles, and the predicted sizes of the PCR products are given in Table I.

VI. Specificity of Primers

It is important to test whether the primers selected specifically identify the mRNAs for the intended genes. This can be accomplished by using templates from the tissues or cells that are known to express these genes. This step also verifies whether the primers recognize RNA-derived templates only, and not genomic DNA. In our example, total RNA was prepared from adult rat spinal cord using the acidic-phenol extraction method (Chomczynski and Sacchi, 1987) and reverse-transcribed into cDNA. All cDNAs were amplified for 35 cycles except β-*actin*, which was amplified for 20 cycles. Each set of primers (except those for *SP*) generates a single PCR product with the size predicted from the known gene sequences (Fig. 2, "+" lanes). Reverse transcription is essential for generation of these expected products, because they do not appear in the samples that are not treated with reverse transcriptase (Fig. 2, "−" lanes). The primers for *SOM* and *TH* presumably also recognize genomic DNA and produce PCR products whose sizes are larger than expected in the samples not treated with reverse transcriptase (Fig. 2, *SOM*, *TH*, "−" lanes). This does not interfere with the assay, however, because these bands disappear when in competition with the proper templates present in the cDNA. The *SP* primers produce three products whose sizes (375, 331, and 276 bp) correspond to the mRNAs for β, γ, and δ-SP, respectively (Krause *et al.*, 1987; Harmar *et al.*, 1990). The assay also yields the

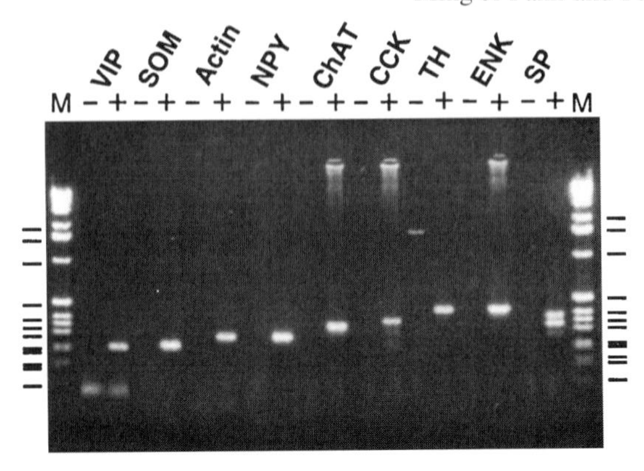

Fig. 2. Specificity of primers. Total RNA was prepared from adult rat spinal cord and either converted to cDNA with reverse transcriptase ("+" lanes), or not treated with reverse transcriptase ("−" lanes). All samples were amplified for 35 cycles in the PCR except actin, which was amplified for 20 cycles. The samples were loaded in ascending order of expected sizes, based on the known gene sequences, as listed in Table 1. A 1-kb DNA marker was used as the DNA molecular weight standard ("M" lanes) and the measured sizes from bottom to top are 75, 134, 154, 201, 220, 298, 344, 396, 506/517, 1018, 1635, 2036 bp. [Modified from Fann and Patterson (1993).]

expected relative levels of these mRNAs; β-SP and γ-SP are more abundant than δ-SP, whereas α-SP comprises less than 1% of the total *SP* mRNAs (Carter and Krause, 1990; Harmar *et al.*, 1990) and is undetectable in our assay. There is primer–dimer formation when *VIP* primers are used, which can be eliminated by diluting the primer concentration twofold and preparing reagents in 4°C before the PCR.

To confirm further the specificity of the primers, PCR products can be checked with restriction enzyme digestion or sequencing. In our example, three PCR products (β-*actin*, *CCK*, and *ENK*) were cloned and sequenced. TA cloning kits (Invitrogen, San Diego, CA) were used to clone PCR fragments. All clones sequenced (at least three clones for each PCR product) contain the sequences expected for these genes. This is expected, because the annealing temperature used in the PCR for each set of primers is close to the predicted melting temperature in an effort to increase the specificity of the PCR. In addition, we have demonstrated that these PCR products appear only when reverse transcriptase is present, that there is only one PCR product for each gene analyzed, and that the size of each PCR product is identical to that predicted from the known sequences. Although the PCR can generate artifacts when unknown sequences are cloned from ambiguous primers, this type of artifact rarely occurs with primers that are identical to the known sequences, especially when the PCR is carried out under high-stringency conditions, as in the present case. Therefore, the PCR products obtained here should represent products of the genes indicated.

VII. Effects of Leukemia Inhibitory Factor, Ciliary Neurotrophic Factor, and Depolarization on Cultured Sympathetic Neurons

To compare the applicability of the RT-PCR assay with conventional methods, we tested effects of cytokines (leukemia inhibitory factor [LIF] and ciliary neurotrophic factor [CNTF]) on neuronal gene expression and further asked whether depolarization modulates effects of these two cytokines. Using methods of metabolic labeling, radioimmunoassay and Northern analysis, it had been shown that LIF and CNTF induce the expression of SP, SOM, VIP, and ChAT, and this induction can been inhibited by depolarization (Nawa *et al.*, 1990, 1991; Rao *et al.*, 1992; Symes *et al.*, 1993). These prior analyses required large numbers of neurons and assay one gene at a time.

Recombinant human LIF and recombinant rat CNTF (both from Dr. B. Samal, Amgen, Thousand Oaks, CA) were added to cultured sympathetic neurons at a concentration of 1.6 ng/ml between days 2 and 7 *in vitro*. Duplicate samples are used for each cytokine to compensate for an occasional failure of amplification (see later). After 7 days of culture, RNA is extracted from each well and reverse transcribed to cDNA. The cDNA derived from a single well was used as template for nine PCRs, in nine separate tubes. β-Actin expression was used as internal control to monitor the amount of RNA in each sample. As shown in Figure 3, the intensity of the β-actin signal is somewhat weaker in samples treated with LIF and CNTF, which indicates some neuronal death triggered by these factors (Kessler *et al.*, 1993). Nonetheless, even before making the adjustment of such RNA differences, the expression pattern of neuronal genes in the presence of LIF and CNTF is similar to that previously reported using conventional methods (Nawa *et al.*, 1990). These cytokines induce expression of *SP*, *ChAT*, and *SOM* mRNAs, while decreasing that of *NPY* and *TH* (Fig. 3). We do not observe induction of *VIP*, which is known to require longer incubation times for induction (Nawa *et al.*, 1990, 1991). Another reason for the lack of *VIP* induction may be the lack of serum in the culture. This RT-PCR assay also demonstrates an induction of *CCK* and *ENK* mRNAs by these two cytokines. These data illustrate that this RT-PCR assay reproduces previous results, as well as provides the sensitivity to uncover new findings.

Membrane depolarization by elevation of extracellular potassium concentration (K^+) changes expression of neuropeptides and neurotransmitters (Fann and Patterson, 1994b). When sympathetic neurons were cultured in medium contain-

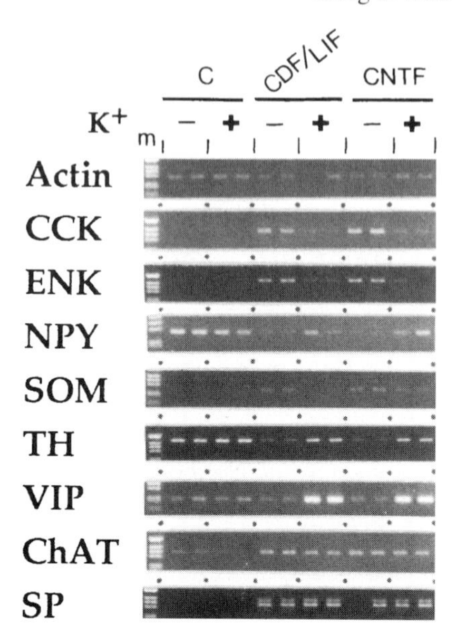

Fig. 3. LIF and CNTF induce expression of *CCK*, *ENK*, *SOM*, *ChAT*, and *SP*, but inhibit *TH* and *NPY*, and depolarization can block these effects. Sympathetic neurons were cultured in the presence of the cytokines, with or without 40 m*M* KCl (K^+) for 6 days, and the RT-PCR assay was used to monitor a variety of mRNAs. Duplicate samples were prepared for each concentration. Depolarization of cells alone (C, +) decreases the expression of *ChAT*. The effects of the cytokines on *CCK*, *ENK*, *NPY*, *SOM*, and *TH* are also diminished. There is, however, potentiation of *VIP* expression in the presence of cytokines and K^+. The 1-kb DNA marker ("M" lane) was used as in Figure 2.

ing 40 m*M* K^+, expression of mRNA for *ChAT* is decreased (Fig. 3). Depolarization also inhibits effects of LIF and CNTF on expression of most genes, except for *VIP*, *ChAT*, and *SP*. If the RNA amounts in each sample are adjusted by normalizing to β-*actin*, the effects of LIF and CNTF on *ChAT* and *SP* expression are also seen to be inhibited by high K^+ medium. Potentiation of *VIP* expression by depolarization is difficult to reconcile with previous findings, but it is not without precedent. It had been shown that expression of *VIP* is enhanced when activin A and high K^+ are added together to sympathetic neurons (Fann and Patterson, 1994b).

This assay does have several minor drawbacks. Great care is needed for pipetting these small samples. In addition, depending on the thermocycler used, there are different degrees of failure of amplification, which is defined in duplicate samples where one lane displays a PCR product and the other does not. Judging from Figure 3, we have a failure rate of 4 of 96 amplifications. Thus, duplicate or triplicate samples are necessary for each condition, and each experiment needs to be repeated several times.

VIII. Closing Comments

Detection of neuronal gene expression can be hampered by low numbers of cells and extreme heterogeneity in neuronal populations. Recent advances in cellular and molecular techniques provide ways to circumvent these problems using small numbers of homogeneous neurons grown in culture. Here we describe a method, based on the RT-PCR, to detect mRNAs for a variety of transmitters and neuropeptides present at low levels in primary sympathetic neuron cultures. Particular primers were designed specifically to detect expression of nine different genes using RNA isolated from about 3000 neurons. Although the assay introduced here is only semiquantitative, quantitative results are easily obtained using competitive PCR or Southern analysis after PCR, and making the adjustment of RNA in each sample by calculating ratio of each gene signal to an internal control (such as β-*actin*).

Different ways of preparing neuronal cultures have been introduced (Banker and Goslin, 1996) and various neuronal stem cells, which are capable of self-replication and generate committed cells, are becoming available (Gage *et al.*, 1995). Combination of these culture systems with the RT-PCR technique should greatly enhance our understanding of gene regulation in the nervous system.

Acknowledgments

We thank Doreen McDowell for help with tissue culture materials and Derek Stemple for advice on designing primers. This project is supported by grants from the NINDS (Javits Neuroscience Investigator Award) to P. H. P., and the National Science Council, Taiwan, R.O.C. to M.-J. F (NSC84-2331-B010-121).

References

Banker, G., and Goslin, K. (1996). "Culturing Nerve Cells," 2nd ed. MIT Press, Cambridge, MA.

Belayavsky, A., Vinorgradova, T., and Rajewsky, K. (1989). PCR-based cDNA library construction: General cDNA libraries at the level of a few cells. *Nucleic Acids Res.* **17,** 2919–2932.

Carter, M. S., and Krause, J. E. (1990). Structure, expression, and some regulatory mechanism of the rat preprotachykinin gene encoding substance P, neurokinin A, neuropeptide K, and neuropeptide γ. *J. Neurosci.* **10,** 2203–2214.

Chomczynski, P., and Sacchi, N. (1987). Single-step method of RNA isolation by acid guanidinium thiocyanate-phenol-chloroform extraction. *Anal. Biochem.* **162,** 156–159.

Darmon, M. C., Guibert, B., Leviel, V., Ehret, M., Maitre, M., and Mallet, J. (1988). Sequence of two mRNA encoding active rat tryptophan hydroxylase. *J. Neurochem.* **51,** 312–316.

Deschenes, R. J., Haun, R. S., Funckes, C. L., and Dixon, J. E. (1985). A gene encoding rat cholecystokinin: Isolation, nucleotide sequence, and promoter activity. *J. Biol. Chem.* **260,** 1280–1286.

Fann, M.-J., and Patterson, P. H. (1993). A novel approach to screen for cytokine effects on neuronal gene expression. *J. Neurochem.* **61,** 1349–1355.

Fann, M.-J., and Patterson, P. H. (1994a). Neuropoietic cytokines and activin A differentially regulate the phenotype of cultured sympathetic neurons. *Proc. Natl. Acad. Sci. U.S.A.* **91,** 43–47.

Fann, M.-J., and Patterson, P. H. (1994b). Depolarization differentially regulates the effects of bone morphogenetic prrotein (BMP)-2, BMP-6, and activin A on sympathetic neuronal phenotype. *J. Neurochem.* **63,** 2074–2079.

Furshpan, E. J., Landis, S. C., Matsumoto, S. G., and Potter, D. D. (1987). Synaptic functions in rat sympathetic neurons in microcultures: I. Secretion of norepinephrine and acetylcholine. *J. Neurosci.* **6,** 1061.

Gage, F. H., Ray, J., and Fisher, L. J. (1995). Isolation, characterization, and use of stem cells from the CNS. *Annu. Rev. Neurosci.* **18,** 159–192.

Garrett, J. E., Collard, M. W., and Douglass, J. O. (1989). Translational control of germ cell-expressed mRNA imposed by alternative splicing: Opioid peptide gene expression in rat tissue. *Mol. Cell. Biol.* **9,** 4381–4389.

Greenlund, L. J. S., Deckwerth, T. L., and Johnson, E. M. (1995a). Superoxide dismutase delays neuronal apoptosis: A role for reactive oxygen species in programmed neuronal death. *Neuron* **14,** 303–315.

Greenlund, L. J. S., Korsmeyer, S. J., and Johnson, E. M. (1995b). Role of BCL-2 in the survival and function of developing and mature sympathetic neurons. *Neuron* **15,** 649–661.

Grima, B., Lamouroux, A., Blanot, F., Biguet, N. F., and Mallet, J. (1985). Complete coding sequence of rat tyrosine hydroxylase mRNA. *Proc. Natl. Acad. Sci. U.S.A.* **82,** 617–621.

Hahn, M., Hahn, S. L., Stone, D. M., and Joh, T. H. (1992). Cloning of the rat gene encoding choline acetyltransferase, a cholinergic neuron-specific marker. *Proc. Natl. Acad. Sci. U.S.A.* **89,** 4387–4391.

Harmar, A. J., Hyde, V., and Chapman, K. (1990). Identification and cDNA sequence of delta-preprotachykinin, a fourth splicing variant of the rat substance P precursor. *FEBS Lett.* **275,** 22–24.

Hawrot, E., and Patterson, P. H. (1979). Long-term culture of dissociated sympathetic neurons. *Methods Enzymol.* **58,** 574–583.

Kessler, J. A., Ludlam, W. H., Friedin, M. M., Hall, D. H., Michaelson, M. D., Spray, D. C., Dougherty, M., and Batter, D. K. (1993). Cytokine induced programmed death of cultured sympathetic neurons. *Neuron* **11,** 1123–1132.

Koh, J.-Y., Gwag, B. J., Lobner, D., and Choi, D. W. (1995). Potentiated necrosis of cultured cortical neurons by neurotrophins. *Science* **268,** 573–575.

Krause, J. E., Chirgwin, J. M., Carter, M. S., Xu, Z. S., and Hershey, A. D. (1987). Three rat preprotachykinin mRNA encode the neuropeptides substance P and neurokinin A. *Proc. Natl. Acad. Sci. U.S.A.* **84,** 881–885.

Larhammar, D., Ericsson, A., and Persson, H. (1987). Structure and expression of the rat neuropeptide Y gene. *Proc. Natl. Acad. Sci. U.S.A.* **84,** 2068–2072.

Lowe, T., Sharefkin, J., Yang, S. Q., and Dieffenbach, C. W. (1990). A computer program for selection of oligonucleotide primers for polymerase chain reaction. *Nucleic Acids Res.* **18,** 1757–1761.

Mahanthappa, N. K., and Patterson, H. P. (1996). Culturing mammalian sympathoadrenal derivatives. *In* "Culturing Nerve Cells" (G. Banker and K. Goslin, eds.), 2nd ed. MIT Press, Cambridge, MA.

McConnell, S. K. (1995). Strategies for the generation of neuronal diversity in the developing central nervous system. *J. Neurosci.* **15,** 6987–6998.

Montminy, M. R., Goodman, R. H., Horovitch, S. J., and Habener, J. F. (1984). Primary structure of the gene encoding rat preprosomatostatin. *Proc. Natl. Acad. Sci. U.S.A.* **81,** 3337–3340.

Nawa, H., Yamamori, T., Le, T., and Patterson, P. H. (1990). Generation of neuronal diversity: Analogies and homologies with hematopoiesis. *Cold Spring Harbor Symp. Quant. Biol.* **55,** 247–253.

Nawa, H., Nakanishi, S., and Patterson, P. H. (1991). Recombinant cholinergic differentiation factor (leukemia inhibitory factor) regulates sympathetic neuron phenotype by alternations in the size and amounts of neuropeptide mRNAs. *J. Neurochem.* **56,** 2147–2150.

Nishizawa, M., Hayakawa, Y., Yanaihara, N., and Okamoto, H. (1985). Nucleotide sequence divergence and functional constraint in VIP precursor mRNA evolution between human and rat. *FEBS Lett.* **183,** 55–59.

Nudel, U., Zakut, R., Shani, M., Neuman, S., Levy, Z., and Yaffe, D. (1983). The nucleotide sequence of the rat cytoplasmic beta-actin gene. *Nucleic Acids Res.* **11,** 1759–1771.

Patterson, P. H., and Nawa, H. (1993). Neuronal differentiation factors/cytokines and sympathetic plasticity. *Cell* **72** (Suppl.), 123–137.

Rao, M. S., Tyrrell, S., Landis, S. C., and Patterson, P. H. (1992). Effects of ciliary neurotrophic factor (CNTF) and depolarization on neuropeptide expression in cultured sympathetic neurons. *Dev. Biol.* **150,** 281–293.

Ray, J., Peterson, D. A., Schinstine, M., and Gage, F. (1993). Proliferation, differentiation and long-term culture of primary hippocampal neurons. *Proc. Natl. Acad. Sci. U.S.A.* **90,** 3602–3606.

Rosen, H., Douglass, J., and Herbert, E. (1984). Isolation and characterization of the rat proenkephalin gene. *J. Biol. Chem.* **259,** 14309–14313.

Symes, A. J., Rao, M. S., Lewis, S. E., Landis, S. C., Hyman, S. E., and Fink, J. S. (1993). Ciliary neurotrophic factor coordinately activates transcription of neuropeptide genes in a neuroblastoma cell lines. *Proc. Natl. Acad. Sci. U.S.A.* **90,** 572–576.

Thoenen, H. (1995). Neurotrophins and neuronal plasticity. *Science* **270,** 593–598.

Wallace, T. L., and Johnson, E. M. (1989). Cytosine arabinoside kills postmitotic neurons: Evidence that deoxycytidine may have a role in neuronal survival that is independent of DNA synthesis. *J. Neurosci.* **9,** 115–124.

Wolinsky, E. J., Landis, S. C., and Patterson, P. H. (1985). Expression of noradrenergic and cholinergic traits by sympathetic neurons cultured without serum. *J. Neurosci.* **5,** 1497–1508.

Wolinsky, E. J., and Patterson, P. H. (1985). Rat serum contains a developmentally regulated cholinergic inducing activity. *J. Neurosci.* **5,** 1509–1512.

Yamamori, T., Fukada, K., Aebersold, R., Korsching, S., Fann, M.-J., and Patterson, P. H. (1989). The cholinergic neuronal differentiation factor from heart cells is identical to leukemia inhibitory factor. *Science* **246,** 1412–1416.

11

Selective Aggregation Assays for Embryonic Brain Cells and Cell Lines

Shinichi Nakagawa, Hiroaki Matsunami, and Masatoshi Takeichi
Department of Biophysics, Faculty of Science
Kyoto University
Kitashirakawa, Sakyo-ku, Kyoto 606-01, Japan

Hiroaki Matsunami
Department of Neurobiology
Harvard Medical School
Boston, Massachusetts 02115

I. Introduction

Animal cells, including embryonic brain cells, have differential adhesivness, and this property of cells is thought to be involved in the patterning of cell assemblies. Cell–cell adhesion mechanisms are complex, and comprose multiple systems, In this chapter, we illustrate how to assay the role of these adhesion mechanisms in selective adhesion of embryonic brain cells and fibroblastic cell lines. The methods to be described include how to dissociate cells while leaving a particular set of adhesion mechanisms intact, how to let them reaggregate, and how to evaluate selective cell adhesion.

Current Topics in Developmental Biology, Vol. 36

0070-2153/98 $25.00

Animals cells are known to exhibit selective adhesiveness. When cells derived from different tissues are mixed, they sort out. This property of cells is thought to play a key role in the organization of complex tissue architecture. Studies suggest that the selective adhesiveness of cells also contributes to regionalization in the embryonic brain (Matsunami and Takeichi, 1995; Götz *et al.*, 1996).

The vertebrate brain has complex regional diversity. This is produced by sequential subdivision of the relatively simple neural tube during development. The anterior part of the neural tube becomes subdivided into three distinct vesicles, the procencephalon, mesencephalon, rhombencephalon, and each brain vesicle becomes further subdivided into smaller units, called "neuromeres." Although the neuromeres were originally defined by their anatomic appearance, studies have revealed that many molecules, including transcription factors and signaling molecules, are also expressed in association with neuromeric organization (Krumlauf *et al.*, 1993; Puelles and Rubenstein, 1993; Rubenstein *et al.*, 1994). The neuromeres in the hindbrain are specifically called "rhombomeres," and each rhombomere is considered as a "compartment" (Lumsden, 1990). Neuroepithelial cells do not cross the boundaries between rhombomeres, and thus cells from adjacent thombomeres do not intermingle with each other (Fraser *et al.*, 1990). Such cell lineage restrictions were also reported at the boundary between the cortical ventricular zone and the lateral ganglionic eminence (Fishell *et al.*, 1993), and among diencephalic neuromeres (Fidgor and Stern, 1993). Thus, the identity of each neuromere seems to be at least in part maintained by the cell lineage restriction.

To elucidate the mechanisms by which cells in different neuromeric regions segregate from each other is an intriguing issue. As one of the mechanisms, selective adhesiveness appears to be essential. For example, neuroepithelial cells derived from the ventral and dorsal thalamus do not randomly intermix with each other in *in vitro* aggregation assays, and this phenomenon was suggested to be regulated by region-specific expression of different cadherin subtypes (Matsunami and Takeichi, 1995). To understand the molecular basis of patterning in the developing brain, therefore, it is critical to accumulate much more information on the adhesion properties of the cells that belong to each compartment.

There are many ways to measure and quantify cell–cell adhesiveness. A classic method is to dissociate cell masses into single cells, and observe their reaggregation process. In this chapter, we present an updated version of these classic methods, illustrating how to dissociate cells, how to allow them to reaggregate, and how to assay their selective adhesiveness. As materials, we use embryonic brains and fibroblastic cell lines as examples. The latter include L-cell lines transfected with cadherin cDNAs, which are widely used for the assay of adhesion molecules. The methods described here can be applied to a variety of different cell types, and also can be modified and improved for more sophisticated ways to measure selective cell adhesion.

II. Principles of How to Disaggregate and Reaggregate Cells

Cells of multicellular solid tissues have complex mechanisms for mutual adhesion. They are connected to each other by various forms of cell–cell junctions (tight junction, adherens junction, desmosome, and gap junction), and by multiple classes of cell surface molecules, each of which has distinctive functions. These complex adhesion mechanisms, however, can operationally be subdivided into two types, Ca^{2+}-dependent and Ca^{2+}-independent ones (Takeichi, 1977, 1988). We can selectively remove molecules of either of these two adhesion systems or of both from cell surfaces by differential trypsin treatment (Fig. 1) (Urushihara *et al.*, 1979; Takeichi *et al.*, 1979). The Ca^{2+}-dependent system (CDS) is highly sensitive to trypsin digestion even at low concentrations of the enzyme, but it can be protected from the protease digestion by Ca^{2+}. On the other hand, the Ca^{2+}-independent system (CIDS) can be removed only by treatment with a relatively high concentration of trypsin, and this proteolysis, in general, cannot be protected by Ca^{2+}. Accordingly, if cells are treated with a high concentration of trypsin in the presence of Ca^{2+} (TC treatment), CDS is left intact but CIDS is removed from the cell surface. However, if cells are treated with a low concentration of trypsin in the absence of Ca^{2+} (LTE treatment), CIDS is left intact whereas CDS molecules are digested and inactivated. These

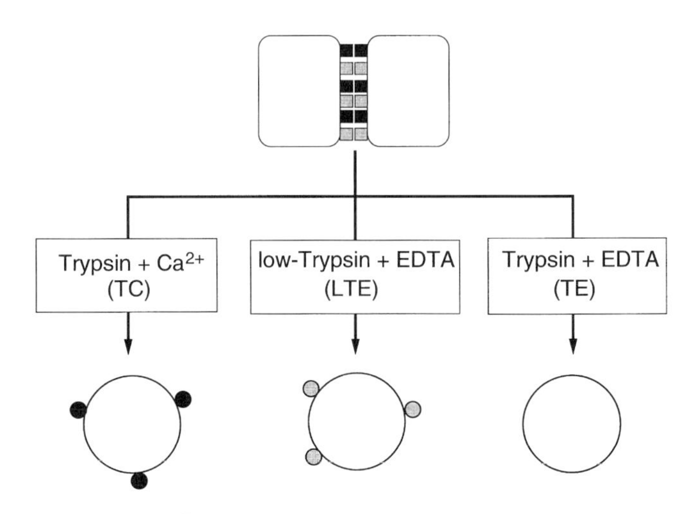

Fig. 1. The cell–cell adhesion systems preserved after different types of trypsin treatment. TC-treated cells have only CDS, LTE-treated cells have only CIDS, and TE-treated cells have neither.

cells disaggregated by TC or LTE treatment can quickly reaggregate if placed in appropriate culture conditions because adhesion molecules belonging to CDS or CIDS are left intact on their surfaces.

As expected from the aforementioned properties of CDS and CIDS, treating cells with a high concentration of trypsin in the absence of Ca^{2+} (TE treatment) inactivates both adhesion systems. Thus, for most cell types, TE-treated cells temporarily lose all cell–cell adhesion activities, although they in general retain cell–substrate adhesiveness. For the TE-treated cells to reaggregate, the adhesion molecules, once removed, must be restored.

Most cell–cell adhesion molecules thus far identified can be grouped into CDS or CIDS. Classic cadherins, Ca^{2+}-dependent cell–cell adhesion molecules, are the major components of CDS, and to our knowledge, no other adhesion molecules are similar to cadherins in their trypsin–Ca^{2+} sensitivity. Some of other cell–cell adhesion molecules require Ca^{2+} for their activity, but they are not preserved during the TC treatment. Therefore, CDS is likely represented by a single family of adhesion molecules, the cadherins. CIDS includes multiple classes of adhesion molecules, especially members of the immunoglobulin superfamily, but its entire biochemical profile remains to be clarified. Adhesion molecules left on the LTE-treated cell surfaces may greatly vary according to the cell type.

By preparing cells by use of the aforementioned methods, we can assay the activity of the two adhesion systems independently. However, each system is still complex. Therefore, for investigating the function of a particular adhesion molecule, the use of specific blocking antibodies is helpful. It should also be noted that some adhesion molecules can still be removed from both TC- and LTE-treated cells. For study of these molecules, some other methods must be considered.

The TC and LTE strategy for cell disaggregation can be applied to many cell types, including early embryonic cells, neuroepithelial cells, and fibroblasts. However, one would experience difficulty in disaggregating certain cell types into single cells by TC treatment, especially differentiated epithelial cells. Some junctional structures, including tight junctions, seem to be resistant to the TC procedure. For these cells, TE treatment should be effective to obtain single cells. In this case, CADS (cadherins) must be sacrificed, but the activity of the cadherin can be studied by having these molecules restored on the TE-treated cell surfaces by a prolonged incubation.

III. Solutions

HEPES-buffered Ca^{2+}-, Mg^{2+}-free salt solution (HCMF): 8 g NaCl, 0.4 g KCl, 0.12 g $Na_2HPO_4 \cdot 12H_2O$, 1.0 g glucose, 2.38 g HEPES in 1000 ml DW. Adjust pH to 7.4 with 1*N* NaOH. This and other salt solutions to be used for cell aggregation assays should be filtered through a Millipore filter with a 0.45-μm pore size to remove any dusty materials.

HEPES-buffered Mg^{2+}-free salt solution (HMF): HCMF supplemented with 1 m*M* $CaCl_2$

HEPES-buffered salt solution (HBSS): HCMF supplemented with 1 m*M* $CaCl_2$ and 1 m*M* $MgCl_2$

0.1% crystalline trypsin solution: Dissolve 10 mg of crystalline trypsin (Sigma T-8253) in 10 ml of HCMF. Divide into aliquots, and store at −20°C.

0.5% soybean trypsin inhibitor solution: Dissolve 50 mg of soybean trypsin inhibitor (Sigma T-6522) in 10 ml of HCMF. Divide into aliquots, and store at −20°C.

1% bovine serum albumin (BSA) solution: Dissolve 1 g of BSA (Sigma A-4503) in 100 ml of HCMF. Adjust pH to 7.4 with NaOH, and sterilize by filtration. Store at 4°C.

0.5 *M* ethylenediamine tetraacetic acid (EDTA) or glycoletherdiamine tetraacetic acid (EGTA): Dissolve in distilled water and adjust pH to 8.0 with NaOH.

Other stock solutions: 1 *M* $CaCl_2$, and 1 *M* $MgCl_2$, prepared in DW.

IV. Preparation of Single-Cell Suspensions

A. Precoating Plastic or Glassware with Bovine Serum Albumin

Cells suspended in protein-free solutions readily attach to plastic or glass surfaces, causing loss of the cells and many other troubles. To avoid this, all plastic or glass dishes, tubes, and pipettes to be used for cell preparation should be precoated with BSA. This can be done by placing 1% BSA into dishes or tubes for at least 1 hour, and rinsing them with HCMF just before use. Pipettes and forceps can be coated by submerging them in the 1% BSA solution.

B. Keys to Preserve or Remove Cadherins

It should be kept in mind that the presence or absence of Ca^{2+} during cell disaggregation treatments greatly affects the preservation or removal of CADS (cadherins) in a temperature-dependent manner. In the presence of physiologic concentrations of Ca^{2+}, cadherins are protected from protease digestion. In Ca^{2+}-free cell suspensions, cadherins tend to be gradually degraded by proteases released from lysed cells. Therefore, when the preservation of cadherins is desired, it is highly recommended to add 1–2 m*M* Ca^{2+} to all solutions used for tissue dissection or cell preparation. However, in the steps to obtain single cells, the cells must be exposed to Ca^{2+}-free conditions to inactivate cadherins. At this time, cadherins can be protected from digestion by low temperatures. Ca^{2+}-free cell suspensions should always be kept at an ice-cold temperature.

If Ca^{2+} is added to suspensions of cells with active cadherins, it induces immediate cell aggregation at physiologic temperatures. Under ice-cold conditions, however, cadherin-mediated aggregation is inhibited. Thus, placing cells in ice-cold, Ca^{2+}-containing medium is the best way not only to preserve their cadherin activity but to keep them dispersed.

Although the presence of Ca^{2+} during trypsin treatment is essential for preserving cadherins, cells of certain tissues cannot be singly dissociated under these conditions, as mentioned previously. If the primary purpose is to obtain single cells in the experiments, Ca^{2+} may be removed with a chelating reagent, EDTA or EGTA, before or during protease treatment (see IV.D). This greatly enhances cell disaggregation.

C. Preparation of Initial Materials for Cell Dissociation

Embryonic brains are dissected in a standard balanced salt solution containing 1–2 m*M* Ca^{2+} or HBSS. If the pia mater cannot be removed easily from the brain with forceps, incubate the brain in ice-cold HCMF supplemented with 1 m*M* EDTA for a few minutes, and then peel off the covering in HCMF. This process should be done as quickly as possible, because the EDTA treatment may inactivate CADS. Cut the tissues into small pieces using a pair of scissors or other tools in HBSS, and store them on ice for further treatments.

When cell lines are to be used, prepare semiconfluent monolayer cultures by seeding cells 1 or 2 days before use. Overcrowded or old cultures should not be used because they are often difficult to dissociate into healthy single cells.

Before protease treatment, the tissues should be washed with a Ca^{2+}-containing balanced salt solution, HMF or HBSS, when TC treatment follows. On the other hand, when they are being subjected to LTE or TE treatment, HCMF should be used for washing. The same rule is applied when monolayer cell cultures are used. Rinse them three times with HMF–HBSS or HCMF, depending on which trypsin treatment is being used in the subsequent step.

D. Trypsin Treatment

Tissues or cell cultures are then subjected to trypsin treatment. Prepare the following trypsin solutions, depending on the purpose of the cell dissociation. The solutions should be freshly prepared from the stock solutions just before use.

1. For *TC treatment*, 0.01% trypsin in HCMF supplemented with 1–10 m*M* $CaCl_2$. Higher concentrations of Ca^{2+} are more effective to preserve cadherins, but could cause more difficulty in cell disaggregation in the following steps. We

usually use 5–10 m*M* Ca^{2+} for embryonic brains, whose cell disaggregation is relatively easy, and 1–2 m*M* for most cell lines. The Ca^{2+} concentration can be reduced to 0.5–0.2 m*M* for certain cells; this enhances cell disaggregation but reduces the number of cadherins left intact on the cell surface. These conditions can be applied to cell layers that are difficult to dissociate. Below 0.1 m*M*, Ca^{2+} fails to protect cadherins. It should also be remembered that more Ca^{2+} is required when higher concentrations of trypsin are used. Ca^{2+} effectively protects cadherins against many types of proteases.

2. For *LTE treatment*, 0.0001–0.0005% trypsin in HCMF supplemented with 1 m*M* EDTA or EGTA. Appropriate trypsin concentrations should be sought for every tissue to ensure that sufficient CIDS activity is preserved. We usually use 0.0005% trypsin for embryonic brains. EDTA and EGTA added to the solution give the same result not only for LTE but for TE treatment.

3. For *TE treatment*, 0.01% trypsin in HCMF supplemented with 1 m*M* EDTA or EGTA. This trypsin solution can be widely used when complete cell dissociation is desired.

Incubate tissues or cell cultures with the desired trypsin solution for 20–30 minutes at 37°C on a rotary shaker. Larger pieces of tissues may require longer incubation periods. We usually use Petri dishes to treat embryonic tissues because we can monitor the dissociation under an inverted microscope. For monolayer cultures, trypsin solution is directly added to them after the rinsing process. We usually add 4 ml of trypsin to a 5-cm dish.

After completion of trypsin treatment, examine the tissues or cells under an inverted microscope. When monolayer cultures are used, one can see clear differences in the appearance of the cells between the different trypsin treatments. TC-treated cells maintain tight clumps (Fig. 2a), LTE-treated cells form loose aggregates (Fig. 2b), and TE-treated cells are virtually dispersed, although small, loose clusters may be still present (Fig. 2c). When the TC treatment results in a complete dispersion of the cells, this means either that the cells used do not have any cadherins or that something was wrong with the treatment procedure. In general, cells spontaneously detach from the dish surface after these trypsin treatments, but some cells might remain adhered to the dish. These cells can be collected by flushing the medium through a pipette. Concerning embryonic tissues, similar differences should be observed if the treatments are successful.

E. Dissociation of Cells

During the trypsin treatments, DNA may be released from dead cells, especially when tissues are used. The DNA forms viscous gels, and traps cells into loose aggregates. These artificial aggregates should be distinguished from those mediated by real cell–cell adhesion molecules. The presence of DNA gels hampers the cell aggregation assay because they nonspecifically clump cells. These mate-

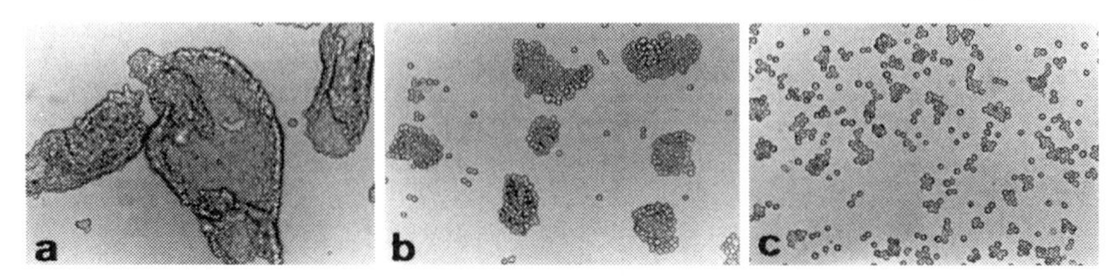

Fig. 2. Cells after TC, LTE, or TE treatment. Semiconfluent cultures of L cells transfected with N-cadherin cDNA (Fujimori *et al.*, 1990) were subjected to (a) TC treatment, (b) LTE treatment, or (c) TE treatment for 15 minutes at 37°C, and the resultant cell suspensions were then immediately photographed. Note that TC-treated cells maintain very tight clumps, LTE-treated cells are in loose aggregates, and TE-treated cells are more dispersed. The cell clumps in the TE-treated cultures can easily be dispersed into single cells by light pipetting. The CIDS activity in L cells is not very high, and larger aggregates would be observed after LTE treatment of other cell types.

rials, therefore, should be removed by DNase treatment before or after the following cell-washing procedure. To digest them before the washing step, add an excess amount of soybean trypsin inhibitor to the cell suspension, and subsequently 10 μg/ml DNase I and 10 m*M* $MgCl_2$ in final concentrations. Then incubate the plates at 37°C until the viscous materials completely disappear, which might take several minutes. This pre-DNase treatment usually is not required when cell lines are used because DNA release is not so extensive for these materials.

Transfer the tissue fragments or cells in the trypsin solution to a 10-ml round-bottomed test tube. All the following procedures are carried out on ice (4°C). Centrifuge the test tubes at 1000 rpm (700 *g*) for 3 minutes, and remove the supernatant. Then, add 100 μl of the soybean trypsin inhibitor stock solution to the cell pellet, if not used in the previous step, and subsequently 5 ml of ice-cold HMF. Resuspend the cells by brief pipetting, and centrifuge again. Then, aspirate the supernatant and add 5 ml of HCMF. Disperse the tissue fragments or cell clumps by gentle pipetting. Centrifuge them again, resuspend the cells in 2–5 ml of HCMF, and then pipette them strongly by use of a Pasteur pipette to obtain completely dissociated cells. It is helpful to polish the tip of the Pasteur pipette in a hot flame before use to prevent cells from being injured. TE-treated tissues are readily dispersed into single cells during these steps; LTE-treated cells require stronger mechanical force to be dispersed. For TC-treated cells, vigorous pipetting is often required for complete cell dispersion. After several pipettings, take a tiny aliquot of the cell suspension onto a slide glass and examine it under an inverted microscope. If cell clumps are still present, repeat the pipetting until most cells become single. To obtain completely dissociated cells is an essential step for the selective cell aggregation assays. After cell dispersion, count cell number, centrifuge the cells again, and finally suspend them in a medium to be used for the cell aggregation assay. Store them on ice.

When the cell suspension still contains large cell clumps or tissue debris, these should be removed by low-speed centrifugation or by use of a filter such as a cell strainer (Falcon 2350). The cell suspension should be as clean as possible, because every noncellular contaminant disturbs natural cell aggregation. For example, the presence of a single piece of cotton fiber could become a core to induce artificial clumping of cells.

As mentioned earlier, certain cell types resist being dissociated into single cells, especially after TC treatment. For such samples, prolonged pipetting might eventually kill them. In this case, other methods for cell dispersion should be used, such as TE treatment.

V. Aggregation of Cells

The dispersed cells suspended in an appropriate culture medium are placed into 24-well dishes and incubated on a rotator to induce reaggregation. Many factors can affect the rate of aggregation, and these factors must be optimized for each experiment. Precoating the dishes to be used for the rotation culture, which prevents the attachment of cells to their surfaces, is an absolute requirement for the success of the experiments. If cells attach to the dish, the cell aggregation process is severely affected. Cell aggregation can be studied by either short- or long-term cultures. Because the methods for precoating dishes and the culture media used in these two systems differ, the procedures for the two types of cultures are explained separately.

A. Short-Term Aggregation Culture

In this system, the cells are incubated for only a few hours in simple balanced salt solutions. This method should be chosen when adhesion molecules are left intact on the cell surfaces after the trypsin treatment and their activity is to be assayed. Physiologic recovery of the trypsinized molecules is not expected in this culture system.

Precoat 24-well dishes (Falcon 3047) with BSA, as described previously. Into each well of the dishes placed on ice, add cells suspended in 0.5 ml of HCMF. Add other substances such as $CaCl_2$, antibodies, and inhibitors from stock solutions, when necessary. If there is a concern about release of DNA from dead cells, add DNase I (10 μg/ml) and 1 m*M* $MgCl_2$ to the medium. The presence of DNase is often critical for accurate measurement of natural cell aggregation. These plates are placed on a shaker (Gyratory shaker model G2, MARYSOL Rotator KS-6300, or equivalent) set to rotate at 80 rpm, and incubated at 37°C. The cells soon become concentrated at the center of each well, and will aggregate at this position. This central accumulation of cells is perturbed by contamination of fibrous materials, such as cotton fibers, in the cultures, and as a consequence the pattern of cell aggregation is affected.

To activate cadherins on TC-treated cells, add 1 m*M* $CaCl_2$ to the cell suspension. The cells start to aggregate within 10 minutes. As a control, prepare wells without Ca^{2+}, and check that no cell aggregation is induced in these wells. If some aggregation is detected in the Ca^{2+}-free wells, one must suspect that adhesion molecules other than cadherins are active in these cells. LTE-treated cells aggregate in the absence of Ca^{2+} (in HCMF). TE-treated cells, in general, do not aggregate in these cultures. However, a low level of aggregation activity is sometimes observed for TE-treated embryonic brain cells, suggesting that some residual adhesion molecules are present in them.

Maximum aggregation of these cells is usually attained within 2 hours under the aforementioned culture conditions. For study of their aggregation properties, 30 to 60-minute incubation is often sufficient. Longer incubations may cause cell death, as well as nonspecific cell aggregation due to various factors.

Concerning cell density, we usually prepare suspensions with $0.5–1 \times 10^5$ cells/ml for cell lines, and $2–5 \times 10^6$ cells/ml for embryonic brain cells. Dishes larger than 24-well dishes can be used. In larger dishes, a larger shearing force is generated by rotation of the medium. Therefore, the rotation speed, medium volume, and other factors must be optimized for each different culture.

B. Long-Term Aggregation Culture

In this case, cells are incubated for several hours to days. To maintain them for such long periods, standard cell culture media are used. As anticipated, cells undergo *de novo* protein synthesis, and therefore replace the adhesion molecules removed earlier. We use these cultures to allow the aggregation of TE-treated cells. As mentioned, we need complete disaggregation of cells, but this cannot be achieved for certain cell types by TC or LTE treatment. TE treatment is the only choice toward this end. TE-treated cells gradually generate new adhesion molecules under the previously described culture conditions. For example, cadherin-mediated aggregation begins a few hours after inoculation. If specific blocking antibodies to those adhesion molecules are available, their functions can be studied by use of this aggregation system. In our experience, when TE-treated L cells transfected with various cadherin cDNAs are cultured under these conditions, their initial aggregation is induced only by these cadherins but not by CIDS, allowing us to study their specific roles in cell aggregation (see VI.B).

For these cultures, BSA coating is not sufficient to prevent cells from attaching to the dishes. We thus use agar to coat the dish bottom. Prepare 1% agar in the same medium as used for the rotation culture. Pour this solution onto the dishes, and then immediately remove it, leaving a small amount. After the gel is set (a few minutes at room temperature), add a culture medium onto the agar plates to prevent the gel from drying.

Cells are suspended in a standard culture medium, such as Dulbecco's minimum essential medium with 10% fetal calf serum. Depending on the purpose of the experiments, select the most appropriate culture medium. Cell density, medium volume, rotation speed, and other conditions can be similar to those used for the short-term aggregation assay (see V.A). One exception is that the rotation culture must be carried out in a CO_2 incubator in this case, if the medium is buffered with a $NaHCO_3$–CO_2 system. Agar is a powerful inhibitor of cell attachment. Therefore, these cultures can be maintained even without shaking if necessary.

C. Mixed Cell Aggregation

To assay whether cells of different origins segregate from each other, one can mix them and allow reaggregation under either of the aforementioned culture systems. For the resultant cell aggregates, the cell segregation pattern is analyzed. To do this, the two cell populations mixed must be distinguished. This can be done in two different ways. One is to stain cells with vital dyes, and the other is to detect some markers that are expressed only by one cell population.

For the dye staining, fluorescence dyes usually are used. These include DiI, DiO, Cell Tracker (Molecular Probes), PKH26 (ZYNAXIS), and Fluoro-gold (Fluorochrome, Inc.). Cells can be labeled before or after the dissociation step. For embryonic cells, the latter may be better because the whole population of the cells can be equally labeled. For details of how to use the dyes, follow the manufacturer's instructions or other literature (Götz *et al.*, 1996). Make sure that all the cells have been equally labeled after the final washing. For cultured cells, it is probably most convenient to label them before the dissociation treatment. Cells in monolayer cultures can be incubated with dyes until satisfactory staining is obtained. The best labeling method should be selected for each cell type and each dye.

The cells thus fluorescently labeled are mixed at a 1:1 ratio with cells unlabeled or labeled with another dye. These are incubated to induce cell aggregation, according to either of the aforementioned methods.

VI. Analyzing the Results

A. Visualizing Markers

To evaluate selective cell aggregation, the samples should be processed for visualizing markers. If the cells are labeled with a fluorescent dye, they can be directly observed under a fluorescence microscope, but for more detailed analysis they should be fixed and examined by the following method.

Add an equal volume of 8% paraformaldehyde to each well of the dish, and fix the cells for 30–60 minutes. Transfer the contents of each well into an Eppendorf tube, and centrifuge the sample (700*g*, 1 minute), if the cells are labeled with a fluorescent dye. The Eppendorf tube should also be coated with BSA; otherwise the cells will stick to the wall of the tube, which results in a tragic end. Discard the supernatant and gently wash the cells three times with HBSS. The cells are then mounted onto a slide glass with a minimum amount of glycerol-based mounting solution containing antifading reagents. We routinely use 90% glycerol containing 1 mg/ml *p*-phenylenediamine (PPDA). The PPDA stock solution is prepared at 10 mg/ml in water, divided into aliquots, and stored at −20°C in the dark.

For immunostaining the cells for certain markers, any standard method can be applied. To handle small aggregates in suspension through the staining and washing processes, we use a hand-made basket for efficient changes of solutions. To make it, cut an Eppendorf tube at its middle with a heated cutter blade (Fig. 3a). Heat the cut end at one side of a flame, and then push it against a nylon mesh with a pore size of 15 μm (Fig. 3b). The baskets should be precoated with BSA before use. These baskets fit nicely into the wells of 24-well plates (Fig. 3c), and the samples are easily washed simply by transferring the basket from well to well. Alternatively, one may use a cell strainer (Falcon 2350) instead of the hand-made basket. In this case, six-well cell culture plates (Falcon 3046) can be used for the processing.

B. Classification of Cell Segregation Patterns

In mixed cell aggregates, three kinds of patterns can be observed for the distribution of the two cell populations: (1) cells of each population form separate

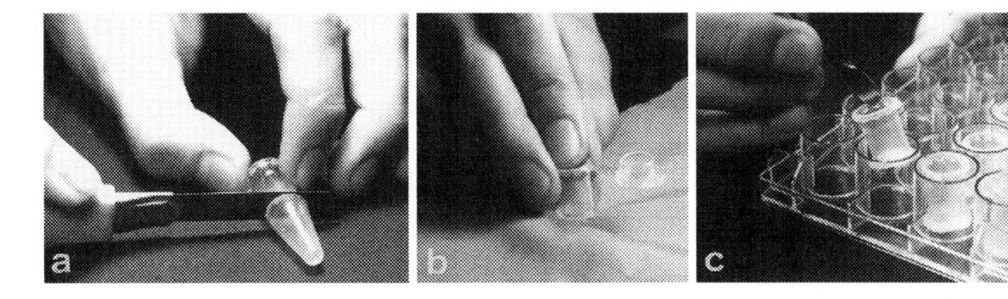

Fig. 3. How to make a basket for cell washing. (a) Cut an Eppendorf tube with a heated knife blade. (b) Push the melted cut end against a nylon mesh. (c) The basket thus made fits into the wells of a 24-well plate.

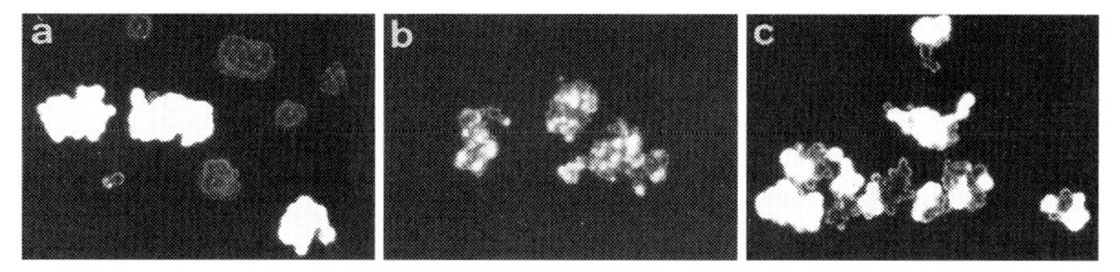

Fig. 4. Aggregates formed by mixing cells expressing different subtypes of cadherin. Cells expressing E-cadherin (Nagafuchi *et al.*, 1987), N-cadherin (Fujimori *et al.*, 1990), cadherin 6B (c-cad6B), and cadherin 7 (c-cad7) (Nakagawa and Takeichi, 1995) were used. Cells expressing one of these cadherins were labeled with DiO and dissociated by TE treatment. They were then mixed at a 1:1 ratio with cells expressing another cadherin that had not been labeled but had been dissociated by the same TE treatment. The cultures were incubated for 3.5 hours. (a) Example of complete cell segregation in a mixture of N-cadherin- and E-cadherin-positive cells. (b) Example of random mixing. In this sample, labeled and unlabeled N-cadherin-expressing cells were mixed in equal numbers. (c) Example of partial cell segregation that occurred in a mixture of labeled cad6B- and unlabeled cad7-positive cells.

aggregates (Fig. 4a), indicating that they can adhere only to like cells; (2) two cell populations randomly intermix with each other (Fig. 4b), suggesting that they share common adhesion mechanisms; and (3) intermediate patterns between the previous two cases can be seen; that is, each aggregate contains the two cell types, but their distribution is not random, and they tend to form clusters segregated from each other within the aggregate (Fig. 4c). By analyzing these patterns, selectivity in adhesiveness of two given cell populations can be assessed.

Using these analyses, we and others have determined the adhesion specificity of different cadherin subtypes (Takeichi, 1995; Murphy-Erdosh *et al.*, 1995) and also have demonstrated that early embryonic brains have region-dependent adhesion specificity based on cadherin activities (Matsunami and Takeichi, 1995). Although we focused on a simple cell aggregation method in this chapter, other ways to measure differential adhesiveness of cells (e.g., Hoffman, 1992; Steinberg and Takeichi, 1994) are also available and should be referred to for complete coverage of the methodology in this field.

References

Figdor, M. C., and Stern, C. D. (1993). Segmental organization of embryonic diencephalon. *Nature* **363,** 630–634.

Fishell, G., Mason, C. A., and Hatten, M. E. (1993). Dispersion of neural progenitors within the germinal zones of the forebrain. *Nature* **362,** 636–638.

Fraser, S., Keynes, R., and Lumsden, A. (1990). Segmentation in the chick embryo hindbrain is defined by cell lineage restrictions. *Nature* **344,** 431–435.

Fujimori, T., Miyatani, S., and Takeichi, M. (1990). Ectopic expression of N-cadherin perturbs histogenesis in *Xenopus* embryos. *Development* **110,** 97–104.

Götz, M., Wizenmznn, A., Reinhardt, S., Lumsden, A., and Price, J. (1996). Selective adhesion of cells from different telencephalic regions. *Neuron* **16,** 551–564.

Hoffman, S. (1992). Assays of cell adhesion. *In* "Cell–Cell Interactions" (B. R. Stevenson, W. J. Gallin, and D. L. Paul, eds.), pp. 1–29. IRL Press at Oxford University Press, Oxford.

Krumlauf, R., Marshall, H., Studer, M., Nonchev, S., Sham, M. H., and Lumsden, A. (1993). Hox homebox genes and regionalization of the nervous system. *J. Neurobiol.* **24,** 1328–1340.

Lumsden, A. (1990). The cellular basis of segmentation in the developing hindbrain. *Trends Neurosci.* **13,** 329–335.

Matsunami, H., and Takeichi, M. (1995). Fetal brain subdivisions defined by R- and E-cadherin expressions: Evidence for the role of cadherin activity in region-specific, cell–cell adhesion. *Dev. Biol.* **172,** 466–478.

Murphy-Erdosh, C., Yoshida, C. K., Paradies, N., and Reichardt, L. F. (1995). The cadherin-binding specificities of B-cadherin and LCAM. *J. Cell Biol.* **129,** 1379–1390.

Nagafuchi, A., Shirayoshi, Y., Okazaki, K., Yasuda, K., and Takeichi, M. (1987). Transformation of cell adhesion properties by exogenously introduced E-cadherin cDNA. *Nature* **329,** 341–343.

Nakagawa, S., and Takeichi, M. (1995). Neural crest cell–cell adhesion controlled by sequential and subpopulation-specific expression of novel cadherins. *Development* **121,** 1321–1332.

Puelles, L., and Rubenstein, J. L. R. (1993). Expression patterns of homeobox and other putative regulatory genes in the embryonic mouse forebrain suggest a neuromeric organization. *Trends Neurosci.* **16,** 472–479.

Rubenstein, J. L. R., Martines, S., Shimamura, K., and Puelles, L. (1994). The embryonic vertebrate forebrain: The prosomeric model. *Science* **266,** 578–580.

Steinberg, M. S., and Takeichi, M. (1994). Experimental specification of cell sorting, tissue spreading, and specific spatial patterning by quantitative differences in cadherin expression. *Proc. Natl. Acad. Sci. U.S.A.* **91,** 206–209.

Takeichi, M. (1977). Functional correlation between cell adhesive properties and some cell surface proteins. *J. Cell Biol.* **75,** 464–474.

Takeichi, M. (1988). The cadherins: Cell–cell adhesion molecules controlling animal morphogenesis. *Development* **102,** 639–655.

Takeichi, M. (1995). Morphogenetic roles of classic cadherins. *Curr. Opin. Cell Biol.* **7,** 619–627.

Takeichi, M., Ozaki, H. S., Tokunaga, K., and Okada, T. S. (1979). Experimental manipulation of cell surface to affect cellular recognition mechanisms. *Dev. Biol.* **70,** 195–205.

Urushihara, H., Ozaki, H. S., and Takeichi, M. (1979). Immunological detection of cell surface components related with aggregation of Chinese hamster and chick embryonic cells. *Dev. Biol.* **70,** 206–216.

12

Flow Cytometric Analysis of Whole Organs and Embryos

José Serna, Belén Pimentel, and Enrique J. de la Rosa
Department of Cell and Developmental Biology
Centro de Investigaciones Biológicas
Consejo Superior de Investigaciones Científicas
E-28006 Madrid, Spain

I. Introduction

Flow cytometry permits sensitive detection and rapid quantification of some intrinsic features of single cells, such as relative size, complexity, and endogenous fluorescence, as well as the quantitative analysis of any cellular compound that can be labeled with a fluorochrome. The general technical basis and a wide review of applications of flow cytometry are beyond the scope of this chapter, and can be found elsewhere (Melamed *et al.*, 1990; Ormerod, 1990). Here, we first discuss the specific technical considerations for performing flow cytometric analysis when the original materials are whole tissues and embryos, and present some selected applications.

Flow cytometry's main achievement, the rapid quantification of large numbers of single events (particles or cells), has generally limited its application to cell lines or primary cells growing in isolation, such as blood cells. It is currently one of the essential techniques in the field of immunology. However, whole organs and organisms, constituted by various heterogeneous populations of cells tightly adherent to each other, present the particular problem of having to be dissociated into a single-cell suspension representative of the tissue or organism of origin. This is requisite to apply the fruitful flow cytometry technique to other fields of biology, such as developmental biology.

Current Topics in Developmental Biology, Vol. 36

0070-2153/98 $25.00

Flow cytometry has been applied occasionally to embryonic development, particularly limb bud development (Tsonis and Walker, 1991), cardiac development (Prados *et al.*, 1992), and to studies in mitochondrial changes during development (López-Mediavilla *et al.*, 1992). Cell cycle analysis is also a major application of flow cytometry. Neoplastic cell or nuclei isolation protocols are found in the literature (Pallavicini, 1987; Pallavicini *et al.*, 1990). These methods can also be applied to developmental biology (Morris and Taylor, 1985). Here we present an appraisal of the limiting steps of flow cytometry applied to whole organs and embryos.

II. Specific Considerations for Whole Organs and Embryos

As stated previously, it is essential to obtain a single-cell suspension representative of the tissue of origin that still maintains its intrinsic and immunologic characteristics. Developing a protocol for the tissue of interest requires optimizing the steps of dissociation, fixation, and permeation.

A. Tissue Dissociation

The protocols most widely used to obtain single-cell suspensions include mechanical disruption, enzymatic digestion, or use of chemicals (Pallavicini, 1987; Melamed *et al.*, 1990). We obtain the best results by combining all of them, enzymatic digestion with trypsin or collagenase and mechanical disruption by pipetting up and down with a glass siliconized Pasteur pipette during digestion. In addition, sometimes we include ethylenediamine tetraacetic acid (EDTA) to prevent Ca^{2+}-mediated cell reaggregation and DNase to avoid clumping of cells by the sticky DNA released from the broken cells.

We present orientative protocols to dissociate the chick embryonic retina, the whole early chick embryo, and the *Drosophila* embryo. Conditions should be worked out for each particular tissue and each batch of proteases. Note that digestion with proteases may affect surface proteins and impair their immunologic detection (see Fig. 4). A detailed overview of this aspect can be found in Pallavicini *et al.* (1990).

To obtain valid results, the cell suspension should be unbiased in cell types, representative of the organism or tissue from which it comes. It is advisable, if possible, to check by another technique that there is no loss of specific cell types during dissociation.

In addition, aggregates alter the experimental results by expressing the summation of all the signals as a single event. They even can clog the cytometer flow system. Therefore, before running samples, big aggregates should be removed through a nylon mesh of 50 μm in pore size, whereas debris and small aggregates can be further excluded from results by means of software analysis, based on their DNA content (see II.C).

B. Cell Fixation and Permeation

Although surface epitopes can be stained in unfixed cells, cell fixation is necessary to maintain cell integrity when the cells need to be permeated to reach intracellular epitopes with antibodies. Fixation can be performed before or after dissociation. Live tissue dissociation yields more and healthier cells and, conversely, cellular debris is also lower with this procedure. On the contrary, some parameters can be altered during the dissociation time, such as the cell cycle or the expression of some antigens, and apoptotic cells are massively lost during dissociation of fresh tissue.

As a general rule, one can use the same fixative for flow cytometry as is used for standard immunocytochemistry or immunohistochemistry. This is also true for the general cell staining protocol. There are two major categories of fixatives, alcohols, which fix the cells by dehydration and denaturation of proteins, and aldehydes, which cross-link cell constituents. Aldehydes can be successfully used before or after dissociation. However, we recommend using alcohols only after dissociation because they remove lipids from the cell membrane and generate pores in the membrane large enough for the digesting enzymes to trespass into the cell and destroy it. For embryos or tissues, we generally use 4% (w/v) paraformaldehyde in 0.1 *M* phosphate buffer, pH 7.1.

Once the cells are properly fixed, they have to be permeated when aldehyde fixation has been used (the pores generated by alcohol are wide enough for the antibody to enter the cell). Aldehyde-fixed cells can be rendered permeable using lipophilic agents or detergents, such as Triton X-100, NP-40, Tween 20, or saponin, or with alcohols. Brief exposure (5–20 minutes) to low concentration of these agents (0.05–0.5%) at 4°C is sufficient to generate pores large enough to allow the antibody molecules, including IgMs, to enter the cells (Bauer and Jacobberger, 1996). It is important to avoid prolonged exposures to these agents because detergent treatments can produce cell debris and vesicles that are counted as events by the cytometer, or may even empty the cells, causing antigen losses.

C. Data Analysis

As a first approach, the population of interest is visualized based on its size (forward-angle light scattered) versus its complexity (side-angle light scattered). Both intrinsic parameters provide an overview of the quality of the cell popula-

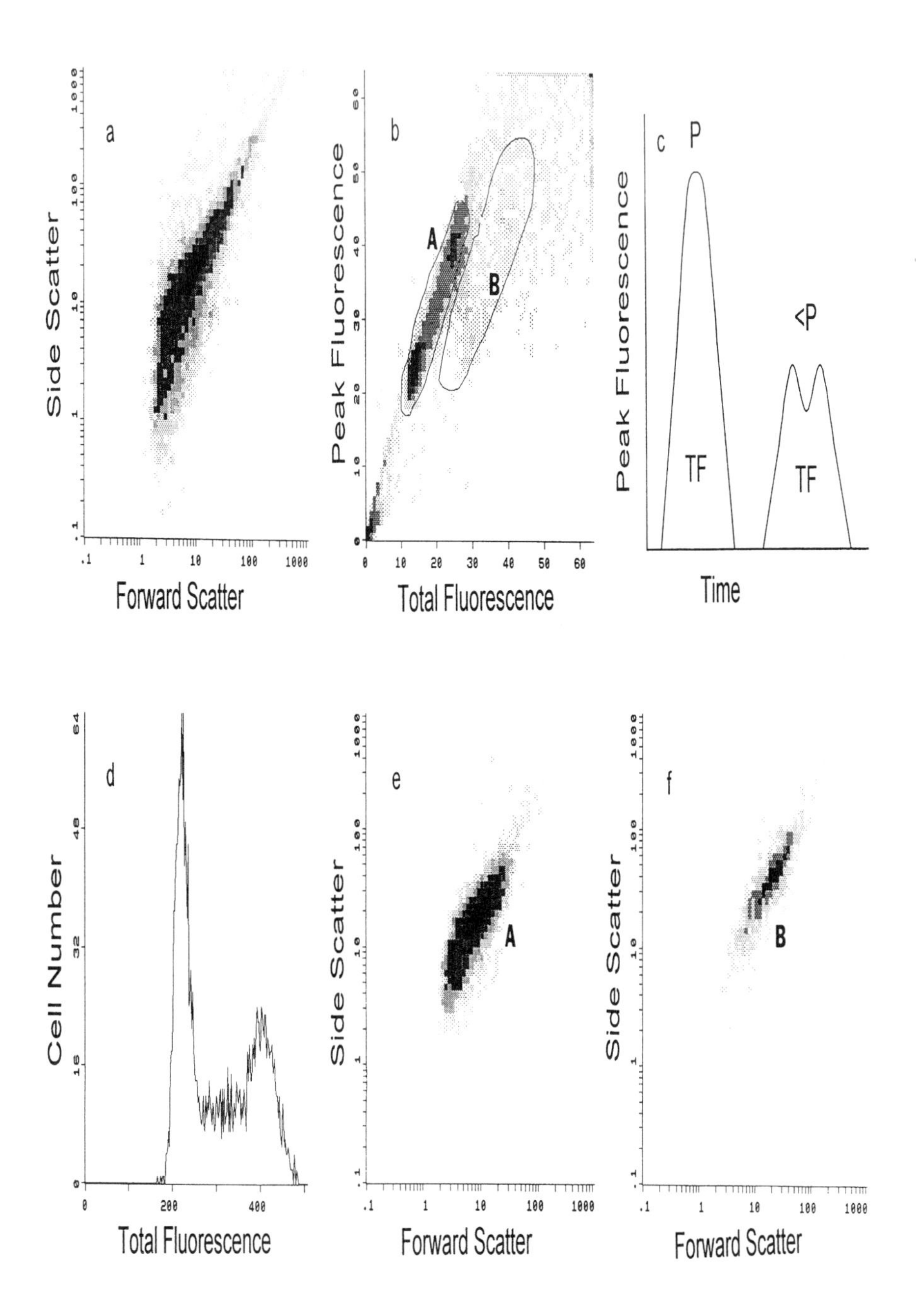
a
Side Scatter
Forward Scatter
b
A
B
Peak Fluorescence
Total Fluorescence
c
P
<P
TF
TF
Peak Fluorescence
Time
d
Cell Number
Total Fluorescence
e
A
Side Scatter
Forward Scatter
f
B
Side Scatter
Forward Scatter

tion obtained. Debris appear smaller and less complex, whereas aggregates are larger and more complex (Fig. 1a).

The next step is to discard the aggregates in the data analysis, as well as the nucleus-free debris. This is easily performed in permeated cells by staining the nuclei with propidium iodide, which gives a fluorescence signal proportional to the DNA content (Fig. 1b,d), where the G0/G1 peak, the S-phase transition, and the G2/M peak are identifiable. Besides the usual analysis of total fluorescence per cell (the integral under the curve obtained when the intensity of the emitted light is plotted as a function of the time that a single event requires to pass through the laser beam; Fig. 1c), the cytometer can also analyze the peak signal per cell (the maximum value of emitted light intensity reached while a single event is crossing the laser beam; Fig. 1c). Whereas the single cells present a linear correlation peak versus total fluorescence, aggregates give lower peak fluorescences than those of single cells with the same DNA content (total fluorescence; Fig. 1c). Therefore, in a total fluorescence versus peak fluorescence representation, the single cells appear, and can be gated for analysis, in a diagonal (Fig. 1b, region A). Aggregates are below this diagonal (Fig. 1b, region B), and nucleus-free debris stay at 0. Projecting back to the size versus complexity representations of the cell populations gated in these two regions (A to Fig. 1e, and B to Fig. 1f), it is possible once again to assess their differences.

Once the single-cell population is gated, the cytometer acquisition parameters need to be set with an appropriate control to study the expression of an antigen. This is the hallmark of flow cytometry applied to tissue or embryos, especially when an intracellular antigen is to be detected. Although in studies of surface antigens in intact cells the absence of primary antibody is an acceptable control, permeabilized cells tend to bind antibodies unspecifically. A homotypic, irrelevant antibody–antiserum should be used with each particular preparation of dissociated cells analyzed (see examples later), setting the parameters to keep the nonspecific fluorescence signal in the lower channels.

Fig. 1. Gating the cells to discard aggregates. An E1.5 chick embryo cell suspension was stained with propidium iodide and analyzed by flow cytometry. (a) Total population visualized by size (Forward Scatter) versus complexity (Side Scatter). (b) Propidium iodide labeled cells are represented by DNA content (Total Fluorescence) versus maximum intensity per cell (Peak Fluorescence). Two regions (A, single events and B, aggregates) are gated for further analysis. Notice that the nucleus-free (DNA-free) events, corresponding to cell debris, stay at value near to 0. (c) Basis of the diagonal gating. Schematic representation of a single cell (left) and of a doublet (right) as detected by propidium iodide fluorescence as they pass the laser beam off the flow cytometer. The total signal is the area (TF, total fluorescence in [b] and [d]) below the curve, corresponding to the DNA content. The maximum signal is the peak (P). A doublet that gives a similar total fluorescence signal (for instance, two G1 cells, compared with a single G2 cell), however, produces a lower peak fluorescence signal. (d) A-region gated cells (in [b]) show the characteristic cell cycle profile, with G0/G1 (first peak), S phase, and G2/M (second peak). (e, f) Size versus complexity representation of the cells gated in (b) (A and B regions, respectively). Compare with the total population in (a), and note that in this representation doublet events from B are not distinguishable from single events from A.

III. Selected Applications

A. Quantification of a Cytoplasmic Marker in Chick Embryonic Retinal Cells

We have explored the use of flow cytometry to quantify a subpopulation of retinal cells. Precursor marker 1 (PM1) is a monoclonal antibody that recognizes a cytoplasmic antigen in a subpopulation of neuroepithelial cells (Hernández-Sánchez *et al.*, 1994). The staining by immunocytochemistry in the retina decreases as development proceeds. However, this decrease has not been quantified in cell numbers because counting in retinal sections is tedious and approximative. Flow cytometric analysis provides a rapid and accurate cell counting. In addition, compared with counting under a microscope, it provides a stable criterion by which to compare different samples. Because PM1 staining shows a gradient of cellular intensity, it is very difficult for a human observer to set the edge for positive–negative or to discriminate different levels of positive signal. Flow cytometry also measures the relative content of the antigen in a quantitative form. The following protocol includes optimized fixation, dissociation, and staining procedures for flow cytometric analysis of PM1 in embryonic chick retinal cells.

1. Neuroretinas are dissected out from the surrounding tissues, including the pigmented epithelium, the lens, and the vitreous humor. A stepwise protocol is presented in Chapter 7.

2. Dissociation of the neuroretina to a single-cell suspension is performed by incubation at 37°C in 2 ml of phosphate-buffered saline (PBS) containing 1.5 mg/ml of bovine serum albumin (BSA) and 0.5 mg/ml of trypsin (freshly prepared; TRL from Worthington Biochemical Corporation, NJ, U.S.A.) for approximately 15 minutes, passing the tissue every 5 minutes through a siliconized Pasteur pipette. After 5 minutes, add 5 μg/ml DNase. After 10 minutes, add 5 m*M* EDTA. Check the cell suspension under phase-contrast microscopy. When a good proportion of single cells is observed, stop the reaction and fix the cells by adding 4 ml of paraformaldehyde 6% (w/v; final concentration 4%) in phosphate buffer, pH 7.1. Keep the cells in fixative for 1 hour with a swinging motion.

3. The cell suspension is filtered through a 50-μm nylon mesh (Falcon) to remove remaining tissue pieces and washed twice by centrifugation at 400*g* for 5 minutes with PBS and twice with 30 mg/ml BSA in PBS to block the remaining paraformaldehyde.

4. Aliquots of the cell suspension, typically 150,000 cells, are transferred to an Eppendorf tube, sedimented by centrifugation at 200*g* for 5 minutes, and resuspended in 500 μl of 0.05% Triton X-100, 30 mg/ml of BSA, and 100 m*M* of glycine in PBS to permeate the cells. The cells are then sedimented again, resuspended in 200 μl of 15% (v/v) normal goat serum in PBS, and transferred to a well of a 96-well microtiter plate, U or V bottom. The cells are again sedimented before adding the primary antibody by centrifugation of the plate at 200*g* for 3 minutes. These conditions are used at each washing step.

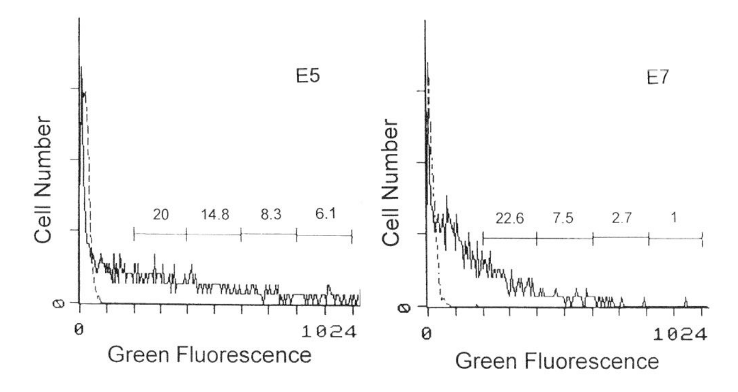

Fig. 2. PM1 staining in chick embryonic retinal cells. Chick retinal cell suspensions from E5 and E7 were stained with monoclonal antibody PM1 and visualized by cyanine-2 (green fluorescence). An irrelevant IgM was used as control (dotted lines). Regions containing cell populations of increasing intensity of fluorescence, corresponding to increasing levels of PM1 antigen, are defined. Numbers above the lines represent the percentage of cells inside the defined region.

5. Resuspend and incubate the cells sequentially in the primary antibody (200 μl, 2 μg/ml of mouse IgM antibody) for 45 minutes at room temperature (RT); then in biotinylated anti-mouse IgG + IgM (Amersham, 1/200 dilution) for 30 minutes at RT; and finally in Cyanine2–streptavidin (Amersham, 1/200 dilution) for 30 minutes at RT. All incubations are done in a swinging platform. Four washing steps between incubations, as well as the dilutions, are done with 30 mg/ml BSA and 100 m*M* glycine in PBS. As a negative control, we use an irrelevant IgM at the same protein concentration.

6. Transfer the cells to a tube containing 1 ml PBS with 0.5 μg/ml propidium iodide. After 5 minutes, they can be analyzed in the cytometer. Set the parameters of the cytometer with the negative control. Gate the single, intact cell population as described in Figure 1. Because PM1 expression shows a gradient, we analyze several regions of different levels of specific signal (Fig. 2).

From E5 to E7, the number of cells expressing PM1 decreases, as well as the quantity of antigen per cell (Fig. 2). This corresponds well with the qualitative observations in tissue sections, and provides a method to estimate the effect of experimental treatments on the level of expression of PM1.

B. Detection of Intracellular Insulin in *Drosophila* Embryonic Cells

Flow cytometric analysis of cells from whole *Drosophila* embryos has been done previously by Krasnow *et al.* (1991), who sorted the cells harboring a β-galactosidase transgene. Here we describe the detection of endogenous insulin immunoreactivity (Pimentel *et al.*, 1996).

Insulin-related peptides have been found in invertebrates, as well as in prepancreatic vertebrate embryos (Smit *et al.*, 1988; Adachi *et al.*, 1989; Lagueux *et al.*, 1990; de Pablo and de la Rosa, 1995). Flow cytometry makes it possible to detect the minute amounts of peptide present in such tissues, even when the peptide is not accumulated in granules.

1. Embryos of the desired developmental stage, in our case 12–15 hours, are washed with cold PBS, dechorionated by immersion in 50% (v/v) sodium hypochlorite for 2 minutes, devitelinated by washing with cold n-heptane–4% (w/v) paraformaldehyde (1:1) for 15 minutes and immersed briefly in cold methanol. The embryos are then washed in cold PBS and fixed in 4% (w/v) paraformaldehyde for 15 minutes. Afterward, 30 mg/ml BSA is added to block the paraformaldehyde and the embryos are centrifuged at 800*g* for 1 minute at RT and resuspended in PBS.
2. The embryos are dissociated to a single-cell suspension by incubation at 37°C with 27 U/ml collagenase (Sigma) and 0.18 mg/ml trypsin (freshly prepared; TRL from Worthington Biochemical Corporation) for 30 minutes. The reaction is stopped by adding a volume of 30 mg/ml BSA. The cells are then collected by centrifugation at 230*g* for 5 minutes at RT, resuspended in PBS containing 0.05% (w/v) Triton X-100, 30 mg/ml BSA, and 100 m*M* glycine, and filtered through a 50-μm nylon mesh (Falcon).
3. Dissociated cells, 10,000 per point, are labeled with 0.1 mg/ml anti-insulin immunoglobulin (Ig; protein A-purified anti-porcine insulin raised in guinea pig, lot 627, Department of Pharmacology, Indiana University, Indianapolis, IN, U.S.A.) by overnight incubation at 4°C in 0.05% (w/v) Triton X-100, 30 mg/ml BSA, and 100 m*M* glycine in PBS, followed by 1-hour incubation at RT with a 1/100 dilution of fluorescein isothiocyanate-labeled anti-guinea pig antiserum (Jackson Immunoresearch Laboratory). After each incubation, the cells are washed three times by centrifugation at 230*g* for 5 minutes at RT and resuspended in 0.05% (w/v) Triton X-100, 30 mg/ml BSA, and 100 m*M* glycine in PBS. The last pellet is resuspended in 1 ml PBS. The negative control to set parameters in the flow cytometer is labeled with 0.1 mg/ml guinea pig Ig (Biotek).

As shown in Figure 3, a small proportion of *Drosophila* embryonic cells presented a specific fluorescence signal (region A). The selected B region represents 7.5% of total cells with a mean fluorescence signal of 77.8, whereas 81% of the cells remain in the 0.1 region.

C. Detection of Insulin Receptor in Chick Embryonic Cells

Insulin receptors in chick embryonic cells have been demonstrated by binding studies (Girbau *et al.*, 1989). The insulin receptor molecule consists of an extracellular, a transmembrane, and an intracellular domain. This example illustrates how the dissociation protocol that uses trypsin digests the extracellular domain of the receptor, because recognition with an antibody against the extracellular do-

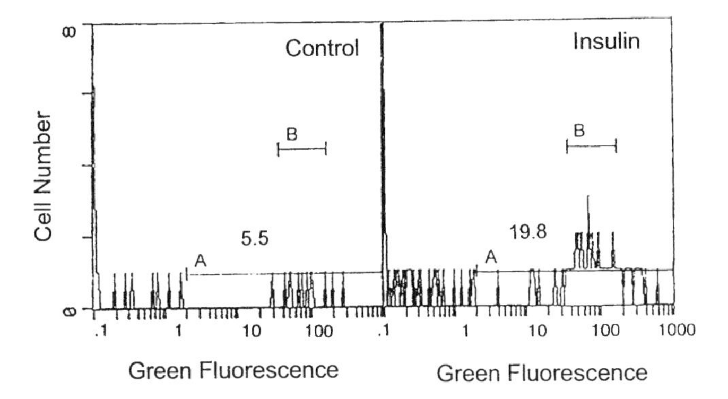

Fig. 3. Insulin immunoreactivity in *Drosophila* embryonic cells. Aliquots of a cell suspension from whole *Drosophila* embryos were stained with anti-insulin Ig and visualized by FITC (green fluorescence) using preimmune guinea pig Ig as a control. The setting of the acquisition parameters was done with control staining and defined A region (fluorescence values over channel 1). Numbers above the line represent the percentage of cells inside the defined region. Region B represents the population of cells presenting high insulin-like immunoreactivitty. (From Pimentel *et al.*, 1996, with permission.)

main disappears when trypsin is used in the dissociation. In addition, this example provides a protocol of fixation before dissociation, which gives better results in the analysis of cell cycle parameters.

1. Chick embryos of the desired stage are dissected out from the surrounding membranes. Embryos are then fixed in 4% (w/v) paraformaldehyde in 0.1 *M* phosphate buffer, pH 7.1 for 1 hour and washed twice with PBS and twice with 30 mg/ml BSA in PBS.

2. Embryonic cells are dissociated in 2 ml of PBS containing 1.5 mg/ml BSA. First, collagenase (Sigma) is added at 37.5 U/ml and incubated at 37°C for 10 minutes. Then, trypsin (freshly prepared; TRL from Worthington Biochemical Corporation) is added at 0.5 mg/ml and incubated for 30 minutes at 37°C. During this incubation, embryos are pipetted up and down every 10 minutes with a siliconized Pasteur pipette. Digestion is stopped by addition of 30 mg/ml BSA in PBS. The cell suspension is filtered through a 50-μm nylon mesh (Falcon) and centrifuged at 300*g* for 5 minutes. The cell pellet is resuspended in 1 ml of 30 mg/ml BSA in PBS.

3. Aliquots of the cell suspension, typically 150,000 cells, are transferred to an Eppendorf tube, sedimented by centrifugation at 200*g* for 5 minutes, and resuspended in 500 μl of 0.05% Triton X-100, 30 mg/ml of BSA, and 100 m*M* of glycine in PBS to permeate the cells for 30 minutes. The cells are then sedimented again, resuspended in 200 μl of 15% (v/v) normal goat serum in PBS as a blocking step, and then transferred to a well of a 96-well microtiter plate, U or V bottom. The cells are again sedimented before adding the primary antibody by

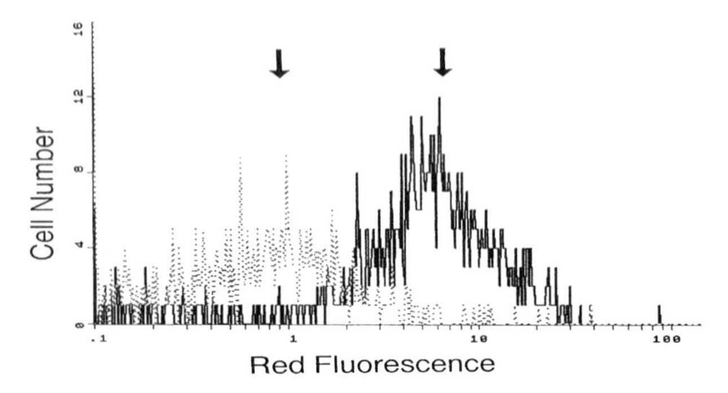

Fig. 4. Insulin receptor staining in chick embryonic cells. Aliquots of an E1.5 whole chick embryo cell suspension were stained with AbP5 (against the intracellular domain of insulin receptor; continuous line) or anti-insulin receptor antiserum from UBI (against the extracellular domain; dotted line) and visualized with cyanine-3 (red fluorescence). Because trypsin digestion during dissociation digests the extracellular domain of the receptor, there is a shift in the labeling to a lower signal (mean fluorescence from 5.8 to 0.9, arrows) when the antibody against the extracellular domain is used.

centrifugation of the plate at 200*g* for 3 minutes. These conditions are used at each washing step.

4. Resuspend and incubate the cells in 200 μl of 10 μg/ml of the primary antibody for 45 minutes at RT. We compared an anti-insulin receptor antiserum (UBI, an antipeptide antiserum raised against the extracellular α-subunit of the human insulin receptor; Fig. 4, dotted line) with AbP5 (Garofalo and Barenton, 1992; an antipeptide antiserum raised against the β-subunit of the human insulin receptor, kind gift of R. S. Garofalo; Fig. 4, continuous line). Afterward the cells are incubated in biotinylated anti-rabbit Ig (Amersham, 1/200 dilution) for 30 minutes at RT, and finally in Cyanine3–streptavidin (Amersham, 1/200 dilution) for 30 minutes at RT. All incubation is done in a swinging platform. Four washing steps between incubations, as well as the dilutions, are done with 30 mg/ml BSA and 100 m*M* glycine in PBS.

5. The cells are then transferred to a tube containing 1 ml PBS with 0.5 μg/ml propidium iodide. After 5 minutes, they can be analyzed in the cytometer. Gate the single, intact cell population as described in Figure 1.

Because trypsin digests the extracellular domain, a shift in the distribution is observed when both antisera are compared, the one recognizing an intracellular peptide (Fig. 4, continuous line) and the other recognizing an extracellular peptide (Fig. 4, dotted line), resulting in an increased mean fluorescence for the AbP5.

IV. Concluding Remarks

The protocols described here could be applicable to other primary tissues after optimization, as well as to organs and embryos experimentally treated *in vitro* and *in vivo*, as described in Chapters 2 and 7.

Flow cytometric analysis permits an accurate and rapid quantification of the level of expression of antigens in single cells. In this way, it is possible to define cell populations with different sets of markers. More interestingly, flow cytometric analysis can determine whether changes in the level of expression of a defined antigen, occurring during development or as a response to an experimental treatment, are affecting the number of cells expressing that antigen or the amount of antigen expressed per cell. In parallel, combined with cell cycle analysis, it is possible to determine if changes in the level of antigens depend on phases of the cell cycle. Besides the major epitope immunocytochemical detection, binding experiments with a labeled ligand are also possible to detect functional receptors (Ziegler *et al.*, 1994).

Despite the major advantage of quantifying developmental processes, all the positional information is lost by the required dissociation step. In some cases, this can be partially overcome by regional subdissections. In our experience, the flow cytometric approach should be complemented by other cellular approaches that may be less quantitative, but that preserve the very important positional information.

Acknowledgments

Participation of Begoña Díaz, who provided Figure 2, Aixa V. Morales, and Flora de Pablo in the establishment of the protocols described here is gratefully acknowledged. We also wish to thank Pedro Lastres and Alberto Orfao for their assistance. Research in our laboratory is supported by grants from DGICYT (PM94-152 and PM96-0003), FIS (94/151), and CAM (AE 376/95).

References

Adachi, T., Takiya, S., Suzuki, Y., Iwami, M., Kawakami, A., Takahashi, S. Y., Ishizaki, H., Nagasawa, H., and Suzuki, A. (1989). cDNA structure and expression of bombyxin, an insulin-like brain secretory peptide of the silkmoth *Bombyx mori*. *J. Biol. Chem.* **264,** 7681–7685.

Bauer, K. D., and Jacobberger, J. W. (1996). Analysis of intracellular proteins. *Meth. Cell Biol.* **41,** 351–376.

De Pablo, F., and de la Rosa, E. J. (1995). The developing CNS: A scenario for the action of proinsulin, insulin and insulin-like growth factors. *Trends Neurosci.* **18,** 143–150.

Garofalo, R. S., and Barenton, B. (1992). Functional and immunological distinction between insulin-like growth factor I receptor subtypes in KB cells. *J. Biol. Chem.* **267,** 11470–11475.

Girbau, M., Bassas, L., Alemany, J., and de Pablo, F. (1989). *In situ* autoradiography and ligand-dependent tyrosine kinase activity reveal insulin receptors and insulin-like growth factor I receptors in prepancreatic chicken embryos. *Proc. Natl. Acad. Sci. U.S.A.* **86,** 5868–5872.

Hernández-Sánchez, C., López-Carranza, A., Alarcón, C., de la Rosa, E. J., and de Pablo, F. (1995). Autocrine/paracrine role of insulin-related growth factors in neurogenesis: Local expression and effects on cell proliferation and differentiation in the chicken retina. *Proc. Natl. Acad. Sci. U.S.A.* **92,** 9834–9838.

Krasnow, M. A., Cumberledge, S., Manning, G., Herzenberg, L. A., and Nolan, G. P. (1991). Whole animal cell sorting of *Drosophila* embryos. *Science* **251,** 81–85.

Lagueux, M., Lwoff, L., Meister, M., Goltzene, F., and Hoffmann, J. A. (1990). cDNAs from neurosecretory cells of brains of *Locusta migratoria* (*Insecta, Orthoptera*) encoding a novel member of the superfamily of insulins. *Eur. J. Biochem.* **187,** 249–254.

López-Mediavilla, C., Orfao, A., San Miguel, J., and Medina, J. M. (1992). Developmental changes in rat liver mitochondrial populations analyzed by flow cytometry. *Exp. Cell Res.* **203,** 134–140.

Melamed, M. R., Lindmo, T., and Mendelshon, M. L. (1990). "Flow Cytometry and Cell Sorting," 2nd ed. Wiley-Liss, New York.

Morris, V. B., and Taylor, I. W. (1985). Estimation of nonproliferating cells in the neural retina of embryonic chicks by flow cytometry. *Cytometry* **6,** 375–380.

Ormerod, M. G. (1990). "Flow Cytometry: A practical Approach." IRL Press, Oxford University Press, New York.

Pallavicini, M. G. (1987). Solid tissue dispersal for cytokinetic analyses. In "Techniques in Cell Cycle Analysis." Humana Press.

Pallavicini, M. G., Taylor, I. W., and Vindelov, L. L. (1990). Preparation of cell/nuclei suspension from solid tumor for flow cytometry. In "Flow Cytometry and Cell Sorting" (M. R. Melamed, T. Lindmo, and M. L. Mendelsohn, eds.), 2nd ed., pp. 187–194. Wiley-Liss, New York.

Pimentel, B., de la Rosa, E. J., and de Pablo, F. (1996). Insulin acts as an embryonic growth factor for *Drosophila* neural cells. *Biochem. Biophys. Res. Commun.* **226,** 855–861.

Prados, J., Fernández, J. E., Garrido, F., Alvarez, L., Hidalgo, R., Muros, M. A. and Aránega, A. (1992). Expression of α-tropomyosin during cardiac development in the chick embryo. *Anat. Rec.* **234,** 301–309.

Smit, A. B., Vreugdenhil, E., Ebberink, R. H., Geraerts, W. P., Klootwijk, J. and Joosse, J. (1988). Growth-controlling molluscan neurons produce the precursor of an insulin-related peptide. *Nature* **331,** 535–538.

Tsonis, P. A., and Walker, E. (1991). Cell populations synthesizing cartilage proteoglycan core protein in the early chick limb bud. *Biochem. Biophys. Res. Commun.* **174,** 688–695.

Ziegler, O., Cantin, C., Germain, L., Dupuis, M., Sekaly, R. P., Drouin, P. and Chiasson, J. L. (1994). Insulin binding to human cultured lymphocytes measured by flow cytometry using three ligands. *Cytometry* **16,** 339–345.

13

Detection of Multiple Gene Products Simultaneously by *in Situ* Hybridization and Immunohistochemistry in Whole Mounts of Avian Embryos

Claudio D. Stern
Department of Genetics and Development
College of Physicians and Surgeons of Columbia University
New York, New York 10032

I. Introduction

Research on the avian embryo has succeeded in successfully combining experimental embryology, for which avian embryos are unrivaled, with the availability of many antibodies and cDNA probes, allowing for the first time the objective identification of tissues and cell types and greatly increasing the level of sophistication possible in the design of experiments. At the same time, methods for

Current Topics in Developmental Biology, Vol. 36

0070-2153/98 $25.00

interfering with the levels of expression of specific genes, including the use of retroviral vectors, have become available for avian embryos, which expands the range of experiments that can be performed targeted at understanding developmental mechanisms at the cellular and molecular levels.

In these experiments, it is often desirable to study the distribution of different mRNAs and proteins at a given stage of development in one embryo. For example, it may be of interest to detect the ectopic expression of a gene introduced with a retroviral vector, at the same time as studying the expression of one or more of its putative target genes. For this, a technique allowing the simultaneous visualization, with single-cell resolution, of two or more different gene products is essential. Good methods have become available only since the early 1990s; this chapter reviews our present state of knowledge, with detailed protocols and instructions for designing appropriate controls and tips for trouble-shooting different problems. It should be borne in mind, however, that most of these techniques are still under development, because improvements are made continuously.

II. Synthesis of Digoxigenin- and Fluorescein-Labeled Riboprobes by *in Vitro* Transcription

When conducting *in situ* hybridization with multiple probes, it is particularly important to remove any unincorporated digoxigenin and fluorescein (especially in the case of the latter, which is "sticky"), which appear to be the main cause of high backgrounds. There are several ways to do this. The most efficient methods include: (1) the use of a Sephadex G25 spin column made in an Eppendorf tube, and (2) repeating the LiCl–ethanol (EtOH) precipitation step. The following protocol includes the latter method. All solutions and equipment for this part of the protocol should be RNase-free, and gloves should be worn. Consult Sambrook *et al.* (1989) for instructions on how to do this. When making up solutions that will contain enzymes (e.g., RNA polymerases, DNase), it is best to use commercial RNase-free water rather than to treat water with diethylpyrocarbonate (DEPC) oneself, because traces of DEPC can inhibit enzyme reactions. If DEPC-treated water is to be used, it helps to autoclave the bottle after treatment for at least 1 hour, with the cap loosely fitted.

First, linearize 3–10 μg of the vector (the most common vector is pBlueScript [Stratagene], which contains promoters for RNA polymerases T3 and T7 at each end of the polylinker) with the appropriate enzyme (Fig. 1A) for 4–5 hours or overnight (after cutting, check the efficiency of cutting in an agarose gel using 1/20th of cutting reaction). Then add an equal volume of phenol (equilibrated to pH 7.2–8.0): chloroform (1:1), vortex hard, and spin in a microfuge for 1–2 minutes. Take the upper layer into a clean Eppendorf and add 10 μl 3 *M* sodium acetate for every 100 μl and 2.5× its volume of absolute alcohol. Vortex briefly and put at −20°C for 2 hours to overnight.

Spin at 4°C (15,000 rpm) for 15 minutes, and take off alcohol with a sterile

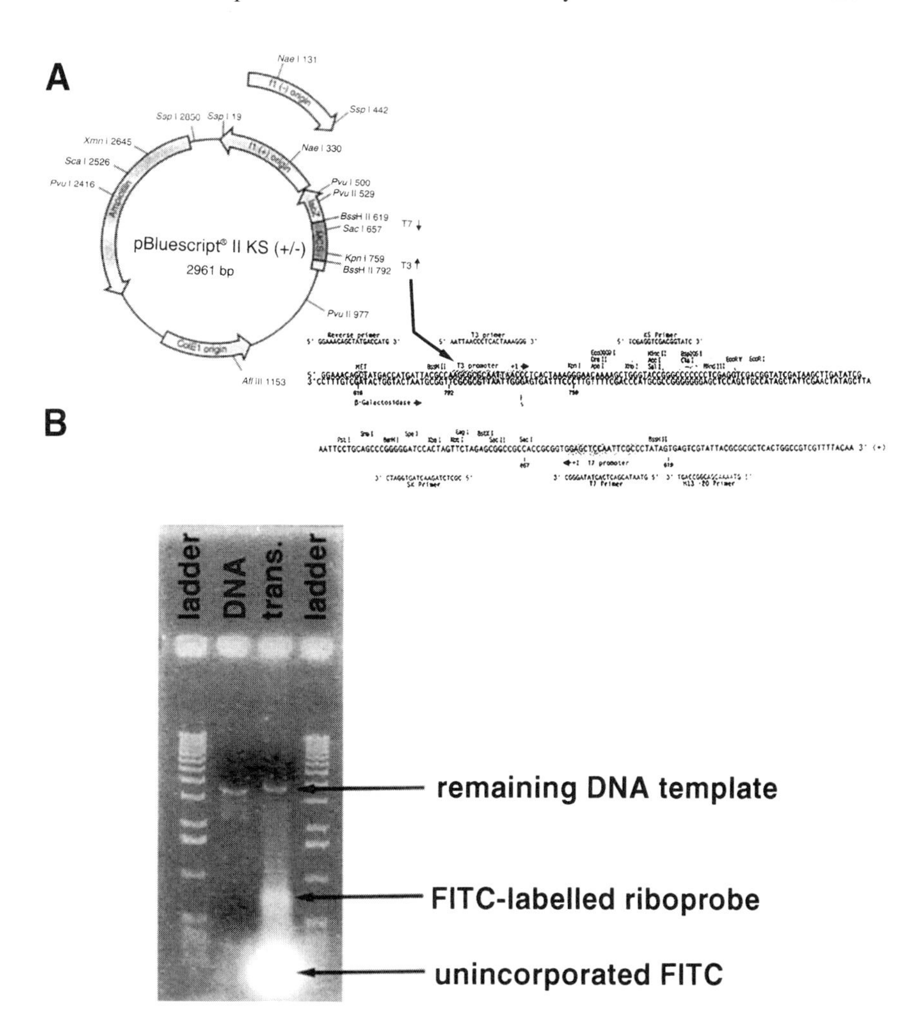

Fig. 1. Linearization of the vector DNA and transcription with fluorescein-labeled UTP. (A) Map of the commonly used vector pBlueScript II KS (Stratagene), and sequence of the polylinker region. The position of the T3 and T7 polymerase sites and direction of transcription from each are shown. (B) Agarose gel, stained with ethidium bromide, showing size markers ("ladder"), linearized plasmid DNA ("DNA"), and transcription reaction before DNAse treatment and precipitation with LiCl ("trans."). Note the large amount of unincorporated fluorescein-labeled UTP remaining, and the relationship between the amount of transcribed riboprobe and the amount of DNA template.

Table I Transcription Mix

Component	For 1 μg DNA	For 3 μg DNA
DNA (1 μg/μl)	1 μl	3 μl
Water	15 μl	22 μl
5× transcription buffer	6 μl	10 μl
Digoxigenin–nucleotide mix	2 μl	5 μl
DTT (10×)	3 μl	5 μl
RNAsin	1 μl	1 μl
Enzyme (T3, T7, or SP6)	2 μl	4 μl
Total	30 μl	50 μl

Pasteur pipette pulled to a fine tip. If the DNA is clean, the pellet will probably be invisible, but it is important not to touch it with the pipette. Wash the pellet with 150 μl 70% ethanol, vortex, and spin again 5 minutes. Take off the alcohol carefully and dry the pellet at 37°C. Add clean (preferably commercial Ultrapure) RNase-free water to give a final concentration of about 1 mg/ml. Allow to dissolve at 37°C for at least 15 minutes, with occasional vortexing or flicking of the tube.

Transcribe with appropriate enzyme (T3, T7, or SP6; see Fig. 1) at 37°C (40°C in the case of SP6) for 2 hours. Proportions of different reagents are as shown in Table I (add in the order shown in Table I).

After transcription, digest the template with RNase-free DNase (1–5 units per microgram DNA): first make up a stock of 1:10 DNase in transcription buffer with 10% DTT (e.g., 6 μl water, 1 μl DTT, 2 μl transcription buffer, 1 μl DNase I), then add this to the tube. It is important that the concentration of glycerol (contained in the polymerase, RNAsin, and DNase solutions) not exceed 10% of the total volume of either the transcription or DNase digestion mix, or activity of the enzymes will be impaired. Incubate for 30 minutes at 37°C. At this point, take an aliquot (e.g., 1/20th of sample) and run an agarose gel to check the probe. The yield of these transcription reactions can be as much as 10× the number of micrograms of DNA template used.

Bring the volume up to 82 μl with water, and add 8 μl ethylenediamine tetraacetic acid (EDTA) to stop the DNAse. Add 10 μl 4 *M* LiCl and 250 μl absolute EtOH. Vortex briefly and put at −20°C as described previously. Precipitate 2 hours to overnight. Spin at 4°C 15 minutes. Take off alcohol (this time the pellet should be clearly visible, particularly when labeling with fluorescein isothiocyanate [FITC] labeled nucleotide). Wash with 300 μl 70% EtOH. Vortex very well. The pellet should be dislodged and should break up into many little pieces. Spin again 5 minutes. Take off alcohol. Rinse carefully with 30 μl absolute EtOH and allow to dry at 37°C. Dissolve the pellet in water at approximately 1 mg/ml (transcription should yield about 8–10× the original weight of the DNA). Leave at 37°C for at least 15 minutes, vortexing occasionally.

When making FITC-labeled riboprobes, repeating the LiCl precipitation helps to remove unincorporated FITC. Bring the volume of riboprobe solution up to 90 μl, add 10 μl LiCl and 250 μl EtOH, and repeat the previous precipitation step. After washing the pellet and dissolving again at 0.1–1 mg/ml in water, check the probe in an agarose gel. In the case of fluorescein-labeled probes, any unincorporated fluorescein will be visible in the transilluminator at the bottom of the gel as a green spot, even in unstained gels (Fig. 1B). If some of this is still present, repeat the LiCl–EtOH precipitation once more or pass through a Sephadex G25 spin column.

Spin briefly, then add ≥ 10× its volume of hybridization buffer (see later) for storage (or make up to about 100–500 ng/ml in hybridization buffer). Store at −20°C.

For *in situ* hybridization with two probes, one probe is transcribed exactly as described previously, and the second probe is transcribed by the same method except that FITC-labeled uridine triphosphate (UTP; Boehringer-Mannheim) mixed with unlabeled nucleotides replaces the digoxigenin–nucleotide mix (the proportion of FITC-labeled to unlabeled nucleotides is described in the sheet supplied by the manufacturers).

Probes synthesized in this way and dissolved in hybridization buffer are stable at −20°C for many months. The heparin, RNA, and formamide in the hybridization buffer act as very powerful RNase inhibitors, and degradation of the probe is virtually completely eliminated. However, fluorescein-labeled probes are probably a little less stable than digoxigenin-labeled ones. In our experience, however, even fluorescein-labeled probes can be stored in this way for at least 9 months and reused at least eight times (which also contributes to reduce background).

III. Preparation of Embryos

When changing solutions for all the procedures involving embryos described here, particularly those involving embryos of 2 days and younger, the integrity of the embryos critically depends on the care used in pipetting the solutions. Always use a Pasteur pipette, not a Gilson pipetman. Tilt the vial and turn it around to remove solutions with the pipette, avoiding sucking the embryos into it. At each step, remove all of the solution from the vial until embryos stay in position attached to the wall, but then add new fluid quickly so they do not dry. When adding new liquid, turn the vial so that the fluid runs down the opposite wall of the vial to that containing the embryos.

Collect embryos in calcium–magnesium-free (CMF) phosphate-buffered saline (PBS). Clean off any adhering yolk, and remove from the vitelline membrane. It is highly desirable to fix the embryos in a flat, stretched-out state. Young embryos up to 1 day's incubation can be laid flat on the lid of a plastic dish, in a small drop of CMF. Just before fixing, carefully remove the CMF and start fixing by adding fixative directly onto the embryo gently using a Pasteur pipette. Older embryos (2–5 days' incubation) are best pinned out on a Silicon rubber- or wax-coated dish using fine insect pins. Once arranged appropriately, the CMF can be sucked out and replaced with fixative. After a few minutes, the pins can be removed and the embryo transferred to a small (5 or 20 ml, depending on the number of embryos) glass scintillation vial. The fixative is freshly made 4% formaldehyde/CMF/EGTA (4% w/v paraformaldehyde powder added to CMF preheated at 65°C, stirring continuously; adjust pH to about 7.5 with 5*N* NaOH. Allow to cool, then add EGTA to final concentration of 2 m*M*). Leave in this 1 hour to overnight at 4°C. Transfer embryos to absolute methanol, and store in this for up to 1 week at −20°C.

On the first day of the *in situ* procedure, start by rehydrating the embryos through 75%, 50%, and 25% methanol in PTW (PTW = CMF with 0.1% Tween-20), allowing embryos to settle between changes. Wash twice with PTW, 10 minutes each. For embryos older than about 2 days, bleach for 1 hour in 6% H_2O_2 (1 ml H_2O_2 + 4 ml PTW from 30% stock). Wash three times with PTW, 10 minutes each. For the last wash, measure the volume of PTW (use 2 or 5 ml, depending on size of tube). Add proteinase K (1:1000; final concentration 10 μg/ml). Incubate at room temperature for 30 minutes regardless of stage of embryos, but reduce this to 15 minutes for very young embryos (less than stage 4–5) or embryos that have been cultured by the New (1955) method, which become very thin. During incubation, gently roll the tube every few minutes to make sure the sides and top of vial get wet with proteinase K (this will remove RNase from the tube and cap).

Take off proteinase K and rinse briefly and very carefully with a very small volume of PTW. Replace PTW with 4% formaldehyde in PTW (made as previously, but does not need to be fresh), containing 0.1% glutaraldehyde. Postfix 20 minutes.

IV. Prehybridization, Hybridization and Post-Hybridization Washes

The following protocol is a modification of a method devised by Drs. David Ish-Horowicz, Domingos Henrique, and Phil Ingham (unpublished). It is based on hybridization conducted under stringent conditions at low pH and low salt concentration, at a high temperature. This simplifies the post-hybridization washes, where the same solution and conditions are used. RNase treatment is not only unnecessary, but actually leads to loss of signal with most probes.

Remove postfixing solution and wash twice briefly with PTW. Remove PTW, and replace with 1 ml hybridization solution (Table II). Remove hybridization mix, and replace with another 1–2 ml (5 ml if using large vials) of the same solution. Place tube upright in a beaker in a water bath at 68–71°C. Incubate 2–6 hours. Remove hybridization mix, and replace with probe in hybridization mix

Table II Hybridization Solution

Component (stock concentration)	Final concentration	Volume to add
Formamide	0.5	25 ml
SSC (20×, pH 5.3 adjusted with citric acid)	1.3× SSC	3.25 ml
EDTA (0.5 *M*, pH 8.0)	5 m*M*	0.5 ml
Yeast RNA (20 mg/ml)	50 μg/ml	125 μl
Tween-20	0.002	100 μl
CHAPS (10%)	0.005	2.5 ml
Heparin (50 mg/ml)	100 μg/ml	100 μl
H_2O		18.4 ml
Total		50 ml

(see section II). For probes ≤400 nucleotides, use a lower hybridization temperature (e.g., 62°C for 150–250 nt, 65°C for 250–400 nt). It is advantageous to prehybridize at the higher temperature, then to lower the temperature when adding the short probe. For *in situ* hybridization with two or more probes, add both probes simultaneously (provided that the probes are labeled differently (e.g., one with digoxigenin-UTP, one with fluorescein-UTP, and the third with biotin-UTP). Allow to hybridize overnight in the water bath.

The next morning, remove the probe (keep at −20°C for reuse up to 6–10 times). Rinse carefully three times with a small volume (1 ml) of prewarmed hybridization solution. Wash twice with 1.5 ml (4 ml if using a large vial) of prewarmed hybridization solution, 30–45 minutes in water bath. Wash 20 minutes with prewarmed 1:1 hybridization solution : TBST (1:10 from the stock in Table III). Rinse three times with TBST at room temperature. Wash three times 30–60 minutes with TBST. Rinses can be made with a very small volume. Washes are best done by filling the vial to the top, replacing the cap, and then laying the vial horizontally on a tilting table.

Table III 10× TBST

NaCl:	8 g
KCl:	0.2 g
1 *M* Tris-HCl, pH 7.5:	25 ml
Tween-20:	11 g
H_2O:	~64 ml
Total:	100 ml

Table IV Preabsorbing Antibody

To preabsorb antibody, proceed as follows:

1. Weigh X mg of embryo powder,[a] where X = 2× number of milliliters of final vol. of Ab. solution needed, into an Eppendorf.
2. Add 500 μl TBST and vortex for 20 seconds.
3. Heat to 70°C for 30 minutes, vortex again 20 seconds, and spin down at low speed for 1 minute—just enough to get the powder to form a loose pellet.
4. Discard the supernatant.
5. Wash the pellet three to five times with 500 μl TBST, spinning low speed each time (this is to remove any fat that may be floating on the supernatant). Repeat until no more fat is seen at the top of the supernatant.
6. Resuspend the pellet in 100 μl blocking buffer (not TBST) for every bottle of embryos, mixing gently (not shaking).
7. Add antibody so that the final concentration will be 1:5000. (For example: if final incubation in antibody overnight will have 1 ml per tube, the embryos will now be sitting in 1 ml blocking buffer. If five bottles will be stained, add 1 μl antibody to the 500 μl blocking buffer with the embryo powder). Leave on a rocking table to absorb for 1–2 hours at room temperature.
8. After absorbing antibody with powder, spin down at high speed (to make a hard pellet) in the microfuge for 3 minutes. Keep the supernatant and discard the pellet.

[a]Embryo powder is made as follows: Homogenize embryos (ideally of the same stage as those being stained) in a minimum volume of ice-cold PBS, through a syringe. Add four volumes of ice-cold acetone, mix and incubate on ice for 30 minutes. Centrifuge at 10,000*g* for 10 minutes, discard the supernatant, and then wash the pellet with ice-cold acetone and spin again. Spread the pellet out and grind it into a fine powder on a sheet of filter paper. Air-dry the powder and store it at 4°C.

V. Antibody Incubations

After TBST washes, embryos are first blocked with a protein solution to lower nonspecific binding of the antibody and then placed in antibody solution overnight. For whole-mount staining to detect two or more labels, each antibody incubation is followed by detection of the appropriate enzyme. Then the enzyme activity is inactivated by one of several possible methods (described in more detail later) before proceeding with incubation in the next antibody. This allows the same enzyme (e.g., alkaline phosphatase) to be detected by a different chromogenic reaction.

Remove the embryos from TBST and place them in 1 ml of blocking solution: 5% heat-inactivated (at 55°C for 30 minutes) sheep serum in TBST with 1 mg/ml bovine serum albumin for 3 hours at room temperature. During this time, preabsorb the antibody (Table IV). Dilute the preabsorbed antibody into the blocking buffer with the embryos so that final concentration of antibody on the embryos is 1:5000. Incubate overnight at 4°C on a rocking platform.

The next morning, remove the antibody from the embryos and keep it for reuse (up to 8–10 times; this also helps to reduce background). Rinse three times with TBST. Wash three times 1 hour with TBST, rocking (fill vial right up to the top and place horizontally on rocker). Embryos older than 2 days should be washed more extensively (five times for 1 hour, and one overnight wash in TBST).

VI. Detection

As mentioned above, each antibody is detected sequentially, which allows several antibodies coupled with the same enzyme (alkaline phosphatase) to be detected in different colours. Procedures for inactivating the enzyme are given later. The paragraphs below describe the chromogenic reactions.

A. Detection of Alkaline Phosphatase with NBT–BCIP (Blue Precipitate)

Remove TBST and wash twice for 10 minutes with NTMT (Table V). Incubate in NTMT containing 4.5 μl NBT (75 mg/ml in 70% dimethylformamide) and 3.5 μl BCIP (50 mg/ml in 100% dimethylformamide) per 1.5 ml, rocking, protected from light, at room temperature. The reaction may take anything between 30 minutes and several days at room temperature. It develops faster at 37°C, but this can speed up the reaction too much and greatly increase background. If color has not quite developed after about 3–4 hours, it is advisable to leave the vials at 4°C room overnight to slow down the process and to take them out to room temperature again the next day if necessary. If color has still not developed after the whole of the next day at room temperature, there is no need to leave them at 4°C again the next night. After color has developed as desired, stop by washing three times for 10 minutes in PBS. The reduction in pH from these washes intensifies the blue color.

Table V NTMT

5 *M* NaCl:	1 ml
2 *M* Tris-HCl, pH 9.5:	2.5 ml
2 *M* $MgCl_2$:	1.25 ml
Tween-20:	500 μl
H_2O:	44.75 ml
Total:	50 ml

B. Detection of Alkaline Phosphatase with Fast Red (Red Precipitate and Fluorescence)

Fast Red TR produces a red precipitate that also fluoresces intensely when viewed with rhodamine optics (green excitation, red emission). Fluorescent detection appears to give higher sensitivity, and if this is to be used, less intense red staining is needed. There are several commercial sources of fast red, which have different properties. Vector Labs supply a kit called Vector Red that contains three solutions. Two drops of each solution are added to 5 ml of Tris-HCl buffer at pH 8.2 to make the staining mixture. Boehringer-Mannheim supply fast red tablets—one tablet is dissolved in Tris-HCl (pH 8.2) by vortexing, and then filtered through Whatman No. 2 paper before use. Sigma supply Sigma *Fast* tablets for the substrate as well as the buffer. With all these sources of fast red, the background can be quite yellow-orange with chick embryos, particularly in the yolky cells of the extraembryonic regions. In our experience, the background problem is worst with the Boehringer-Mannheim tablets. The Vector kit gives the lowest backgrounds, provided that the kit is fairly new. Older kits ($\geq$ 1 year after purchase) reduce the signal and increase background considerably. It is best to use the fast red substrate to detect only the most abundant transcripts (mRNAs encoding transcription factors are usually good for this).

First remove the TBST from the embryos. Then rinse twice for 10 minutes in Tris-HCl (pH 8.2). Finally, add the substrate as specified by the manufacturer. With the Vector kit, color should develop within 0.5–3 hours. With the Boehringer tablets, it may take several days. After staining, stop the reaction by washing several times with PBS. The reduction in pH slightly improves the signal-to-background relationship, but in our experience the background is too high with this chromogen (from whichever source) for it to be useful in two- or three-color detection protocols.

C. Detection of Alkaline Phosphatase with ELF (Yellow-Green Fluorescence)

Molecular Probes, Inc. has introduced the fluorescent substrate ELF for alkaline phosphatase, which they supply in kit form. Two kits are available: E-6604 allows direct detection of alkaline phosphatase activity, whereas E-6605 also contains streptavidin–alkaline phosphatase conjugate, allowing detection of biotinylated molecules. We find that avian embryos contain too much endogenous biotin and avidin and therefore backgrounds can be too high for use of the biotin–streptavidin system. However, the E-6604 kit can be used successfully and has also been used for mouse (Bueno *et al.*, 1996) and zebrafish (Jowett and Yan, 1996) embryos. Kit E-6604 contains wash buffer (10×), blocking buffer, developing buffer, substrate and two additives, Hoechst 33342 dye (as a counterstain if desired), mounting medium, and some plastic coverslips.

To stain whole mounts of embryos with this kit, begin by washing the embryos three times for 10 minutes in PBS containing 0.5% Triton X-100 (because Tween-20 appears to cause the ELF crystals to be too large; see Jowett and Yan, 1996), and then follow the manufacturer's instructions supplied with the kit. Just before developing, place one or more embryos in a cavity slide (a glass slide with a concave depression or a home-made chamber made by attaching several layers of electrical tape to a glass slide and cutting out a hollow using a scalpel) in the prereaction solution. Then remove this solution and replace with 200–500 μl substrate working solution. Do not cover with a coverslip because this prevents convection and mixing of the substrate components. During incubation (at room temperature), check under a fluorescence microscope with ultraviolet (UV) illumination (350–380 nm excitation; >500 nm emission; DAPI/Hoechst filters are suitable) until the desired level of signal is attained. Molecular Probes state that signal should develop within 10 minutes to 1 hour, and recommend not exceeding a 2-hour incubation. However, weak signals may take up to 4–6 hours to develop. If no signal is visible after 2 hours at room temperature, it appears to be advantageous to place the embryos in a humid chamber at 37°C and check occasionally under fluorescence (it is best not to check too frequently because continuous illumination may lead to photobleaching of the fluorochrome). After the desired incubation, stop the reaction by washing in PBS containing 25 m*M* EDTA and 0.05% Triton X-100 (pH 7.2). Then postfix the embryos in 2% paraformaldehyde in PBS for 20 minutes and mount them with the materials supplied with the kit. Always perform ELF detection as the last step in a series.

D. Detection of Alkaline Phosphatase with Other Chromogens (Magenta, Pink, Green, and Brick-Red Precipitates)

An analog of BCIP, 5-bromo-6-chloro-3-indolylphosphate (B8409 from Molecular Probes or MagentaPhos from Biosynth AG; Avivi *et al.*, 1994) produces a magenta–mauve precipitate when reacted with alkaline phosphatase. The resulting color can be fairly similar to that produced by NBT–BCIP, so it is not ideal for two-probe detection where the expression patterns overlap. It is also less sensitive than most of the other methods described here, so it can be used only to detect very abundant molecules. However, we have found that using tetrazolium red (Sigma) as a coprecipitant together with MagentaPhos makes the color less violet and more red and slightly improves the sensitivity of the reaction. To use, simply replace MagentaPhos for BCIP and tetrazolium red for NBT in the protocol given earlier for the blue (NBT–BCIP) substrate.

A pink precipitate can be obtained using 6-chloro-3-indolyl phosphate (C8413 from Molecular Probes or SalmonPhos from Biosynth AG; Avivi *et al.*, 1994). It is possible that tetrazolium red can enhance this substrate by acting as a coprecipitant, but we have not experimented with this.

Another variation is to omit the NBT in the NBT–BCIP reaction, which gives a more greenish reaction product, but there is significant loss of sensitivity. This can therefore be used only to detect very strongly expressed molecules, but its color provides better contrast than NBT–BCIP when using either MagentaPhos or fast red to detect the other probe. Finally, INT [2-(4-iodophenyl)-3-(4-nitrophenyl)-5-phenyltetrazolium chloride; I6496 from Molecular Probes] can be used instead of NBT together with BCIP to give a brick-red precipitate.

E. Detection of Horseradish Peroxidase with Diaminobenzidine–H_2O_2 (Brown Precipitate)

We have never succeeded at detecting anti-digoxigenin antibodies coupled to horseradish peroxidase in whole mounts, for reasons that remain mysterious. We therefore reserve this reaction for detection of embryonic antigens with antibodies. It is important to remember that it is not usually possible to combine *in situ* hybridization with whole-mount detection of embryonic antigens if the latter are integral membrane proteins or anchored to the cell membrane, because the strong detergent washes and proteinase treatments used for detection of mRNAs lead to loss of these embryonic antigens. However, most intracellular antigens survive and retain antigenicity after the *in situ* procedure. When combining *in situ* hybridization with immunocytochemistry for embryonic antigens, the hybridization must be performed first because antibody solutions usually contain RNases that degrade single-stranded (but not hybridized, double-stranded) RNA. However, after hybridization, it is possible to detect embryonic antigens and hapten-labeled probes in any order. It appears advantageous to detect embryonic antigens immediately after hybridization and before detecting the haptens, for the following reasons: (1) embryonic antigens are probably more labile and may degrade after several enzyme reactions and fixations, and can be affected by the acid glycine treatments used to inhibit alkaline phosphatase; and (2) the diaminobenzidine (DAB) precipitate, once developed, is completely stable and is not removed by solvents or detergents, unlike many of the precipitates generated by alkaline phosphatase reactions.

To stain embryonic antigens in whole mounts in combination with *in situ* hybridization, take the embryos up to the post-hybridization washes and block them in blocking solution as described previously. Then place them in an appropriate dilution of the primary (anti-embryonic antigen of choice) antibody in blocking buffer at 4°C, rocking, for 3–4 days (see Stern, 1993). After this, rinse three times and wash three to four times for 1 hour in TBST (Table III) and place in appropriate dilution of the correct secondary antibody (e.g., anti-mouse IgG, coupled to peroxidase) overnight at 4°C, rocking. It is generally advantageous to preabsorb this secondary antibody against chick embryo acetone powder as described in Table IV. The next day, rinse three times and wash three to four times for 1 hour in TBST and then place embryos into a measured volume of 500 μg/ml DAB* in Tris-HCl (0.1 *M*, pH 7.4) for 1 hour in the dark. Then add H_2O_2 to a final dilution of 1:10,000 from a 100 VOL (30%) solution (first make a 1:100 dilution of the peroxide in Tris buffer, and then dilute this 1:100 in the DAB solution containing the embryos). The brown color should begin to develop within 5–10 minutes and the reaction should be complete in less than an hour. At the end of this, rinse several times with distilled water or TBST and postfix for 20 minutes in 4% paraformaldehyde in PBS.

Antibody staining of cell surface or other labile antigens is possible before the *in situ* hybridization procedure, provided that RNase activity is minimized. One way to do this is to include 1 *M* LiCl in the initial fixation, blocking, antibody, and washing solutions before staining with DAB–H_2O_2. We are just starting to experiment with this with some encouraging results, but further development is needed.

F. Combining the Colors

When performing detection of several molecules, the combination of colors chosen is very important. One criterion for choosing the colors is that they should contrast with each other so that they are easily distinguished. Another criterion, particularly when it is important to distinguish regions of expression that overlap with one another, is that the colors should not obscure one another. Clearly, these two criteria are partly contradictory. The best solution is to use fluorescent detection with fluorochromes emitting at two different wavelengths (see Jowett and Yan, 1996), because it is then easy to examine each of the two signals separately as well as to photograph them together.

In our experience, the blue product produced by the NBT–BCIP reaction is not sufficiently different from the magenta product generated by MagentaPhos to allow regions of overlap to be resolved, but this combination is good when expression is in adjacent regions (Fig. 2A,C). The green product produced by BCIP alone is sufficiently different from MagentaPhos when the latter is used with tetrazolium red (see earlier), but both substrates lack the sensitivity to detect weakly expressed molecules. NBT–BCIP is sufficiently different from fast red to allow the simultaneous use of both of these substrates in ordinary light microscopy (however, the backgrounds produced by fast red can be quite orange-yellow and distracting; see Fig. 2B). At present, the best choice for detecting regions of overlap appears to be either a combination of MagentaPhos–tetrazolium red for the first probe with NBT–BCIP for the second probe for the light microscope (Fig. 2A,C), or a combination of fast red as a fluorescent substrate with either

*Unreacted DAB is a potent carcinogen, but can be inhibited by treatment with bleach (sodium hypochlorite), which causes it to precipitate. Wear gloves while handling DAB and place any solutions or objects that have come into contact with it in a 1:50 dilution of household bleach in tap water to inactivate.

NBT–BCIP (viewed by conventional light microscopy), if the second signal is weak, or ELF (viewed by fluorescence microscopy with UV excitation), provided that the signal is strong.

VII. Timetable for Multiple Color Detection

Table VI gives a timetable for performing *in situ* hybridization with two probes (using digoxigenin- and fluorescein-labeled riboprobes and alkaline phosphatase enzyme reactions with fast red and ELF fluorescence) and immunocytochemical detection of one intracellular antigen (using horseradish peroxidase) in the same embryos, as a guide to designing other experiments of this type. This table is only a guide, and it should be remembered that when using chromogenic reactions for

Table VI Timetable for Simultaneous Detection of Multiple Molecules

Day 1:	Linearize DNA for both probes (4–6 hr) and check on agarose gel
	Phenol–chloroform extract and ethanol–precipitate overnight
Day 2:	Wash and redissolve pellet in water
	Transcription of FITC- and digoxigenin-labeled riboprobes (2.5 hr)
	DNAase digestion (15–30 min) and check on agarose gel
	LiCl–ethanol precipitation overnight
	Fix embryos and store at 4°C overnight
Day 3:	Transfer embryos to methanol and place at −20°C
	Wash and redissolve riboprobe pellets in water
	Repeat LiCl–ethanol precipitation (2 hr to overnight)
	Wash and redissolve pellet in water; check on gel
	Dilute probes in hybridization buffer and store at −20°C
Day 4:	Hydrate embryos, proteinase K digestion, postfix (2 hr)
	Prehybridization of both riboprobes simultaneously (4–6 hr)
	Add probes and leave overnight
Day 5:	Post-hybridization washes (4–6 hr); incubation in blocking buffer (2–3 hr)
	Place in first (anti-embryonic antigen) antibody (2–3 days at 4°C)
Day 8:	Wash off primary antibody (4–6 hr)
	Put in secondary (HRP-labeled) antibody overnight
Day 9:	Wash off secondary antibody (4–6 hr)
	DAB preincubation (1–3 hr); develop DAB reaction (1 hr)
	Wash (1 hr), postfix 20 min, and wash overnight
Day 10:	Wash 1–2 hr; incubation in blocking buffer (2–3 hr)
	Place in anti-FITC antibody (alkaline phosphatase-labeled) overnight at 4°C
Day 11:	Wash off antibody (4–6 hr); develop with fast red (1 hr)
	Wash (1–2 hr); inactivate alkaline phosphatase in glycine buffer
	Wash 1 hr, incubate at 70°C overnight
Day 12:	Postfix 20 min; wash 2–3 hr; place in blocking buffer (2–3 hr)
	Place in anti-digoxigenin antibody overnight
Day 13:	Wash off antibody (4–6 hr)
	Develop with ELF (1 hr)
	Wash (1–2 hr), postfix 20 min, and wash again.
	Mount and examine under bright-field (DAB) and fluorescence optics (probes).

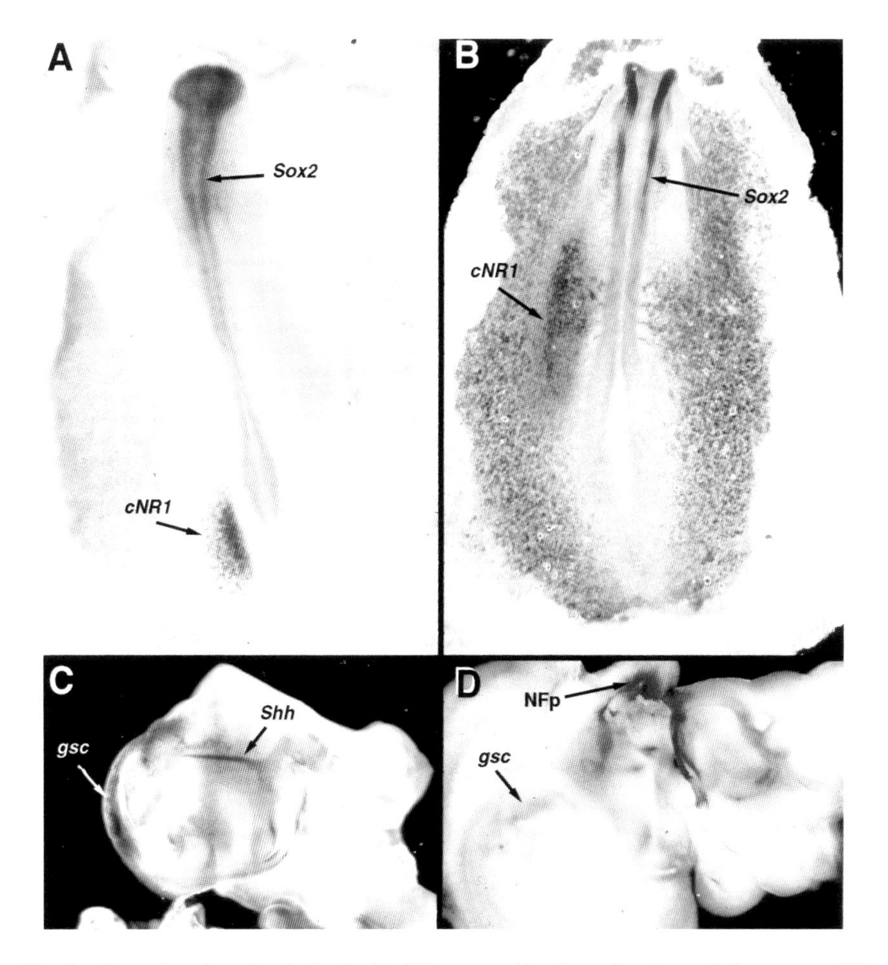

Fig. 2. Examples of results obtained with different combinations of probes and chromogens. (A) Stage 10 chick embryo hybridized with a probe for the transcription factor *Sox2* (expressed throughout the neural tube; Kamachi *et al.*, 1995; Uwanogho *et al.*, 1995; Streit *et al.*, 1997) that had been labeled with digoxigenin and detected with MagentaPhos and tetrazolium red, and with a probe for the secreted transforming growth factor-β–superfamily member *cNR1*(=*nodal*; expressed in the left lateral plate; Levin *et al.*, 1995) that had been labeled with fluorescein and detected with NBT–BCIP. The two colors can be distinguished easily, particularly because the domains of expression do not overlap. Bright field optics. (B) Embryo at stage 8, hybridized with the same probes as in A, but this time *Sox2* was visualized with fast red and *cNR1* with NBT–BCIP. Note the disturbing yellowish background from the fast red, particularly in extraembryonic regions. Dark-field optics (which makes the problem more acute). (C) Here, an embryo at stage 5 was injected with a retroviral vector encoding the full-length *goosecoid* gene (Izpisúa-Belmonte *et al.*, 1993). After 3 days incubation, the embryo was fixed (about stage 23) and hybridized with a fluorescein-labeled probe for *goosecoid* (detected with NBT–BCIP) and with a digoxigenin-labeled probe for *Sonic hedgehog* (Riddle *et al.*, 1994), detected with MagentaPhos and tetrazolium red. Dark-field optics. (D) Embryo overexpressing *goosecoid* as for C, hybridized with a *goosecoid* probe labeled with digoxigenin (detected with NBT–BCIP) and then processed by immunoperoxidase (with DAB/H_2O_2) using an antibody against a neurofilament-associated protein (NFp) (antibody 3A10; Yamada *et al.*, 1991).

the light microscope (e.g., NBT–BCIP, BCIP alone, MagentaPhos), the reactions may take several hours or even several days to develop. The timetable shown here uses horseradish peroxidase for the embryonic antigen, but this can be replaced by a secondary antibody coupled to Cascade Blue (Molecular Probes, Inc.), which gives a bright blue signal that is clearly separated from both the ELF and fast red emission wavelengths, so that three-color fluorescence detection can be done. In this case, do not use the Hoechst counterstain provided with the ELF kit because this is not clearly separable from Cascade Blue.

VIII. Photography

Whole mounts that have been stained with chromogens suitable for light microscopy often benefit from being photographed in a dissecting microscope fitted with dark-field optics (Fig. 2B–D). If such a microscope is not available, the embryos can be mounted in a cavity slide under a coverslip, placed over a matte black surface (such as a piece of black paper) and illuminated obliquely with a fiberoptics source (taking care to avoid reflections from the coverslip). Other specimens benefit from being photographed in a conventional compound microscope, either with normal transmitted light (Köhler illumination; see Stern, 1993c) or with Nomarski differential interference contrast optics. It is a good idea to experiment with different microscopes and illumination techniques for each combination of specimens, probes, and chromogens. In our experience, the best film to use is Fuji 64T (tungsten balanced), but it is important to adjust the illumination so that the color temperature of the light source matches the sensitivity of the film to avoid excessively bluish, greenish, or orange backgrounds (Stern, 1993c).

Specimens stained using fluorescent chromogens, or with a combination of fluorescent and visible light substrates, are best photographed on more sensitive film (e.g., Fuji Provia 1600) by taking multiple exposures onto the same frame. These may combine different fluorescence excitation filter sets for each fluorescent chromogen and transmitted light. However, it is important to take into account that in older microscopes, which do not have infinity-corrected optics, the path length of the light through different optical filter sets may differ, which affects the alignment of the images. A possible solution for this in some microscopes is to photograph the bright-field image through a fairly colorless dichroic mirror and filter sets (Stern, 1993c).

IX. Histological Sectioning of Whole Mounts

Of the chromogens used for light microscopy, only one (DAB) is completely resistant to organic solvents and can therefore tolerate dehydration, clearing, wax embedding, and sectioning by conventional methods. Most of the other chromogens (NBT–BCIP, BCIP alone, MagentaPhos, and fast red) are somewhat sensitive to organic solvents and therefore benefit from an accelerated dehydration and clearing process before embedding. After postfixing the embryos, incubate for 5 minutes in absolute methanol (this intensifies some of these chromogens) and 10 minutes in propan-2-ol (2-isopropanol), followed by clearing in absolute tetrahydronaphthalene for 30 minutes. Then infiltrate in 1:1 tetrahydronaphthalene:paraffin wax for 30 minutes at 60°C and follow with three to four changes (30 minutes) of pure wax at 60°C before placing in a mold to set. After sectioning, the sections can be dewaxed in xylene or Histoclear as normal and mounted in Canada Balsam, Permount, or DePeX.

By contrast, some other substrates (notably ELF and some other substrates marketed by Vector Labs) are very soluble in organic solvents and therefore cannot be wax embedded and sectioned. There are two possible solutions to this problem: (1) embed the stained embryos in gelatin and cut frozen sections in a cryostat or vibratome (see Stern, 1993a), and mount these in an aqueous medium; and (2) perform all staining with other chromogens first, wax embed and section as described in the previous paragraph, and then stain the sections with the ELF kit as described by Bueno *et al.* (1996). The first of these solutions has the advantage that embryos can first be photographed as whole mounts and then sectioned for further analysis.

X. Designing Appropriate Controls

With all *in situ* detection methods, it is important to be certain that the methods used are specific and sensitive enough to give a true image of the distribution of the molecules being studied. This is particularly important and difficult in the case of simultaneous detection of multiple molecules. Sense probes are often used as controls for *in situ* hybridization with riboprobes, but the nucleotide sequence of these is so different from that of the corresponding antisense probe that they are not really adequate controls. For example, sense probes do not allow the investigator to determine whether the sense probe is revealing overlapping expression between different members of a family of genes. Some workers check the specificity of the probe in Northern blots, but again this is not ideal because of the different hybridization and other conditions used *in situ* and in the blots. When studying the distribution of a new molecule in the embryo, it is best to start with a single probe and a well characterized and sensitive detection method (NBT–BCIP is probably the best). Ideally, two or more different antisense riboprobes should be synthesized, covering different portions of the mRNA, and these detected in separate batches of embryos to check that they give an identical distribution. If they do not, this could indicate cross-reaction with another gene product or the possibility of alternatively spliced transcripts (which should be visible in Northern blots).

When performing *in situ* detection of several molecules simultaneously, it is important to check that each detection method gives the same result as when the same molecules are detected singly. In particular, persistent alkaline phosphatase activity from detection of the first molecule could account for a false appearance of regions of overlap between two gene products (see Section XI). To avoid this, it is valuable to do the following: (1) in some embryos, omit the antibody against the second probe–this should not produce the second color; and (2) perform reciprocal experiments with all combinations of colors (e.g., fast red and ELF, to detect an FITC- and a DIG-labeled probe, respectively, and then vice-versa) and labeling haptens (fluorescein, DIG, biotin) (see Jowett and Yan, 1996, for more details).

XI. Trouble Shooting

Several problems may arise from the protocols given previously. The following sections give some guidelines on how to diagnose and solve those most commonly encountered.

A. Embryos Disintegrated or Folded

The former is most likely to be due to rough pipetting when changing solutions. Avoid pipetting solutions directly onto the embryos, do not use a Gibson pipetman but rather a Pasteur pipette with a good rubber teat, and avoid sucking up the embryos into the pipette. If this fails, reduce the proteinase K treatment time but do not lengthen the postfixation after this step because this leads to some reduction of signal.

Folding of very young embryos (2 days or less) occurs if they are not initially fixed flat. Follow the instructions given previously and in Stern (1993a) for fixing embryos. If this is not the problem, it is most likely that they have been allowed to dry out too much between removing a solution from the vial and filling the vial again with the next solution. Change the solutions one vial at a time.

B. Background Problems

It is often difficult to diagnose these accurately because there are many possible causes. Some of the most common are listed here.

1. Unincorporated Hapten in the Riboprobe

The most common source of background signals in *in situ* hybridization experiments is the presence of unincorporated hapten (fluorescein, digoxigenin, or biotin) in the probe solution. These haptens appear to be "sticky" and can therefore cause significant background problems. Several ways to reduce the amount of free hapten have already been discussed (see Section II), including the use of spin columns or reprecipitation in LiCl–EtOH to clean the riboprobe, and reuse of the probe solution several times. Unincorporated fluorescein is particularly troublesome, but other haptens can also cause background.

2. "Trapping" within Cavities

An intensely colored precipitate can often develop within organs that contain internal cavities (e.g., the brain vesicles in 2 to 4-day embryos, heart, eye, gut). This seems to be a particularly serious problem with the NBT–BCIP substrate. I am unaware of the causes of this, but we have found that it can be alleviated substantially by perforating these cavities many times with a fine pin or a miniature knife during fixation of the embryo. This appears to improve the exchange of solutions during the washing steps and virtually eliminates this problem. If puncturing is not enough, the dorsal midline of the neural tube may be opened with a microscalpel in two or three places.

3. Cross-Reactivity with Other mRNAs

This is also a common problem, particularly when detecting a member of a large family of genes, and also when using short ($\leq$ 300 nt) riboprobes where the stringency of hybridization has to be reduced (see earlier). It is difficult to control for this problem in every case, but it is recommended to determine the distribution of specific gene products by comparing results obtained with at least two different riboprobes derived from nonoverlapping regions of the same gene (see Section X). Use of the 3′ untranslated end of the cDNA is often a good way to produce very specific probes (even ones that do not cross-react between closely related avian species, which is useful for analysing chick–quail chimeras; see Izpisúa-Belmonte *et al.*, 1993), but one has to be aware that there may be several alternatively spliced forms of the mRNA affecting this region, with different expression patterns.

4. Nonspecific Binding of Antibodies

The commercially available anti-fluorescein and anti-digoxigenin antibodies purchased from Boehringer-Mannheim give very low levels of nonspecific binding to avian tissues provided that they have been preabsorbed against chick embryo powder as described in Table IV, and that the embryos have been blocked efficiently before addition of antibody. If nonspecific binding of the antibody remains a problem despite it having been preabsorbed, the cause should be determined empirically by omitting one step of the protocol at a time.

5. Endogenous Peroxidase or Phosphatase Activity

The protocol given in this chapter has been successful in our hands in eliminating all endogenous phosphatase and peroxidase activity that may interfere with the detection procedures. The bleaching step in 6% H_2O_2 after fixation of older embryos is designed to eliminate residual peroxidase, particularly in erythrocytes, as well as reducing eye pigmentation, which may obscure signals in this region. The proteinase K treatment and the high temperature of hybridization both contribute to reduce or eliminate phosphatases. If endogenous phosphatase remains a problem, the situation may improve by incubating the embryos in a 1-m*M* solution of levamisol for 1–2 hours before incubation in each alkaline phosphatase-coupled antibody.

C. No Signal or Low Sensitivity

Lack of signal, despite knowledge by the investigator that a particular gene is expressed at a given stage of development, probably indicates contamination with RNases (provided that other steps in the protocol have been carried out correctly and that the probe is good). We have found that the initial fixation step is important, but do not know the reasons for this. If no signal is detected or if only weak reactions are obtained, it is useful to start trouble shooting by reducing the duration of the initial fixation of the embryos to just 30 minutes. In our experience, contamination with RNases during subsequent handling of the embryos is not a problem. If the steps follow one another swiftly, RNase-free solutions are needed only in the hybridization solution itself, and gloves need be worn only between the proteinase K digestion and the end of the post-hybridization washes (up to the first TBST wash). Furthermore, the vials containing the embryos need not be RNase free because the proteinase K digestion should remove these from the inner surfaces of the vial.

When detecting mRNAs expressed at low levels, the most sensitive detection techniques should be used (e.g., NBT–BCIP on digoxigenin-labeled probes), and detection of this mRNA should be carried out first. It is important to remember that signals will be stronger if the riboprobe is longer (and therefore contains more molecules of hapten per riboprobe molecule), so it is useful to design longer riboprobes for the more weakly expressed molecule.

In the case of embryonic antigens, the use of a polyclonal antiserum for weakly expressed antigens is preferable to a monoclonal antibody. Polyclonal secondary antibodies recognizing several regions of the primary immunoglobulin can also be advantageous.

D. Persistent Alkaline Phosphatase Activity from the First Antibody

This can be an important problem when using several alkaline phosphatase-labeled antibodies directed against different haptens, and detected sequentially. No method of inactivation is foolproof. Those described to date include: (1) heating at 70°C overnight (this intensifies the blue color of NBT–BCIP and the magenta of Magenta Phos, but can remove fast red and ELF reaction products); (2) fixation in 4% paraformaldehyde (not very effective if used alone); (3) incubation in 1m*M* levamisol (again not very effective if used alone); (4) two 15-minute washes in absolute methanol (this will intensify NBT–BCIP and Magenta Phos as well as fast red to some extent, but will remove ELF); and (5) incubation twice for 10 minutes in 100 m*M* glycine, pH 2.25 at room temperature.

In our hands, the acid glycine incubation, followed by incubation overnight at 70°C and then by fixation in 4% paraformaldehyde (as well as a short wash in methanol if alkaline phosphatase has been revealed by NBT–BCIP) works best. However, if the order of these steps is altered, the procedure becomes much less effective. Occasionally, we have found that the acid glycine breaks down double-stranded RNA with resulting loss of the labeling hapten. We do not yet know the reasons for the variability between experiments.

Acknowledgments

I am most grateful to Drs. David Ish-Horowicz, Juan Carlos Izpisúa-Belmonte, Randy Johnson, Chris Kintner, Hermann Rohrer, Jonathan Slack, and Cliff Tabin for numerous tips and valuable advice over many years. Our research involving these methods is currently funded by the National Institutes of Health, the Muscular Dystrophy Association, and the Human Frontier Science Program.

References

Avivi, C., Rosen, O., and Goldstein, R. S. (1994). New chromogens for alkaline phosphatase histochemistry: Salmon and magenta phosphate are useful for single- and double-label immunohistochemistry. *J. Histochem. Cytochem.* **42,** 551–554.

Bueno, D., Skinner, J., Abud, H., and Heath, J. K. (1996). Double in situ hybridization on mouse embryos for detection of overlapping regions of gene expression. *Trends Genet.* **12,** 385–387.

Izpisúa-Belmonte, J. C., De Robertis, E. M., Storey, K. G., and Stern, C. D. (1993). The homeobox gene *goosecoid* and the origin of the organizer cells in the early chick blastoderm. *Cell* **74,** 645–659.

Jowett, T., and Yan, Y.-L. (1996). Double fluorescent in situ hybridization to zebrafish embryos. *Trends Genet.* **12,** 387–389.

Kamachi, Y., Sockanathan, S., Liu, Q., Breitman, M., Lovell-Badge, R., and Kondoh, H. (1995). Involvement of SOX proteins in lens-specific activation of crystallin genes. *EMBO J.* **14,** 3510–3519.

Levin, M., Johnson, R. L., Stern, C. D., Kuehn, M., and Tabin, C. J. (1995). A molecular pathway determining left–right asymmetry in chick embryogenesis. *Cell* **82,** 803–814.

New. (1955). A new technique for the cultivation of the chick embryo in vitro. *Journal of Embryology and Experimental Morphology* **3,** 326–331.

Riddle, R. D., Johnson, R. L., Laufer, E., and Tabin, C. (1993). Sonic hedgehog mediates the polarizing activity of the ZPA. *Cell* **75,** 1401–1416.

Sambrook, J., Fritsch, E. F., and Maniatis, T. (1989). "Molecular Cloning: A Laboratory Manual," 2nd ed. Cold Spring Harbor Laboratory Press, Cold Spring Harbor, NY.

Stern, C. D. (1993a). Avian embryos. In "Essential Developmental Biology: A Practical Approach" (C. D. Stern and P. W. H. Holland, eds.), pp. 45–54. IRL Press at Oxford University Press, Oxford.

Stern, C. D. (1993b). Immunocytochemistry of embryonic material. In "Essential Developmental Biology: A Practical Approach" (C. D. Stern and P. W. H. Holland, eds.), pp. 193–212. IRL Press at Oxford University Press, Oxford.

Stern, C. D. (1993c). Simple tips for photomicrography of embryos. In "Essential Developmental Biology: A Practical Approach" (C. D. Stern and P. W. H. Holland, eds.), pp. 67–78. IRL Press at Oxford University Press, Oxford.

Streit, A., Sockanathan, S., Perez, L., Rex, M., Scotting, P. J., Sharpe, P. T., Lovell-Badge, R., and Stern, C. D. (1997). Preventing the loss of competence for neural induction: Roles of HGF/SF, L5 and Sox-2. *Development* **124,** 1191–1202.

Uwanogho, D., Rex, M., Cartwright, E. J., Pearl, G., Healy, C., Scotting, P. J., and Sharpe, P. T. (1995). Embryonic expression of the chicken Sox2, Sox3 and Sox11 genes suggests an interactive role in neuronal development. *Mech. Dev.* **49,** 23–36.

Yamada, T., Placzek, M., Tanaka, H., Dodd, J., and Jessell, T. M. (1991). Control of cell pattern in the developing nervous-system: Polarizing activity of the floor plate and notochord. *Cell* **64,** 635–647.

14

Cloning of Genes from Single Neurons

Catherine Dulac
Department of Molecular and Cellular Biology
Harvard University
Cambridge, Massachusetts 02138

I. Introduction: The Cloning of Mammalian Pheromone Receptor Genes

Perception through sensory systems is essential to the knowledge of the external world and constitutes the basis for animal learning, memory, and behavior. Unlike other sensory modalities, olfaction in mammals employs two independent neuronal networks each involved in a distinct chemosensory perception. Olfactory cues defined as odorants elicit signals ultimately transmitted to higher cortical centers and generate various cognitive and emotional responses, measured thoughts, and behaviors. In contrast, pheromones lead to innate and stereotyped behaviors that are likely to result from a nonconscious perception of this class of odors. Odorant and pheromone olfactory perceptions are thought to be mediated by two anatomically and functionally distinct olfactory sensory organs, the main olfactory epithelium (MOE) and the vomeronasal organ (VNO).

The initial step in olfactory discrimination requires the specific interaction of odorant and pheromone molecules with their receptors expressed on the dendrites

Current Topics in Developmental Biology, Vol. 36

0070-2153/98 $25.00

of olfactory sensory neurons in the MOE or on the microvilli of VNO neurons. In the MOE, the repertoire of odorant receptor genes consists of about 1000 genes (Buck and Axel, 1991), each encoding a distinct seven-transmembrane-domain protein. Discrimination among odorants requires that the brain determine which of numerous receptors has been activated. Because individual sensory neurons in the MOE are likely to express only one of the thousand receptor genes, the problem of distinguishing which receptors have been activated reduces to a problem of distinguishing which neurons have been activated. Recent experiments demonstrate that neurons expressing a given receptor, and therefore responsive to a given odorant, although randomly distributed in domains of the epithelium, project their axons to one or a small number of discrete loci or glomeruli in the olfactory bulb (Vassar *et al.*, 1994; Ressler *et al.*, 1994; Mombaerts *et al.*, 1996). The positions of specific glomeruli are topographically fixed, and are conserved in the brains of all animals within a species. These data provide physical evidence that the olfactory bulb defines a two-dimensional map that identifies which of the numerous receptors have been activated in the sensory epithelium. Such a model is in accord with previous experiments demonstrating that different odors elicit defined patterns of glomerular activity in the olfactory bulb. Thus, the quality of an olfactory stimulus would therefore be encoded by the specific combination of glomeruli activated by a given odorant.

Analysis of the patterns of expression of receptor genes in the main olfactory system has provided significant insight into mechanisms for the diversity and specificity of odor recognition in mammals. The vomeronasal system shares anatomic and physiologic features with the main olfactory system, suggesting that the isolation of the genes encoding the pheromone receptors from VNO neurons might similarly provide insight into the chemical nature of the pheromones themselves, the logic of olfactory coding in the VNO, and the way in which perception of this class of odors leads to innate behaviors.

Our initial efforts to identify the genes encoding the pheromone receptors were based on the assumption that the main olfactory epithelium and the vomeronasal organ might share a common evolutionary origin such that DNA sequence homology may exist between the two receptor families. However, *in situ* hybridization experiments with olfactory receptor probes showed very rare positive cells that could not account for all VNO neurons, suggesting that MOE odorant receptors and VNO pheromone receptors might be different (Dulac and Axel, 1995). Moreover, low-stringency hybridization of MOE receptor probes to rat vomeronasal cDNA libraries, as well as polymerase chain reactions (PCR) using conserved motifs from both the family of odorant receptor genes and from the superfamily of known seven-transmembrane-domain receptors were consistently unsuccessful. In addition, the components of the olfactory signal transduction cascade in the main olfactory epithelium (the olfactory-specific G-protein,

G_{olf}, the olfactory-specific adenylate cyclase, and one subunit of the cyclic nucleotide-responsive ion channel) were not detectable in rat and mouse VNO neurons (Dulac and Axel, 1995; Berghard *et al.*, 1996). Furthermore, in contrast to results obtained with olfactory sensory neurons from the MOE, patch-clamp recordings of rodent VNO neurons show no response to cyclic nucleotides (Liman and Corey, 1996). These observations suggested that the pheromone receptors and the signal transduction pathways they activate might have evolved independently and that the VNO and the MOE might represent two extremely divergent olfactory systems. Therefore, to isolate mammalian genes encoding pheromone receptors, we had to develop a cloning procedure that made no assumptions concerning the structural class of the receptor molecules (Dulac and Axel, 1995).

We assumed that the expression of the family of pheromone receptors would be restricted to the vomeronasal organ, and that, by analogy with the MOE, individual neurons within the VNO were likely to express different receptor genes. In the main olfactory epithelium, 0.1 to 1% of the mRNA in a given sensory cell encodes a given receptor. However, the thousand different receptors are each expressed in different neurons such that the frequency of a specific receptor RNA will be diluted to 0.0001 to 0.001% of the message population, making it impossible to isolate by classic differential or subtractive screening. Differential screening of libraries from single neurons provided an experimental solution to the problem of detecting any given receptor mRNA in a heterogeneous population of neurons.

Reverse transcription PCR (RT-PCR) was therefore used to generate double-stranded cDNA, as well as cDNA libraries from individual vomeronasal sensory neurons. By analogy with the MOE, we expected that the frequency of a specific receptor cDNA in libraries from single neurons would be 0.1 to 1%, thus permitting the isolation of pheromone receptor genes by simple differential screening.

The cloning strategy used to isolate this gene family, differential screening of a cDNA library constructed from a single neuron, may be more broadly applicable to the analysis of the specific gene expression in diverse populations of cells. In the nervous system, for example, functionally distinct neurons each expressing different genes, and each projecting to different targets, are often interspersed. It has therefore been difficult to isolate RNA species unique to functionally distinct subsets of neurons within a heterogeneous cell population. The ability to generate cDNA libraries from individual cells in a diverse population of neurons may permit the identification of that subset of genes that afford a cell a unique identity.

Three steps lead to differential cloning using cDNA libraries prepared from individual neurons (Fig. 1): isolation of selected individual neurons from a dissociated neuroepithelium, synthesis and amplification of single-cell cDNA, and construction and screening of single-cell cDNA libraries. The strategies and

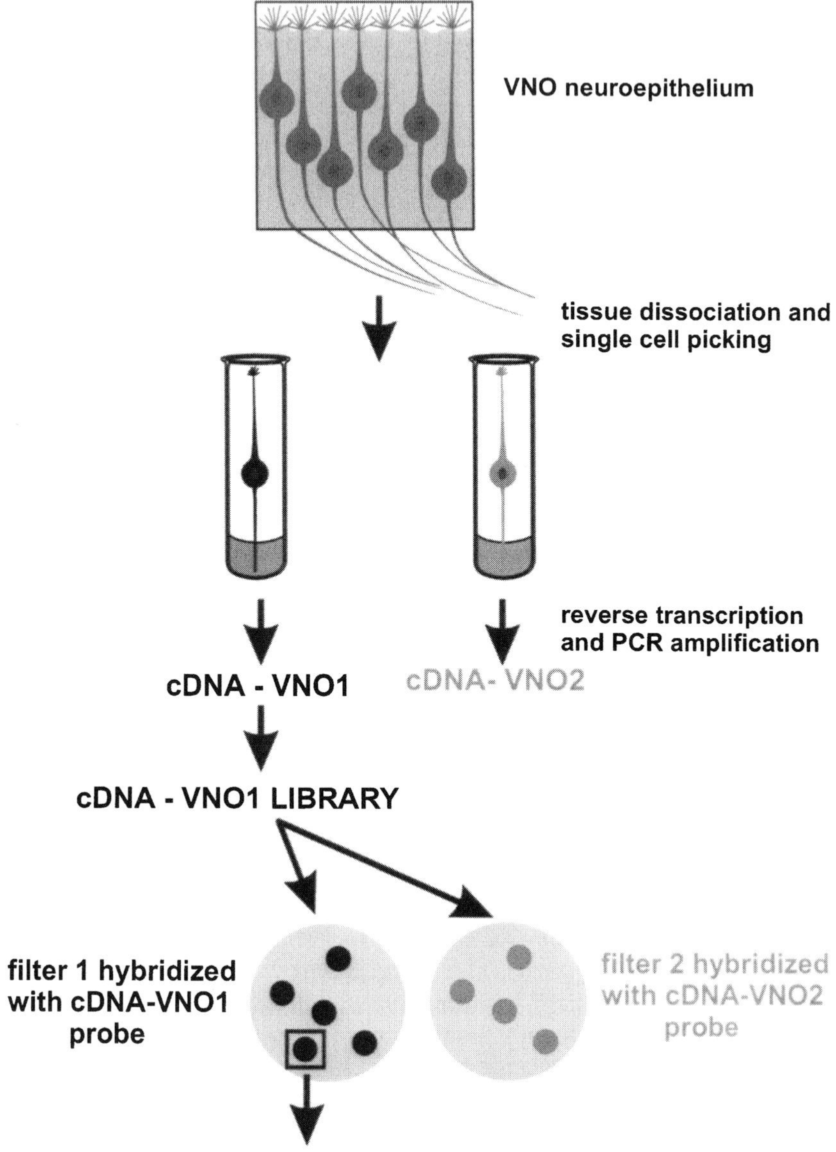
CLONING OF GENES FROM SINGLE NEURONS
VNO neuroepithelium
tissue dissociation and
single cell picking
reverse transcription
and PCR amplification
cDNA - VNO1
cDNA- VNO2
cDNA - VNO1 LIBRARY
filter 1 hybridized
with cDNA-VNO1
probe
filter 2 hybridized
with cDNA-VNO2
probe
Putative pheromone receptor of VNO1 neuron

Fig.1.

protocols we used to identify differentially expressed genes in the olfactory system are detailed and discussed in the next sections.

II. Tissue Dissociation and Isolation of Single Neurons

A. Preliminary Controls and Precautions

Preliminary experiments should be designed to assess the ability to identify and isolate the cell of interest. It is difficult and rather expensive to synthesize more than 30 to 40 single-cell cDNAs per experiment; moreover, to confirm the specificity of clones identified in the differential screen, it is important to obtain several cell cDNAs from the same cell type. Therefore, and if possible, the cell of interest should not represent less than 10% of the cells that are picked and further processed. In the olfactory system, a very mild trypsin dissociation allows a single-cell suspension to be obtained in which neurons still bear their axon and dendrites and can therefore be picked based on their morphology. Most of the cells we picked expressed the olfactory marker protein, a marker of mature olfactory neurons. In most cases, however, particularly in neural or embryonic tissues, different cell types of similar morphology are intermingled. The cell of interest has therefore to be defined by the expression of one or more molecular markers.

Therefore, and before starting any single-cell cDNA synthesis, it is crucial to spend some time confirming the representation of markers on dissociated cells. There are indeed huge discrepancies between reports in the literature indicating the presence of a given marker in a particular tissue and its actual expression by a significant number of individual cells. I have been aware of several groups questioning endlessly the quality of their single-cell cDNAs that were lacking an important transcript, and who finally discovered by a quick antibody staining or an *in situ* hybridization on dissociated cells that a very small percentage of cells were in fact positive. If this is indeed the case, one should try either to choose a more appropriate marker or to enrich the cell population into the cells of interest.

Fig. 1. Experimental strategy to clone pheromone receptor genes. After mild dissociation of a rat vomeronasal organ (VNO), isolated neurons were identified under the microscope, picked with a microcapillary, and directly seeded into PCR tubes. Subsequent steps of reverse transcription and PCR amplification allowed the synthesis of several micrograms of cDNA from each VNO cell. The cDNA prepared from the VNO neuron VNO1 was ligated into phage arms to construct a single cell cDNA library. Differential screening of the cDNA-VNO1 library with ^{32}P-labeled cDNA probe originating from VNO1 and VNO2 neurons led us to the isolation of a first transcript specifically expressed by a subpopulation of VNO neurons, and then of an entire family of novel genes encoding putative seven-transmembrane-domain receptors unrelated to the receptors expressed in the main olfactory epithelium. These genes are likely to encode mammalian pheromone receptors.

The dissection procedure can sometimes be improved, or living cells can be labeled with dyes or with antibodies recognizing surface antigens. An elegant approach allowed Amrein and Axel (1997) to isolate rare neurons from the fly brain using lines carrying the green fluorescence protein as a reporter gene.

Because the cell dissociation just precedes a 50-cycle PCR step, it is crucial to be free of any possible DNA contaminant: avoid the proximity of phages or recombinant bacteria. Use solutions from unopened bottles, and tips with filters; if possible, take a set of "pipetmen" that is not regularly used for molecular biology, or clean yours extensively, autoclave it, or leave it under ultraviolet overnight. Everything has to be RNase free as well.

B. Protocol

The protocol described here was used to dissociate brain and olfactory tissues in mouse and rat. By simply adapting the duration of the enzymatic treatment, we get consistently good results both with adult and embryonic neuroepithelia. The same procedure was successfully used by other colleagues in various species and with many developing tissues. It is important to use high-quality trypsin or dissociating enzymes.

The neuroepithelium is dissected under the microscope, placed in a 35-mm Petri dish, and rinsed several times in phosphate-buffered saline (PBS) without Ca^{2+} and Mg^{2+}. The tissue is then fragmented into many small fragments with fine forceps, microscalpels, or microscissors. The PBS is then removed with a "pipetman" or a Pasteur pipette and replaced by 2 ml of PBS without Ca^{2+} and Mg^{2+} containing 0.025% trypsin, 0.75 m*M* ethylenediamine tetraacetic acid (Low Trypsin–High EDTA solution from Specialty Media) prewarmed at 37°C. Tissue and trypsin are mixed very gently by pipetting up and down two or three times with a 2-ml plastic pipette.

The Petri dish containing the dissociating tissue is kept in a 37°C incubator for 15 to 30 minutes. After 15 minutes, pipette very gently two or three times as before and observe under an inverted microscope. The dissociation is stopped when cells at the periphery of the big clumps start to dissociate and some (but not too many) fully dissociated cells can be seen at the bottom of the dish. If the clumps are still very cohesive after 20 to 30 minutes, remove the trypsin with a pipette, again add 2 ml of prewarmed trypsin, and keep 10 more minutes at 37°C.

To stop the trypsinization, transfer with a pipette the 2 ml of trypsin and tissue into a 10-ml solution of prewarmed Dulbecco's modified Eagle's medium + 10% fetal calf serum. Do not triturate at this stage. Centrifuge 10 minutes at 2000 rpm, remove all supernatant, and add 5 ml of cold PBS without Ca^{2+} and Mg^{2+}. Triturate very gently by pipetting up and down four to five times with pipettes and pipetman tips of gradually smaller diameters: 2-ml plastic pipette, 1-ml plastic pipette, then 1 ml followed by a 200-μl-tip pipetman. Keep on ice.

Under the microscope, the cell suspension should still show a lot of clumps as well as isolated neurons retaining intact axonal and dendritic processes. Decant for 10 minutes to get rid of the clumps.

An appropriate dilution of the cell suspension is observed on a Leitz inverted microscope and neurons are identified by their round cell body and long axonal and dendritic processes. Cells have to be quite sparse; otherwise, additional cells are likely to be picked at the same time or stick to the outside of the pipette. Isolated neurons can be picked with a Leitz micromanipulator fitted with a pulled and beveled microcapillary, or directly with a mouth pipette connected to a pulled 25-μl microcapillary. Successful picking of individual cells requires only a few hours of training.

I suggest the use of a four-well Multidish (Nunc) with 500 μl of PBS in each, so the focus of the microscope does not have to be changed from one well to the other, and to transfer the candidate neuron from the well containing the cell suspension to the adjacent well containing no cell. Then rinse the microcapillary several times in a dish containing PBS, or even take a new one, repick the cell, and seed it in a PCR tube.

Single cells or groups of 10 to 20 cells are seeded in a volume of 0.2 to 0.5 μl into thin-walled PCR reaction tubes containing 4 μl of ice-cold lysis buffer (see later). These PCR tubes are transparent enough so the tip of the microcapillary can be seen reaching the solution. Spin immediately for 30 seconds to make sure the cell went to the bottom of the tube and does not stick to the wall. Keep the collected cells on ice. The whole procedure should not exceed a couple of hours.

III. cDNA Synthesis and Amplification

A. Summary and Concerns

Individual neurons are picked with a microcapillary and directly seeded in PCR reaction tubes containing cell lysis buffer. Lysis is subsequently performed at 65°C, and oligodT-primed first-strand cDNA synthesis is achieved with the addition of a mixture of reverse transcriptases at 37°C, then of reagents allowing the synthesis of a poly(A) tail in 5′ of the first-strand cDNA. The 5′ poly(A) and 3′ poly(T) tails allow PCR amplification to be performed using a unique primer containing a poly(T) sequence. This protocol, modified from Brady *et al.* (1990), allows more than 50 μg of PCR-amplified cDNA to be synthesized from individual neurons in a single tube (Dulac and Axel, 1995). The reverse transcription is performed in limiting conditions to generate cDNA uniform in size (between 500 bp and 1 kb), which are then likely to be equally amplified. In this manner, and despite the PCR step, the amplified cDNA maintains an accurate representation of the different cell RNAs. This cDNA synthesis can be done on single cells or groups of cells, as well as on very small amounts of RNA purified from several hundred cells by the RNAzol procedure.

It is then crucial to control both the quality of the cDNA generated and the nature of the cell picked. This can easily be done by Southern blot analysis with several cell-specific and ubiquitous, rare and abundant transcripts. This study determines the choice of those cells one wishes to make into cDNA probes and cDNA libraries.

Finally, I would suggest first trying this protocol with fibroblasts or a cell line and quickly checking the quality of the amplified cDNA with a tubulin probe to become familiar with the whole procedure and control the quality of all reagents.

B. Equipment and Reagents

37°C and 65°C waterbaths, thin-walled reaction tubes (Perkin Elmer), DNA Thermal cycler (Perkin Elmer)
NP40 (USB)
Prime RNAse inhibitor (3′5′ incorporated)
RNAguard (Pharmacia), RNAse, and DNAse free H_20 (Specialty Media)
deoxyadenosine triphosphate (dATP), deoxycytidine triphosphate (dCTP),
deoxyguanosine triphosphate (dGTP), deoxythymidine triphosphate (dTTP) 100 m*M* (Boehringer)
pdT12-18 or 25-30, 5 units, (Pharmacia)
Moloney muzine leukemia virus (MMLV) reverse transcriptase + buffer 5× (Gibco-BRL)
Avian myelo blastosis virus (AMV) reverse transcriptase (Gibco-BRL)
Terminal transferase (Boehringer)
Terminal transferase buffer 5× (BRL)
Amplitaq, 10× PCR buffer II and 25 m*M* $MgCl_2$ (Perkin Elmer)
Bovine serum albumin (BSA) molecular grade 20 mg/ml (Boehringer)
Triton X-100, RNAse and DNAse free (Sigma)
AL1 primer: ATT GGA TCC AGG CCG CTC TGG ACA AAA TAT GAA TTC (T)24 (0.1 μmole scale, Oligo etc.)
Mineral oil, Molecular biology grade (Sigma)

The reagents purchased from the companies indicated here give consistent good results. However, some batches of MMLV-RT seem not to work for single-cell cDNA synthesis, although they work fine for other, less sensitive applications. Therefore, for each new batch, it is necessary to compare the cDNA prepared from a few cells with AMV-RT alone, with MMLV-RT alone, and with AMV-RT combined with MMLV-RT as indicated in the protocol.

C. Protocol

During cell dissociation, thaw reagents on ice and prepare the cDNA lysis buffer. For 100 μl, mix on ice 20 μl of MMLV buffer 5×, 76 μl of H_2O, 0.5 μl of NP40,

1 μl of PrimeRNase inhibitor, 1 μl of RNAguard, and 2 μl of a freshly made, 1/24 dilution of the stock primer mix. The stock primer mix, kept aliquoted at −20°C, consists of 10 μl each of 100 m*M* dATP, dCTP, dGTP, and dTTP solutions (12.5 m*M* final); 10 μl of 50 OD/ml pd(T)12-18; and 30 μl H_2O.

Aliquot 4 μl of this buffer in thin-walled PCR reaction tubes (Perkin Elmer) for as many samples or cells you wish to collect. Do not forget a zero control tube with no cell in it; it is also useful to have a couple of tubes with clumps of 10 to 20 cells in them as positive controls. Keep on ice.

When all the single cells are collected in the PCR tubes, lyse the cells at 65°C for 1 minute, then keep the rack of tubes for 1 to 2 minutes at room temperature for the oligodT to anneal to the RNA. Put the PCR tubes back on ice and spin quickly at 4°C to remove the condensation. Then add 0.5 μl of a 1:1 (vol:vol) mix of AMV and MMLV-RT and incubate for 15 minutes at 37°C (no longer). Inactivate the enzymes for 10 minutes at 65°C, put the tubes back on ice, and spin 2 minutes at 4°C.

On ice, add 4.5 μl of 2× tailing buffer containing 10 U of terminal transferase. The stock 2× tailing buffer contains 800 μl of 5× BRL terminal transferase buffer, 30 μl of 100 m*M* dATP, and 1.17 ml H_2O. Incubate at 37°C for 15 minutes, then inactivate the enzyme 10 minutes at 65°C, put back on ice, and spin 2 minutes at 4°C.

Add to each tube 90 μl of ice-cold PCR mix (keep all reagents and PCR mix on ice to avoid primer dimers); 90 μl of PCR mix contains 10 μl of 10× PCR buffer II, 10 μl of 25 m*M* $MgCl_2$, 0.5 μl of 20 mg/ml BSA, 1 μl of each 100 m*M* deoxynucleotide triphosphate, 1 μl of 5% Triton, 5 μg of AL1 primer, H_2O qs 90 μl, 2 μl of AmpliTaq, and 2 drops of mineral oil.

On a Perkin Elmer DNA Thermal Cycler perform 25 cycles: 94°C for 1 minute, 42°C for 2 minutes, 72°C for 6 minutes, with a 10-second extension time at each cycle. When these 25 first cycles are finished, add 1 μl of AmpliTaq directly to each tube and perform 25 more cycles with same program as before but without the extension time at each cycle, but add a 30-minute extension at 72°C at the end of the 25th cycle. I strongly recommend starting this second PCR as soon as the first one is finished; I noticed that it gives higher yields. This implies that the cell dissociation is done in the morning and that the cDNA synthesis starts in the early afternoon.

Extract in phenol–chloroform and ethanol precipitate and freeze half of the sample at −80°C as a stock, so you do not have to thaw and freeze repeatedly the whole cDNA while analyzing it.

To check the quality of the cDNA obtained, run two agarose gels 1.5% with 5 μl of cell cDNA in each well. There should be a very intense smear of DNA (around 500 ng) from 0.4 to 1.2 kb. It is not unusual to find a similar result with the zero control (this probably results from some minor bacterial contaminants present in the enzyme solutions, but no specific probe should hybridize to that lane in further controls). Then transfer the DNA to 4 Hybond N+ membranes in two double-sandwich Southern blots and hybridize them with ubiquitous genes (e.g., tubulin and a riboprotein) to check the quality of the single-cell cDNA, and with gene probes specifically expressed in the cell of interest. The cDNAs generated are mostly shorter than 1 kb, so make sure the probes contain the 3′ untranslated region and do not forget that there are usually no cross-hybridizations between different animal species at the 3′ untranslated region, even between rat and mouse, even for very conserved genes like tubulin.

Signals for tubulin and riboprotein should be extremely intense and appear after less than 1 hour of exposure. Dividing cells might have reduced levels of these markers. It is worth checking the cDNAs with as many markers as you can think of, and start the difficult step of differential screening only with cDNA of the best quality obtained from well identified cells. Single-cell cDNA can be reamplified at will with 1 μl of the original cDNA into 100 μl of PCR buffer indicated earlier, amplifying for 30 cycles at 94°C for 1 minute, 42°C for 1 minute, and 72°C for 2 minutes. The quality of this reamplified cDNA should be indistinguishable from the original.

IV. Differential Screening of Single-Cell cDNA Libraries

A. Materials and Reagents

Hybond N+ membranes and 137-mm filters from Amersham (Hybond N+ appears to be the most sensitive membrane and gives very clean results when hybridized at 65°C with the suggested hybridization buffer)
65°C shaking waterback or hybridization oven
Phenol and chloroform
Low-melting-point agarose
PCR prep DNA purification resin (Promega)
Lambda ZapII arms EcoR1 cut and dephosphorylated (Stratagene)
Gigapack II gold extract (Stratagene)
High-concentration ligase, 2000 U / μl (NEB)
Push columns (Stratagene)
Hybridization buffer: 0.5*M* of PO_4, pH 7.3, 1% BSA, 4% sodium dodecyl sulfate (SDS) (for 1 liter: 342 ml of 1 *M* Na_2HPO_4, 158 ml of 1 *M* NaH_2PO_4, 10 g of BSA, 200 ml of 20% SDS, 300 ml ultrapure water)

B. Construction of the Libraries and Screening Protocol

Extract with phenol–chloroform and ethanol precipitate 10 μg of single-cell cDNA and resuspend into 25 μl of H_20. Digest 2 to 4 hours with EcoR1 and run the whole sample on a 2% low-melting agarose gel. Cut everything above 500 bp and purify with the PCR preps DNA purification resin from Promega. Ethanol-precipitate the DNA in the presence of glycogen and resuspend in 10 μl H_2O. Quantify the amount of DNA on an ethidium bromide–agarose plate by comparing with 0 to 200 ng/μl DNA standards.

To ligate into EcoR1-cut and -dephosphorylated λ ZapII arms, mix on ice 1 μl phage arms, 30 to 100 ng EcoR1-cut cell cDNA, 0.5 10× ligase buffer, H_2O qs 5 μl, and 0.3 μl ligase. Perform the ligation overnight at 12°C or 2 days at 4°C. Do not use more than 100 ng or multiple inserts will ligate in the same phage arms. Package 1 μl as recommended by the manufacturer and titer it (should be around 5.10^4 to 5.10^5 ufp/ml). Do not use the X-gal-IPTG system suggested by Stratagene because the inserts are too small and will appear blue.

To control the library, amplify the inserts from 10 to 20 plaques with the T3 and T7 primers; 90% of the phages should have inserts. Digest the largest PCR inserts with EcoR1 to confirm none gives a doublet. If everything looks fine, package the rest and amplify part of the obtained library. Check the representation of the library by screening 1000 to 5000 plaques with control genes. Tubulin, for example, should be present at 0.2 to 0.5%.

For the differential screening, plate the library overnight at very low density (maximum 1000 pfu/plate) to obtain large plaques (2-mm diameter). In the morning, leave the plates for 1 hour at 4°C, then take two duplicate filters. Prehybridize for at least 6 hours at 65°C in the hybridization buffer containing 10 μg/ml salmon sperm. Prepare the cell cDNA probes by reamplifying for 10 cycles 1 μl of the original cell cDNAs into 50 μl total reaction with the AL1 primer, in the absence of cold dCTP and with 100 μCi of newly received ^{32}P-dCTP. Purify the probes on Stratagene push columns. Hybridize for at least 16 hours with 1 to 5×10^7 cpm/ml in the hybridization buffer plus salmon sperm. Note that if the library is prepared from cell A, the first filter (with most of the DNA) has to hybridize with the cell B probe and the second with the cell A probe.

The next day, wash three times at 65°C in 0.5% SDS, 0.5× SSC. Expose for 3 days.

Compare very carefully the positive plaques in the two duplicates. Note that there are very strongly positive plaques (probably ribosomal or bacterial RNA contaminants) and plaques that appear gray to very light gray. The genes you are looking for are very likely to belong to this category (less than 0.5%) and should appear clearly after 3 days' exposure. Make sure cell A cDNA probe hybridizes to most of the plaques of cell A library, but not all of them! If you cannot find any negative plaques, the background is too high, you can try to expose for a shorter period or do the procedure again with less probe. If cells A and B are similar enough, less than 1% of the plaques should hybridize differentially. Pick these plaques. Because it is sometimes difficult to distinguish a very faint positive from the background, do not hesitate to pick too many phage plaques; the next stop is designed to eliminate quickly all the nonspecific cDNAs.

C. Analysis of the cDNA Clones

The level of sensitivity of hybridization on phage DNAs is very low. It is therefore necessary to confirm the specificity of the cDNA clones identified previously in a more sensitive assay. The inserts of all the phages picked in the screening are amplified with T3 and T7 primers. These amplified inserts are run on a 1.5% gel and transferred to nylon membranes. Each membrane is then hybridized with the cDNA probes prepared from cell A and cell B using the same conditions as for the library screening. Because of the high sensitivity of this test compared with the differential screen of phage plaques, most of the inserts should appear nonspecific after 6 hours to overnight exposure. At this point, it is useful to prepare four or six of these blots by double-sandwich Southern and to hybridize them with other cell cDNAs. They will confirm (or reject) the specificity found for the first two blots, and might very quickly give some interesting information on the nature of the cDNAs isolated.

When we screened the cDNA library from the neuron VNO1 with cDNA prepared from VNO1 and VNO2, about 2% of the cDNA clones screened showed specific hybridization with cDNA probes from neuron VNO1, but not with probes from neuron VNO2 (Fig. 1). The inserts from these cDNA clones were amplified by PCR, and the DNA products were hybridized on Southern blots with cDNA probes from VNO neuron 1, VNO neuron 2, or from an MOE sensory neuron (Fig. 2). Of 20 clones initially isolated from the VNO1 cDNA library, only two (clones indicated by stars in Fig. 2) appeared to be specific to neuron VNO1 in this more sensitive screen. These two clones represented independent isolates of an identical cDNA sequence present in the cDNA library of VNO1 at a frequency of 0.5%.

If some cDNA inserts display an interesting pattern as judged from the preceding Southern blots, the next step is to analyze their sequence and distribution after *in situ* hybridization. Because the clones obtained from the single-cell cDNA are usually short, the sequence is usually informative only when it matches to an already known sequence. Full-length clones are isolated by screening a cDNA library with larger inserts prepared from milligram amounts of tissue. *In situ* hybridizations were performed as described in Schaeren-Wiemers and Gerfin-Moser (1993) using cRNA probes labeled with digoxigenin. This protocol combines high sensitivity comparable to S35 *in situ* and very low background. The absence of 3′,5′ orientation of the cDNA inserts obtained from single-cell libraries makes it necessary to prepare cRNA probes from both T3 and

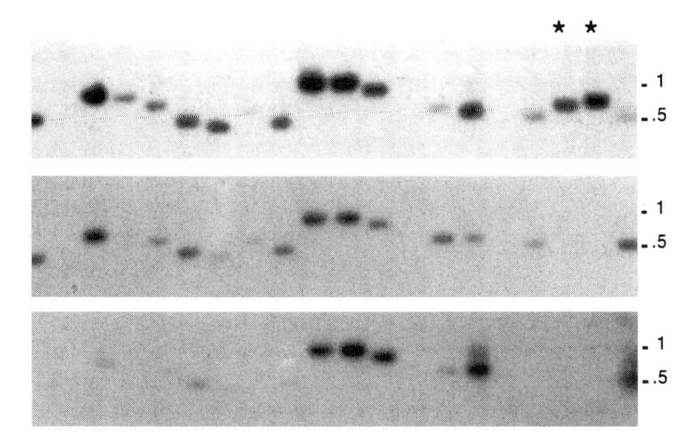

Fig. 2. Identification of cDNA clones specifically expressed in an individual vomeronasal organ (VNO) neuron. Twenty cDNA clones initially identified by differential screening of a cDNA library from a single VNO neuron were isolated. The inserts were amplified by PCR, electrophoresed on 1% agarose gels, and blotted to nylon filters. Blots were annealed with ^{32}P-labeled cDNA probe from VNO neuron 1 (higher panel), VNO neuron 2 (middle panel), or a neuron from the main olfactory epithelium (lower panel). Two cDNA clones (stars) anneal only with cDNA prepared from VNO neuron 1.

T7 orientations for each insert. Although this makes it necessary to synthesize twice as many RNA probes, it also provides an internal negative control.

V. Concluding Remarks

Differential screening of single-cell cDNA libraries prepared from VNO neurons indeed led us to the isolation of candidate pheromone receptors (Dulac and Axel, 1995). The identification of these genes should permit us now to explore the mechanisms by which pheromone signals are encoded and processed to generate innate and stereotyped fixed-action behaviors. Although several groups pioneered the technology to generate amplified cDNA from few cells (Belyavsky *et al.*, 1989; Brady *et al.*, 1990; Eberwine *et al.*, 1992; Froussard, 1992) this is, to my knowledge, the first example of successful cloning of new genes using single-cell cDNA and single-cell cDNA libraries.

The techniques and strategies presented here have broader implications for the study of heterogeneous cell populations, specifically in the olfactory system, and more generally for the study of the acquisition of cell identity in the brain or during embryonic development. Despite the obvious limit in sensitivity of differential screening, the absence of dilution of cell-specific transcripts when using single-cell cDNA and single-cell libraries allows the detection of very rare RNAs at the level of a whole tissue. In addition, the large amount of single-cell cDNA obtained and the possibility to reamplify it at will makes it very easy to adapt PCR-based subtraction protocols (see Bouillet *et al.*, 1995).

References

Amrein, H., and Axel, R. (1997). Genes expressed in neurons of adult male Drosophila. *Cell* **88,** 459–470.

Belyavsky, A., Vinogradova, T., and Rajewsky, R. (1989). PCR based cDNA library construction: General cDNA libraries at the level of a few cells. *Nucleic Acids Res.* **17,** 2919–2933.

Berghard, A., Buck, L. B., and Liman, E. R. (1996). Evidence for distinct signalling mechanisms in two mammalian olfactory sense organs. *Proc. Natl. Acad. Sci. U.S.A.* **93,** 2365–2369.

Bouillet, P., Oulad-Abdelghani, M., Vicaire, S., Garnier, J. M., Schuhbauer, B., Dolle, P., and Chambon, P. (1995). Efficient cloning of retinoic acid-responsive genes in P19 embryonal carcinoma cells and characterization of a novel mouse gene, Stra 1 (mouse LERK-2/Eplg2). *Dev. Biol.* **170,** 420–433.

Brady, G., Barbara, M., and Iscove, N. N. (1990). Representative *in vitro* cDNA amplification from individual hemopoietic cells and colonies. *Methods in Molecular and Cellular Biology* **2,** 17–25.

Buck, L., and Axel, R. (1991). A novel multigene family may encode odorant receptors: A molecular basis for odor recognition. *Cell* **65,** 175–187.

Dulac, C., and Axel, R. (1995). A novel family of genes encoding putative pheromone receptors in mammals. *Cell* **83,** 195–206.

Eberwine, J., Yeh, H., Miyarisho, K., Cao, Y., et al. (1992). Analysis of gene expression in single live neurons. *Proc. Natl. Acad. Sci. U.S.A.* **89,** 3010–3014.

Froussard, P. (1992). A random-PCR method to construct whole cDNA library from low amount of RNA. *Nucleic Acids Res.* **20,** 2900–2911.

Liman, E. R., and Corey, D. P. (1996). Electrophysiological characterization of chemosensory neurons from the mouse vomeronasal organ. *J. Neurosci.* **15,** 4625–4637.

Mombaerts, P., Wang, F., Dulac, C., Chao, S., Edmonson, J., Nemes, A., Mendelsohn, M., and Axel, R. (1996). Visualizing an olfactory sensory map. *Cell* **87,** 675–686.

Ressler, K. J., Sullivan, S., and Buck, L. B. (1994). Information coding in the olfactory system: Evidence for a stereotyped and highly organized epitope map in the olfactory bulb. *Cell* **79,** 1245–1255.

Schaeren-Wiemers, N., and Gerfin-Moser, A. (1993). A single protocol to detect transcripts of various types and expression levels in neural tissue and cultures cells: *In situ* hybridization using digoxigenin-labeled cRNA probes. *Histochemistry* **100,** 431–440.

Vassar, R., Chao, S. K., Sitcheran, R., Nunez, J. M., Vosshall, L. B., and Axel, R. (1994). Topographic organization of sensory projections to the olfactory bulb. *Cell* **79,** 981–991.

15

Methods for Detecting and Quantifying Apoptosis

Nicola J. McCarthy and Gerard I. Evan
Imperial Cancer Research Fund Laboratories
London WC2A 3PX, United Kingdom

I. Introduction

The importance of programed cell death (PCD) as a fundamental process that, together with cell proliferation and differentiation, shapes and crafts tissues and organisms is now well accepted. It is most obviously evident in tissues such as central nervous system and the immune system, tissues that self-assemble through the process of iterative matching that selects for productive neural pathways or competent immunocytes and requires disposal (by apoptosis) of unmatched or autoreactive clones. In such tissues, up to 90% of all cells born may undergo PCD. However, apoptosis is essential in the modeling and maintenance of almost all tissues in metazoans. Indeed, in vertebrates, there is substantial evidence supporting the notion that somatic cells require continuous survival signaling to forestall their spontaneous suicide, so providing an innate mechanism to remove excess or misplaced, infected, or neoplastic located cells.

In vertebrates, the word "apoptosis" is used to describe the characteristic phenotypic and biochemical changes that accompany PCD. During apoptosis, cells shrink and their surfaces bleb: cytoplasm becomes vacuolated and chromatin becomes condensed and marginalized within the nucleus. This is followed by degradation of nuclear DNA, first into large 50- to 300-kb fragments and ultimately into the oligonucleosomal debris that was originally a diagnostic hallmark of apoptosis. At the level of the individual cell, each apoptotic event is rapid: cells go from onset of apoptosis (characterized by membrane blebbing and cell

Current Topics in Developmental Biology, Vol. 36

0070-2153/98 $25.00

shrinkage) to fragmented apoptotic bodies in under an hour, and in somatic tissues the apoptotic remnants are cleared with equal rapidity. Apoptosis is, therefore, an ephemeral event whose extent is difficult to determine from "snapshots" of cell populations taken at various times.

An early preoccupation of those studying apoptosis was the need to distinguish apoptosis, an active process, from cellular necrosis, the passive rupture and dissolution of the cell through direct physical damage. By and large, however, cell necrosis seems to be rare and may be of relatively little physiologic significance. Thus, cell death arising from ischemia, radiation, cytotoxic agents, and most toxins is apoptotic rather than necrotic. However, if large numbers of cells undergo apoptosis at one time in one place, the remaining viable cells and professional phagocytes often are insufficient to remove the apoptotic cells. The resulting apoptotic debris then degenerates, owing to secondary necrosis, which is often assumed to be the result of primary necrotic cell death.

One of the principal reasons that the role of PCD or apoptosis has hitherto been systematically undervalued in developmental biology is that it is difficult to detect. *In vitro*, apoptosis has, until recently, been largely ignored and the offending cells discarded in the trash. Successful cell culture is predicated on the establishment of *living* cells, with the result that cell death in tissue culture is usually regarded as the consequence of technical inadequacy. *In vivo*, detection and quantitation of apoptosis is bedeviled by the speed of the apoptotic process itself, the speed of clearance of apoptotic debris by neighboring cells and professional phagocytes, and the sporadic nature of apoptosis in somatic tissues. Thus, tissue sections seldom intersect with many individual apoptotic events and so tend to underrepresent the extent of PCD in the tissue.

II. The Problems of Measuring Apoptosis

The rapidity of apoptotic events generally makes measurement of apoptosis difficult. *In vitro*, it is a relatively trivial matter to measure cell viability and identify cells in the process of undergoing apoptosis. However, although a host of factors are known that influence apoptosis of cell populations, we know virtually nothing about how each individual cell decides to activate its apoptotic program. Once activated, each apoptotic event is very rapid, and during the 20–60 minutes between the onset of apoptosis (typically the start of membrane blebbing) and the indisputable end point of apoptosis (cell and nuclear fragmentation), a host of biochemical processes is implemented—cytoskeletal reorganization, alterations in membrane structure, transglutamination, proteolytic cleavage, DNA degradation, cell shrinkage, and fragmentation into a variable number of apoptotic bodies. Presumably, many of these processes operate only transiently within the already short duration of each apoptotic event. Thus, long-dead apoptotic cells are unlikely to share all the same properties as cells either "caught in the act" of apoptosis or only recently dead. To make matters worse still, apoptotic cell fragments are readily phagocytosed by neighboring cells with an efficacy that depends on the density and lineage of the live cells remaining. For this reason, counting apoptotic bodies in a culture gives little reliable information as to the number of cells that have died. Moreover, it is easy to confuse apoptotic bodies with apoptotic cells, leading to an overestimate when determining numbers of dead cells. Finally, in many cell populations apoptosis occurs side by side with cell proliferation. Thus, apoptotic cells are being continuously replaced (and eaten) by new cells, so that the end fate of the culture depends on relative rates of cell proliferation versus apoptosis. In such cultures, it is clearly insufficient to monitor the rate of apoptosis by determining live cell numbers. *In vivo*, measurement of apoptosis is complicated by the same problems as *in vitro*, but the processes of detection and measurement are even more difficult for obvious practical and technical reasons.

One last issue concerns confusion of the frequently used term "rate of apoptosis." This confusion can perhaps best be demonstrated using the example of acquisition of drug resistance by tumor cells. During the evolution and progression of cancers, particularly after treatment of the primary tumor with radiation or cytotoxic drugs, recurring tumors frequently become resistant to therapy. In part, this can arise because sufficient numbers of tumor cells evolve a reluctance to activate their apoptotic program when in receipt of an appropriate cytotoxic insult. The expectation would therefore be that the "rate of apoptosis" of drug-resistant tumor cells might be less than for their drug-sensitive forebears. However, there are three interrelated parameters underlying the notion of rate of apoptosis. One concerns the rate of each apoptotic event, the second the probability with which cells undergo apoptosis, and the third concerns the overall fate (death or survival) of the cell population being examined. Currently, the only way of examining the duration of individual apoptotic events is to monitor individual cells over time by time-lapse video microscopy and, by and large, the duration of the apoptotic events in each drug-resistant cell is the same as that in each drug-sensitive cell (McCarthy *et al.*, 1997). With respect to the probability of cells undergoing apoptosis, data indicate that sensitive cells respond to cytotoxic drugs because they activate their apoptotic program more frequently than do resistant cells. Thus, more apoptosis occurs within the cell population per unit time, and this would be reflected in the presence of greater numbers of apoptotic cells at any time. From the stance of growth of a tumor and the outcome for the patient, however, the critical parameter is the relative rate of apoptosis versus that of cell proliferation within the cell population. If the probability of a cell undergoing apoptosis exceeds that of a cell dividing, the cell population will involute; if the reverse obtains, the cell population will expand. Quite modest and subtle alterations in these two dynamics could have catastrophic consequences for the host. The overall fate of a cell population is most effectively monitored by simply counting viable cells over time.

III. Methods for Identifying and Measuring Apoptosis

There are three main classes of technique for detecting and measuring apoptosis, each with its advantages and disadvantages. The first class comprises biochemical methods that make use of some generalized characteristics of apoptosis that can be detected in mixed populations of cells. Examples of this are DNA fragmentation, immunoblotting analyses to detect cleavage of caspase substrates, and viability assays based on mitochondrial function such as the MTT assay. The second class of technique examines individual cells for evidence of processes that characterize apoptosis: examples include altered cell morphology, chromatin condensation, nuclear pyknosis, DNA content, *in situ* labeling of fragmented DNA, and the surface expression of phosphatidylserine (PS). These attributes can be determined by light or electron microscopy and, in some cases, flow cytometrically. The third type of technique follows the fates of individual cells over time; currently, the only way of doing this is to use time-lapse video microscopy.

Biochemical methods that assay cell populations are easy to carry out and robust. However, they suffer from the major drawback that apoptosis is very rapid and almost always asynchronous in cell populations. Thus, any population of cells in which apoptosis is occurring consists of a few cells in the process of apoptosis but, depending on the time point examined, most cells in the population are either alive or already dead. Assays that examine individual cells provide good information about the manner of each cell's demise but suffer from the same limitation that they effectively take a "snapshot" of an asynchronous cell population. Time-lapse video microscopy allows for the systematic morphologic analysis of individual cells over time; however, it requires specialized equipment and acquisition of data is slow and limited to study of only a few cells. Almost all techniques for studying apoptosis are based on the original morphologic criteria by which the process is defined. A molecular mechanical basis for apoptosis, however, has emerged in the form of the caspases—cysteine proteases that are activated during apoptosis and that probably act as key effectors of the process (Alnemri *et al.*, 1996; Takahashi and Earnshaw, 1996). The first human capsase to be identified was the interleukin-1β converting enzyme (ICE) that processes interleukin-1 (IL-1) to its active form (Cerretti *et al.*, 1992; Thornberry *et al.*, 1992). The involvement of ICE-like enzymes with apoptosis became apparent on the discovery of homology between ICE and the protein encoded by the nematode "death" gene *ced*-3 (Yuan *et al.*, 1993). Some 16 members of the ICE–Ced-3 family are now known in humans. All cleave their substrate's C-terminal to an aspartate residue, hence their present name "caspase" (for cysteine–asp protease), and a number have been directly implicated in the execution of the apoptotic process. Some substrates of caspases are known, although in no case is it yet clear that cleavage of that substrate is either necessary or sufficient for apoptosis to occur. Nonetheless, techniques to measure apoptosis are emerging that make use of caspase activity as a measure of activation of the cell suicide program.

In this chapter, we discuss the main methods available to the biologist for the analysis and quantitation of apoptosis. In each case, the principal advantages and disadvantages of each method are discussed. The methods are grouped according to the type of property of apoptotic cells being assayed.

IV. Practical Methods for Detecting and Measuring Apoptosis

A. Standard Methods of Determining Cell Viability

These are all methods that measure viability of cells but cannot distinguish the mode of cell death. Vital dye exclusion examines individual cells in a population, whereas chromium release and MTT assay provide information about the proportion of cells alive or dead in a population.

1. Vital Dye Exclusion

Cell viability can be easily assessed by the exclusion of vital dyes such as Nigrosin, eosin Y, naphthalene black, fast green, neutral red, and trypan blue. Cell populations are typically sampled at various time points, mixed with the vital dye, and live versus dead cells counted in a hemocytometer by light microscopy (Mishell and Shiigi, 1980). Dead cells loose membrane integrity, take up the vital dye, and appear dark under the microscope, whereas viable cells exclude the dye and remain bright in appearance.

An example of this method using trypan blue is as follows:

Trypan blue stock is best kept as 0.2% in water and diluted to isotonicity just before by mixing 4 volumes of trypan blue stock with 1 volume of 4.25% NaCl—this prevents growth of micro-organisms in the stock. Suspension cells are resuspended in growth medium and an equal volume of diluted trypan blue added and mixed. Adherent cells can be quantitated by either direct microscopic inspection, which severely underestimates the numbers of dead or apoptotic cells because dead cells do not adhere to tissue culture plastic, or by first trypsinizing the cells and then analyzing them with trypan blue. After trypsinization, cells should be resuspended in balanced salt solution or growth medium containing at least 1% serum albumin or serum.

Advantages: Vital dye exclusion is very useful for rapidly assessing numbers of dead cells in a culture at any given time point, allowing decisions to be made about optimal time points for assessing apoptosis by more informative methods.

Disadvantages: The major problem with vital dye exclusion assays is that loss of membrane permeability occurs very late in apoptotic cells; indeed, a hallmark of apoptotic cell death is that cellular debris is packaged in membrane-bound vesicles that exclude vital dyes until these particles break down. Moreover, dye exclusion does not distinguish apoptosis from cell necrosis. Trypsinization of adherent cells must be carefully regulated: excess enzyme causes loss of viability, whereas insufficient treatment with enzyme leads to failure to recover all the adherent cells.

2. Chromium Release Assay

Viable cells of most types spontaneously take up and retain $(Na)_2{}^{51}CrO_4$. As a consequence, release of ^{51}Cr by such labeled cells can then be used to assess cell viability radioisotopically. Details of techniques for ^{51}Cr release assays can be found in most standard text on methods in immunology (e.g., Mishell and Shiigi, 1980). In brief, cells are incubated at high density (10^7/ml) in complete growth medium containing 100 μCi/ml ^{51}Cr. After 45 minutes at 37°C, cells are washed three times in cold phosphate-buffered saline (PBS) or complete growth medium. Assays are typically conducted on adherent cells seeded onto TC grade microtiter wells (round bottom) in 100 μl growth medium or, if the targets are suspension cells, on 10^5 cells in about 100 μl medium. After the cytotoxic insult, cells are pelleted (if suspension cells), and 50 μl of the supernatant removed and counted. Important controls are total release (cells treated with 100 μl of 2*N* NaOH and 50 μl counted) and spontaneous ^{51}Cr release (no cytotoxic insult).

Advantages: ^{51}Cr release assays are routine in many immunology laboratories and give consistent and objective data.

Disadvantages: ^{51}Cr release is rather an outdated type of assay, with all the complications of handling radioisotopes thrown in. As with dye exclusion assays, ^{51}Cr release provides no information as to the manner of death (apoptotic vs. necrotic). Not recommended.

3. Determination of Cell Viability by Assessing Mitochondrial Function—the MTT Assay

Cell death eventually results in a collapse of mitochondrial function. One hallmark of functional mitochondria is their ability to convert soluble 3-(4,5-dimethyl thiazol-2-yl)-2,5 diphenyltetrazolium bromide (MTT) into an insoluble blue–black formazan product through the action of their dehydrogenases (Mosmant, 1983), giving an intense blue color in viable cells. Formazan crystals can be dissolved by addition of acid propan-2-ol and the color intensity measured by a colorimetric plate reader at 570 nm. The MTT assay is another rapid method for assessing viability, but, again, it does not distinguish between apoptotic or necrotic death.

An example of the MTT assay is as follows:

Cells are seeded in 96-well plates at 3000 cells per well in complete growth medium. The next day, cells are subjected to the appropriate cytotoxic insult and, at various time points, 10 μl MTT (Sigma) from a stock of 5 mg/ml in PBS is added to each well and the plates incubated at 37°C for 4 hours. Two hundred microliters of acid propan-2-ol (10 m*M* HCL) is then added to each well and mixed. The plates are left for 15 minutes at room temperature and then read at OD (570–630 nm) using a microtiter plate reader. Blank well controls contain growth medium plus MTT but no cells. Percentage cell survival is defined by the formula:

$$\{(\text{Experimental blank})/(\text{Control blank})\} \times 100.$$

Advantages: A very rapid, easy, and robust assay for cell viability.

Disadvantages: Cannot distinguish mode of death (apoptosis vs. necrosis). Problems can arise when using the MTT assay if the number of apoptotic cells at any time is quite small (as is usual) and the target the cell population is cycling. In this situation, cell numbers in the culture may stay constant or even increase as cell division replaces cell loss.

B. Analysis of DNA in Apoptotic Cells

1. Agarose Gel Electrophoresis

One of the earliest defined diagnostics of apoptosis was the degradation of cellular DNA into oligonucleosome fragments. These fragments are typically resolved into the eponymous "nucleosome ladder" by agarose gel electrophoresis. At one time, demonstration of a nucleosome ladder was a mandatory part of proving that cell death occurred by apoptosis. However, there are several reasons why it is unwise to place too much emphasis on the requirement for demonstration of a nucleosome ladder. First, we have no real idea which nucleases are responsible for degradation of cellular DNA in apoptotic cells or, indeed, whether there are any such nucleases dedicated to that role. More recent studies suggest that initial cleavage of DNA during apoptosis generates very large (50–300 kb) fragments by some unknown mechanism. It is possible that the further degradation of these large chromatin fragments to oligonucleosomes represents a rather nonspecific and nonessential scavenging action of liberated nucleases. For technical reasons, it can also prove very difficult to generate a convincing nucleosome ladder. Typically, DNA is extracted from a large population (at least 10^6) of cells of which relatively few are likely to be undergoing apoptosis at any one time. The large excess of intact chromosomal DNA from live cells in the population will, of course, not be resolved by conventional agarose gel electrophoresis and, moreover, will distort and compromise resolution of the minor degraded component from any apoptotic cells. It is also unclear how long "laddered" DNA persists in cells once apoptosis is completed, so even the collection of overt apoptotic debris may not provide a reliable source of degraded DNA. Furthermore, artefactual DNA degradation is not uncommon during cell fractionation or lysis because of the presence of adventitious nucleases.

Despite all of the preceding caveats, many peer reviews and journals still insist on demonstration of nucleosome ladders as evidence of apoptosis. For this pragmatic reason, it is always worth attempting to determine the status of cellular DNA in an apoptotic population. Typically, 10^6 cells are used as starting material. Adherent cells are best scraped into cold PBS containing 1% bovine serum albumin (BSA), and pelleted and washed ×3 in cold PBS. Suspension cells are pelleted directly before washing in cold PBS. The pellet is thoroughly drained and resuspended in 0.2% sodium dodecyl sulfate–50 m*M* Tris–HCl, pH 8.0 containing 1 mg/ml proteinase K. After 30 minutes at 37°C, the solution is extracted with phenol–chloroform and the DNA pelleted with ethanol. The dried pellet is resuspended in 50 μl of tris borate EDTA (TBE) (Sambrook *et al.*, 1991) buffer containing 1 mg/ml RNase A and fractionated on a 2% agarose gel in TBE buffer (Sambrook *et al.*, 1991).

As outlined previously, in many cases the crude genomic DNA analysis of bulk cell populations is not sensitive enough to detect DNA fragmentation because of the low ratio of oligonucleosomal DNA to uncleaved DNA. In such circumstances, low-molecular-weight nucleosomal fragments of DNA can be isolated and separated from the uncleaved high-molecular-weight DNA (Wyllie and Morris, 1982). The low-molecular-weight DNA is again analyzed on 2% agarose gels.

Advantages: Useful for convincing referees and editors that apoptosis is occurring.

Disadvantages: Nonquantitative, difficult to reproduce, unknown mechanistic basis, prone to artefacts that generate false-positive and false-negative results. Does not reveal apoptotic initial cleavage of DNA into large fragments.

2. TUNEL *in Situ* Staining of Cellular DNA

The TUNEL method (TdT-mediated dUTP–biotin nick end labeling) exploits the specific binding of terminal deoxynucleotidyl transferase (TdT) to 3′-OH ends of DNA produced by the active endonuclease. TdT attaches a polydeoxynucleotide to this end, and for the purposes of this technique this is biotin-labeled deoxyuridine triphosphate (dUTP). Thus, the DNA can be labeled *in situ*, revealing the DNA breaks in individual nuclei. The ideal substrate for TdT is single-stranded DNA or 3′-OH ends of double-strand breaks: blunt ends and single-strand breaks are poor substrates.

Nuclear DNA in histologic sections is prepared by exposing the sections to proteolytic enzymes. TdT is added together with biotinylated dUTP, which is incorporated into the sites of strand breaks. The signal is detected by binding to biotin of avidin-conjugated peroxidase, allowing visualization of the signal by light microscopy. The method can be used on fixed or frozen sections. Problems that have been encountered with this technique are high background levels of staining in "viable" controls. However, this can be overcome by pretreatment with DNA ligase, which does not effect the apoptotic strand breaks. "Positive controls" can be created by treating the sections with DNase I.

Advantages: Easy to apply to fixed cells and tissue sections. Many kits available.

Disadvantages: Prone to false positives (necrotic and live cells occasionally score positively in TUNEL; the assay is meaningless if any contaminating nucleases are present during preparation) and false negatives (some apoptotic cells fail to score, presumably if DNA fragmentation is not extensive or has proceeded too far). There is no firm mechanistic basis for the end point being assayed.

3. Flow Cytometric Analysis of Apoptosis

Analysis of apoptotic cells by flow cytometry provides simultaneous data on cell cycle distribution and cell death, together with expression of any interesting immunologic markers. Cycling cells are distributed between those with 2*N* DNA content (G0–G1 cells), 2*N*–4*N* content (S-phase cells), and 4*N* (G2 cells). Many cell lines also exhibit tetraploid (8*N*) and higher ploidy cells. A flow cytometer uses a laser to excite a DNA-binding fluorescent dye such as propidium iodide (PI) and the intensity of signal indicates the DNA content in each cell analyzed. Apoptotic cells are characterized by their sub-G1 DNA content (a population of cells that used to be routinely gated out of flow cytometric analyses because such debris was considered merely as noise; see Fig. 1). Verification of the sub-G1 peak as apoptotic is most easily done by collecting the cells using a fluorescence-activated cell sorter and analyzing the DNA by agarose gel electrophoresis for oligonucleosomal fragments (Telford *et al.*, 1994). In addition to their fluorescent profile, however, a flow cytometer also analyzes the scattered light from the cells passing through its laser beam. Forward scatter and side scatter provide indications of cell size and density, respectively. Because apoptotic cells shrink in size from condensation of both DNA and cytoplasm, flow cytometry is able to identify apoptotic cells as less fluorescent and smaller particles.

For routine flow cytometric analysis, cells are fixed and subsequently stained with PI. A very simple method (Ormerod, 1990) is to take a cell suspension in 200 μl buffered saline at any density from 10^4 to 10^7 ml and, while mixing thoroughly, add 2 ml of ethanol. Fixed cells are pelleted at 800g and resuspended in 1 ml buffered saline plus 1% BSA. The fixed suspension can be kept for several days at 4°C. Before flow cytometric analysis, preboiled RNase is added to

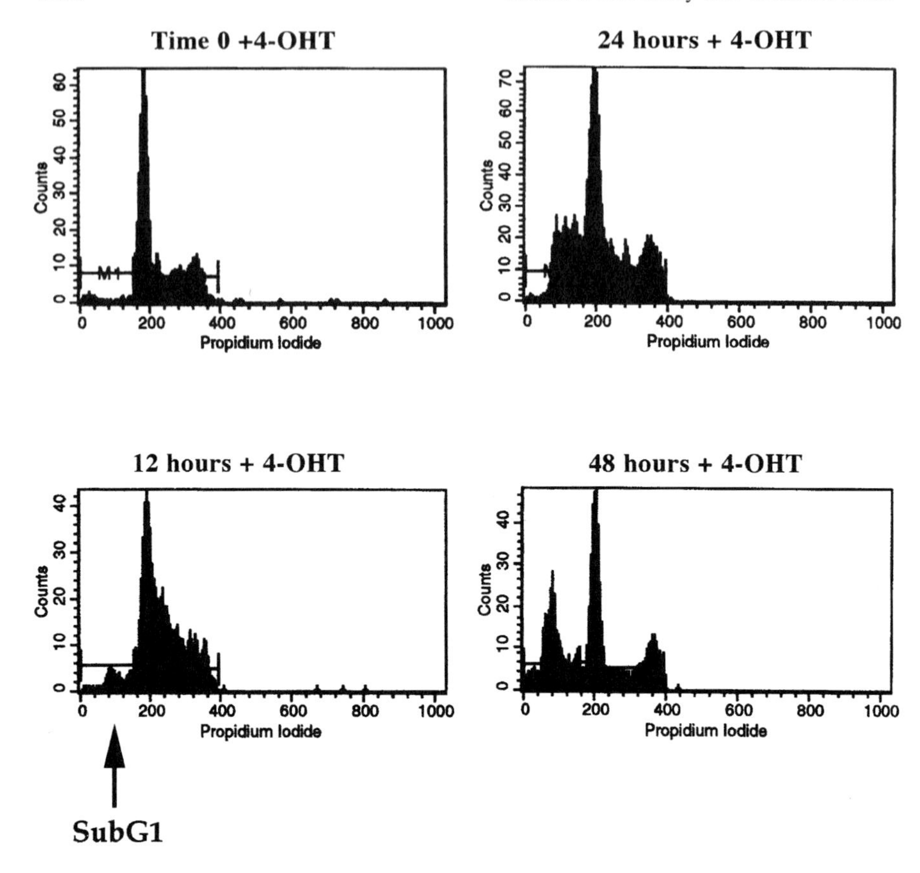

Fig. 1. Flow cytometric analysis. A 48-hour serum-starved culture of Rat-1 cells expressing the 4-hydroxytamoxifen (4-OHT)-regulatable MycER protein was induced to undergo apoptosis by the addition of 70 n*M* 4-OHT. At the time indicated above, cells were trypsinized, washed in PBS, and then resuspended and fixed in 70% ethanol. Before flow cytometric analysis, cells were stained with propidium iodide. Cells at time 0 show a typical G1 arrest profile. On activation of MycER, the cells re-enter the cell cycle and undergo apoptosis, as shown by the appearance of the sub-G1 peak visible at 12 hours.

a final concentration of 100 μg/ml and PI to 40 μg/ml, and the suspension incubated at room temperature for 15 minutes.

Flow cytometry is particularly useful if more than one parameter of apoptotic cells is to be analyzed at one time. For example, thymocytes undergoing apoptosis can be simultaneously analyzed for DNA content with PI, and typed by their expression of CD4 and CD8 antigen using anti-CD4 and anti-CD8 fluorescent conjugated antibodies. The only limitation is that each of the fluorochromes used can be simultaneously excited by the available lasers yet emit at different wavelengths (Fraker *et al.*, 1995). Flow cytometry can also be used to assess the expression in individual cells of specific proteins, such as Bcl-2, p53, and oncoproteins, that modulate apoptosis. Another emerging technique is the flow cytometric monitoring of surface expression of PS in apoptotic cells. PS is an inner membrane leaflet phospholipid that acts as a receptor for macrophages during phagocytosis of apoptotic cells. PS is maintained in the inner leaflet in living cells by a phospholipid flippase, but when cells undergo apoptosis, the flippase becomes inactive and PS flips to the outside of the membrane. The protein annexin V specifically binds PS and is now commercially available in FITC-conjugated form that can be used to identify later-stage apoptotic cells (Koopman *et al.*, 1994).

C. Analysis of Caspase Activity in Apoptotic Cells

1. Western Immunoblot Analysis of Caspase Substrates

Caspases are cysteine proteases that share homology with the Ced-3 killer protein of the nematode worm *Caenorhabditis elegans* and human ICE. Caspases are critical effectors of apoptosis. Not surprisingly, therefore, they are activated in response to diverse apoptotic stimuli and cleave a variety of substrates, including themselves and other caspases. All caspases cleave C-terminally to aspartate residues, usually with a preferred upstream tripeptide substrate consensus that differs between the various different capsases (Takahashi and Earnshaw, 1996). Caspases all share a conserved active site sequence of QACR/QG and all are synthesized as inactive proenzymes that are activated by cleavage at sites that are themselves caspase consensus cleavage sites. Activation requires abscission of the prodomain and cleavage of the remaining protein into a large (~20 kDa) and small (~10 kDa) fragments that assemble to form a mature $(p20{:}p10)_2$ enzyme.

The activation of caspases during apoptosis, together with the discovery of several caspase substrates, has fostered the notion that measurement of cleavage of caspase substrates can be used as a monitor of apoptosis. Unfortunately, it is not clear if cleavage of any of the known caspase substrates is required for apoptosis. Moreover, caspase activation does not necessarily imply apoptosis; for example, the caspase ICE is required to process IL-1β, yet IL-1β is secreted by many cells that are palpably not apoptotic. Measurement of cleavage of caspase substrates in cultures of cells is also fraught with all the problems attending analyses of mixed cell populations, only some of which are apoptotic at any one time: cleaved substrates may be present only briefly in dying cells and therefore difficult to detect.

The most common method for studying caspase activity is by Western immunoblotting of cell extracts with antibodies specific for known caspase substrates and looking for a characteristic signal fragment generated after caspase cleavage (see Fig. 2). Known caspase substrates are other capsases, nuclear lamins (Lazeb-

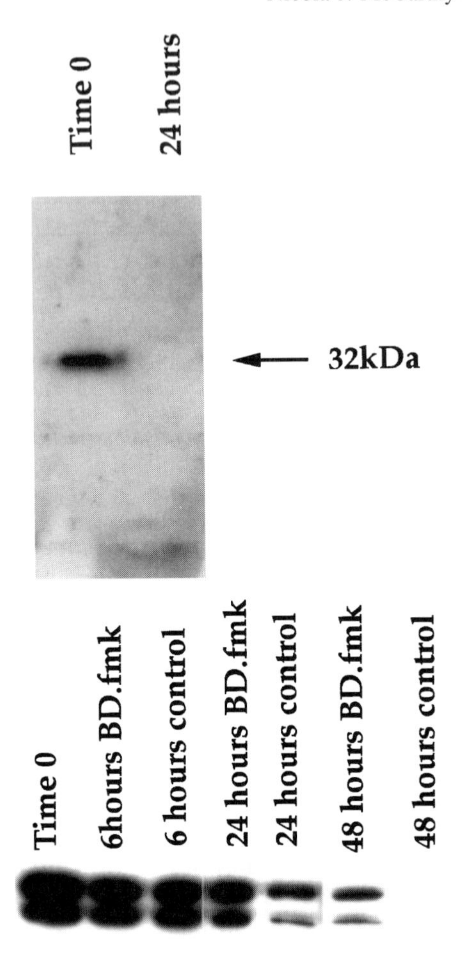

Fig. 2. Lysates from Rat-1 MycER cells with and without 4-hydroxytamoxifen were harvested over 48 hours and analyzed by Western blot to determine caspase activity. (A) Activation of caspase-3 (CPP32) as determined by Western blot. Time 0 control shows the inactive proform of CPP32. Cells harvested 24 hours after Myc activation do not have the 32-kDA form, suggesting CPP32 has been cleaved into its active subforms. Here we have used our own rabbit polyclonal antibody to CPP32 that recognizes only the proform of this protein. (B) Cleavage of lamin A as determined by Western blot. Lamin A is cleaved over a period of 48 hours in cells expressing Myc in the absence of serum. If cells are incubated in the presence of the caspase inhibitor 50 μM BD.fmk, lamin A cleavage is suppressed.

nik *et al.*, 1995), poly (ADP-ribose) polymerase (PARP; Gu *et al.*, 1995; Lazebnik *et al.*, 1994), DNA-dependent protein kinase (Casciola Rosen *et al.*, 1995), the sterol regulatory element-binding proteins SREBP-1 and SREBP-2 (Wang *et al.*, 1995), the 70-kDa protein of the U1-snRNP (Casciola Rosen *et al.*, 1994),

PKCδ (Emoto *et al.*, 1995), and various components of the cytoskeleton such as actin (Mashima *et al.*, 1995) and Gas2, a component of the microfilament system (Brancolini *et al.*, 1995). Several substrates are preferentially cleaved by different caspase family members. For example, lamin A is cleaved by caspase-6 (Mch2), producing a characteristic 40-kDa band, and PARP is cleaved by caspase-3 (CPP32β), producing an 85-kDa band from the 110-kDa PARP protein.

Advantages: Commercial availability of antibodies to both caspase and caspase substrates makes the analysis of caspase activity in any given cell population relatively straightforward.

Disadvantages: Western blot analyzes the activity of proteins in the cell population and not in individual cells. Asynchrony of death in cell populations can complicate analyses of caspase substrate cleavage, which indicates only that the point at which sufficient cells in the population have acquired the caspase activity of interest.

2. Cell-Free Models for Apoptosis

Cell-free systems have been successfully used to reconstitute complex changes within cells, such as mitosis, *in vitro.* A number of cell-free systems have been able to recapitulate aspects of apoptosis (Cosulich *et al.*, 1996; Lazebnik *et al.*, 1993; Newmeyer *et al.*, 1994)—in the main, collapse and fragmentation of the nucleus. The usual source of apoptotic extract is cells that have been primed to undergo apoptosis but that as yet exhibit no apoptotic morphology. The targets are typically interphase nuclei from viable cells. *Xenopus* egg extracts have also been successfully used to study caspase activity, and in some respects represent a more powerful system because they show very clear time windows when different antiapoptotic (Bcl-2) and proapoptotic (caspase-3) proteins function (Cosulich *et al.*, 1996).

Nuclei added to "apoptotic" lysates clearly undergo the normal morphologic changes associated with apoptosis, such as chromatin condensation and fragmentation. Such changes are easily observed by harvesting nuclei from the lysates at given time points, mixing with either Hoechst or acridine orange, and analyzing by fluorescence microscopy. Nuclei in apoptotic cells appear as condensed, pyknotic, and fragmented. Fragmentation of cell DNA can also be verified by agarose gel electrophoresis. Western blotting can again be used to analyze caspase activity. However, in a cell-free system caspase activity can be more directly and specifically monitored by cleavage of fluorescent peptide substrates (Pennington and Thornberry, 1994).

Advantages: Potentially controllable, biochemically defined system enabling use of reagents that are cell impermeable. Allows the possibility of systematic dissection of apoptosis through addition of purified agents to the extract and removal of species by inhibitors or through immunodepletion.

Disadvantages: Looks only at nuclear apoptotic events: inhibition of caspase activity does not necessarily inhibit commitment of cells to death, even though it suppresses nuclear changes (McCarthy *et al.*, 1997; Xiang *et al.*, 1996). Differing cell-free systems exhibit differing levels of commitment to apoptosis; for example, some are amenable to modulation by the antiapoptotic protein Bcl-2, whereas others are not. However, none as yet can be regulated by antiapoptotic survival cytokines, which are a major mechanism for control of cell viability.

D. Measurement of Apoptosis by Microscopic Analysis of Cell Morphology

1. Analysis of Apoptosis by Light Microscopy

Cells stained with histologic dyes observed using standard light microscopy techniques clearly demonstrate the highly condensed chromatin seen in apoptosis compared to necrosis. For example, Giemsa staining shows the differing states of the DNA in viable and apoptotic cell populations. Apoptotic nuclei, which appear characteristically condensed, stain a deep purple color, whereas viable nuclei stain pink, with purple nucleoli often visible.

A particular clear assessment of apoptosis using light microscopy can be achieved by staining ultrathin sections prepared from a resin block used for electron microscopy. When stained with toluene blue, these sections give a clear overall low-power image of the extent of apoptosis occurring in the culture. One drawback with histologic stains is that, to the inexperienced eye, apoptotic and necrotic nuclei can appear similar. Thus, a good first step is to use acridine orange staining and view the cells by fluorescence microscopy (see next section).

2. Fluorescence Microscopy

Acridine orange, an intercalating fluorescent DNA-binding dye, identifies apoptotic cells because of their characteristic condensed nuclear DNA morphology. DNA in viable cells appears as a diffuse yellow–green fluorescence, whereas apoptotic DNA appears as bright green and condensed. Differing DNA states in viable and apoptotic cells (i.e., open and accessible compared with condensed and inaccessible, respectively) result in differing access for acridine orange and the difference in color between viable and apoptotic nuclei. RNA in both viable and apoptotic cells appears red. Acridine orange staining is particularly simple because cells in suspension can be mixed vol:vol with a 25-μg stock of acridine orange in water or PBS, smeared onto a slide, and visualized directly under blue–ultraviolet light. Hoechst is another fluorescent DNA dye that is routinely used to identify condensed DNA in apoptotic cells.

3. Electron Microscopy

Electron microscopy can reveal most clearly DNA condensation in the nucleus and other morphologic changes characteristic of apoptosis. Standard methods for electron microscopy are used and 70-nm sections are stained with uranyl acetate and Reynolds lead citrate. Electron microscopy is very effective in determining unambiguously whether apoptosis is occurring in a cell population. Condensed DNA is very obvious in the nucleus of apoptotic cells, as is general shrinkage of the cytoplasm, whereas internal organelles such as mitochondria remain morphologically normal. In contrast, necrotic cells are distinguished by the lack of condensed DNA and by the general organelle degradation within the cytoplasm. In addition, other, more subtle aspects of apoptosis are evident under the electron microscope. For example, phagocytosed apoptotic cells are quite often visible in electron micrographs, as are the detailed changes within the nucleolus of the apoptotic cell.

For an excellent demonstration of the differences between apoptotic and necrotic cells at the electron microscopic level, the reader is referred to Wyllie (1980).

Advantages: Direct microscopic analysis of cells gives fairly unambiguous data as to the manner of death of individual cells and the technique is easily applicable to cells in culture or to tissue sections.

Disadvantages: It is difficult and tiring to generate data on large numbers of cells by microscopic examination. Microscopic analysis is also another "snapshot" technique that is good at revealing the number of apoptotic cells at any one time, but is less useful in determining *rates* of apoptosis in the cell culture as a whole.

E. Time-Lapse Video Microscopy

In time-lapse video microscopy, a field of cells is viewed microscopically over an extended period. Animation control devices are used to activate a recording device at preset intervals and also to control necessary shutters and light sources on the microscope. Typically, a single video frame is captured at each interval, either on videotape or using a frame storage card on a personal computer. When the collected data are viewed at normal video rates (25 frames per second), a greatly time-accelerated view of the microscopic field is obtained that shows the fates of many individual cells over several hours or days. The resultant films can be viewed and subjected to various forms of analysis.

Time-lapse video microscopy is the only method available that permits repeated and sustained observation of individual cells over time. Cell deaths can be unambiguously diagnosed as either apoptotic or necrotic by their morphology. Moreover, data concerning individual cell divisions, migration, and differentia-

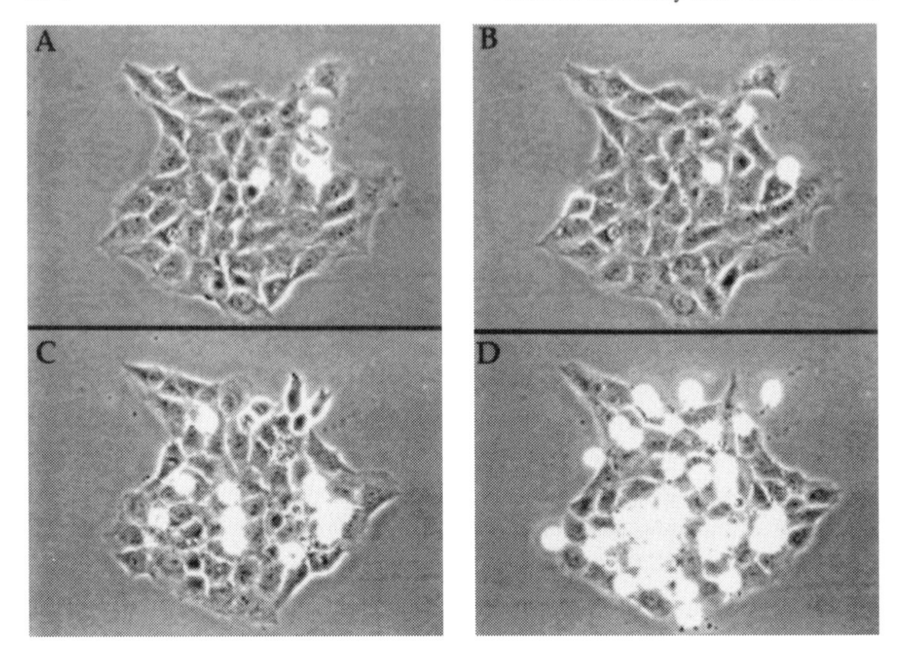

Fig. 3. Time-lapse video microscopy. (A–D) Stills taken from a time-lapse video microscopy film showing approximately 50 cells under normal magnification (×20). The cells were serum starved for 48 hours before addition of 4-hydroxytamoxifen (4-OHT) to activate MycER. The cells were then filmed for 24 hours with one frame taken every 3 minutes. The stills show the asynchronous onset of death in the culture over a 10-hour period. The dead cells appear small, rounded, and phase-bright. (E) Hoffman optics can be used to show clearly the cytoplasmic blebbing that occurs in apoptosis. The Rat-1 MycER cells shown here have been treated with 4-OHT to induce Myc, and with 100 μ*M* of the caspase inhibitor zVAD.fmk, which causes the cells to remain in the blebbing stage for several days (McCarthy *et al.*, 1997).

tion are also acquired at the same time. Thus, time-lapse video microscopic analysis can provide unique information concerning the duration of individual apoptotic events and the relationship between the different cell fate decisions made by each cell. In addition, the technique provides information on the effects on apoptosis of factors such as cell density and direct contact with neighboring cells, and allows visualization of phagocytosis of apoptotic debris.

In a typical time-lapse video microscopic study, a representative number of cells from the culture, typically ~100, is filmed over a designated time period (e.g., one in which 50–90% of the cells are likely to die; see Fig. 3 A–D). One frame is captured at selected intervals; to observe populations of cells for rates of cell death and division, a capture rate of one frame every 3–5 minutes is suitable (time acceleration of 4500–7500); to examine individual apoptotic events of around 20–30 minutes' duration, a capture rate of one frame every 10–30

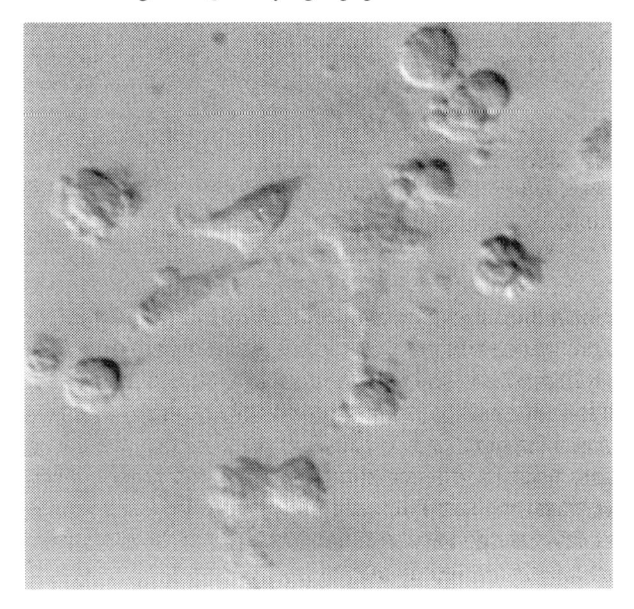

Fig. 3E.

seconds is useful (time acceleration of 250–750). At the end of the filming period, the tape is played back at normal video rate (25 frames per second) to give a dynamic film of cells undergoing apoptosis *in vitro*. By careful analysis, it is possible to assess how long each individual cell takes to die, whether death is stochastic in the culture in response to different stimuli, the effects of neighboring cells' proximity and contact on cell death, relationships between sibling cells, the sequence and consistency of ultrastructural changes during apoptosis, and, with the use of fluorescence time-lapse microscopy, dynamic aspects such as rate and pattern of condensation of the DNA. Moreover, it is straightforward to examine suppression of cell death by genes such as *bcl*-2, *bcl*-x_L or inhibitors of caspases, and assess whether the gene or inhibitor in question affects the proportion of cells that die, or the rate of onset of apoptosis, or the kinetics of each individual apoptotic death. By this technique, for example, we have been able to demonstrate that *bcl*-2 and antiapoptotic survival factors like insulin-like growth factor I suppress the onset of apoptosis in cells but do not affect the kinetics of individual apoptotic events, whereas inhibitors of caspases have no affect on the onset of apoptosis, as determined by membrane blebbing, but prevent each apoptotic death from going to completion (McCarthy *et al.*, 1997).

In our own institute, we have established six independent time-lapse video microscopic systems. In four of them, the first to be established, an animation controller (Eos, Wales, U.K.) is used to activate the light source of an Olympus inverted microscope and to co-ordinate this with acquisition of an image from a monochrome charge-coupled device (CCD) camera and storage on BetaCam high-capacity videotape. Satisfactory results can also be obtained using an sVHS editing video recorder. With the advent of high-capacity storage systems on microcomputers, however, setting up time-lapse video microscopy has become much easier. Our later time-lapse installations make use of either analog video LaserDisc (Sony) to store images (videodisc allows random retrieval of each frame, which can be useful if the images are derived from a Z-series of focal planes and are being used for three-dimensional reconstruction) or write the images from the frame store directly onto rewritable 640-MByte or 1.2-GByte magneto-optical discs using appropriate software (e.g., Improvision, U.K.). The cells under observation are maintained in a small gassed chamber humidified by passing 10% CO_2 through a dilute sodium dichromate solution, and the *whole* microscope is maintained at 37°C in a Perspex box. In our experience, it is vital not to use the common type of commercially available 37°C microscope chamber that sits on the stage. Diurnal changes in temperature, even in climate-controlled buildings, are sufficient to move the image out of focus during the course of the night, especially when observing at high magnification. Heating is supplied by a solid-state heating block regulated by a standard optical switching box, and the air temperature maintained using a low-vibration fan. Cells are observed under phase-contrast or Hoffman optics (DIC/Nomarski optics are not possible with plastic dishes; see Fig. 3E). Further variations can be added at will. For example, metalicized glass cavity slides are now available in which the temperature of the sample is maintained by direct application of electrical current to the coated glass.

Advantages: Time-lapse video microscopy provides unique and indisputable information about the fates of individual cells. Differences between apoptosis and necrosis are unambiguously clear. The technique allows analysis of individual apoptotic events.

Disadvantages: Time-lapse video microscopic systems are expensive and require dedicated equipment and space. Moreover, they can be used only for one experiment at a time—a major problem for experiments that can each last days. Time-lapse analysis can sometimes prove very difficult with suspension cells or cells (such as some keratinocytes) that migrate rapidly (and so move in and out of the visual field). Data acquisition is limited to the relatively small number of cells within each visual field. Data analysis must be carried out entirely by eye. There are unique quantitation difficulties that emerge with time-lapse video analysis and that are ignored or discarded in other methods of measuring apoptosis. Although in principle, it should be straightforward merely to count cell divisions and cell deaths, the number of events observed will obviously depend on the number of cells within the visual field. It is therefore best to standardize experiments to a fixed number of starting cells (e.g., 100) at the start of the video and follow only those cells. However, some cells migrate out of the visual field and are lost to analysis. A more taxing conundrum concerns those cells that divide before they die: are the daughter cells the same as the parent, or are they completely "new" cells? The answer is unknown because it is unclear to what extent cells retain memory of their status and preconfiguration from one cycle to the next.

A useful new emerging technique is time-lapse fluorescence microscopy, which should allow continuous observation of fluorescent probes such as fluorochromatic caspase substrates, fluorescein isothiocyanate labeled annexin V for binding surface PS, vital chromatin stains, and so on. However, fluorescence microscopy requires a number of expensive modifications over the standard time-lapse video microscopy system. Incident excitatory light sources must be heat filtered and of very low intensity, to avoid direct damage to living cells. To pick up such weak signals requires specialized cooled CCD (for optimal resolution) or intensified (for optimal sensitivity) cameras. A controllable shutter also needs to be fitted between the exciting light source and the sample to avoid bleaching, and this must be synchronized with image acquisition.

References

Alnemri, E., Livingston, D., Nicholson, D., Salvesan, G., Thornberry, N., Wong, W., and Yuan, J. (1996). Human ICE/CED-3 protease nomeclature. *Cell* **87,** 171.

Brancolini, C., Benedetti, M., and Schneider, C. (1995). Microfilament reorganization during apoptosis: The role of Gas2, a possible substrate for ICE-like proteases. *EMBO J.* **14,** 5179–51790.

Casciola Rosen, L. A., Anhalt, G. J., and Rosen, A. (1995). DNA-dependent protein kinase is one of a subset of autoantigens specifically cleaved early during apoptosis. *J. Exp. Med.* **182,** 1625–1634.

Casciola Rosen, L. A., Miller, D. K., Anhalt, G. J., and Rosen, A. (1994). Specific cleavage of the 70-kDa protein component of the U1 small nuclear ribonucleoprotein is a characteristic biochemical feature of apoptotic cell death. *J. Biol. Chem.* **269,** 30757–30760.

Cerretti, D. P., Kozlosky, C. J., Mosley, B., Nelson, N., Van, N. K., Greenstreet, T. A., March, C. J., Kronheim, S. R., Druck, T., Cannizzaro, L. A., Hubner, K., and Black, R. A. (1992). Molecular cloning of the interleukin-1 beta converting enzyme. *Science* **256,** 97–100.

Cosulich, S., Green, S., and Clarke, P. (1996). Bcl-2 regulates activation of apoptotic proteases in a cell-free system. *Curr. Biol.* **6,** 997–1005.

Emoto, Y., Manome, Y., Meinhardt, G., Kisaki, H., Kharbanda, S., Robertson, M., Ghayur, T., Wong, W. W., Kamen, R., Weichselbaum, R., and Kufe, D. (1995). Proteolytic activation of protein kinase C delta by an ICE-like protease in apoptotic cells. *EMBO J.* **14,** 6148–6156.

Fraker, P., King, L., Lill Elghanian, D., and Telford, W. G. (1995). Quantification of apoptotic events in pure and heterogenous populations using the flow cytometer. *Methods Cell Biol.* **46,** 57–76.

Gu, Y., Sarnecki, C., Aldape, R. A., Livingston, D. J., and Su, M. S. (1995). Cleavage of poly(ADP-ribose) polymerase by interleukin-1 beta converting enzyme and its homologs TX and Nedd-2. *J. Biol. Chem.* **270,** 18715–18718.

Koopman, G., Reutelingsperger, C. P., Kuijten, G. A., Keehnen, R. M., Pals, S. T., and van Oers, M. H. (1994). Annexin V for flow cytometric detection of phosphatidylserine expression on B cells undergoing apoptosis. *Blood* **84,** 1415–1420.
Lazebnik, Y., Cole, S., Cooke, C., Nelson, W., and Earnshaw, W. (1993). Nuclear events of apoptosis *in-vitro* in cell-free mitotic extracts: A model system for analysis of the active phase of apoptosis. *J. Cell Biol.* **123,** 7–22.
Lazebnik, Y., Kaufmann, S., Desnoyers, S., Poirier, G., and Earnshaw, W. (1994). Cleavage of poly(adp-ribose) polymerase by a proteinase with properties like ICE. *Nature* **371,** 346–347.
Lazebnik, Y. A., Takahashi, A., Moir, R. D., Goldman, R. D., Poirier, G. G., Kaufmann, S. H., and Earnshaw, W. C. (1995). Studies of the lamin proteinase reveal multiple parallel biochemical pathways during apoptotic execution. *Proc. Natl. Acad. Sci. U.S.A.* **92,** 9042–9046.
Mashima, T., Naito, M., Fujita, N., Noguchi, K., and Tsuruo, T. (1995). Identification of actin as a substrate of ICE and an ICE-like protease and involvement of an ICE-like protease but not ICE in VP-16-induced U937 apoptosis. *Biochem. Biophys. Res. Commun.* **217,** 1185–1192.
McCarthy, N., Whyte, M., Gilbert, C., and Evan, G. (1997). Inhibition of Ced-3/ICE-related proteases does not prevent cell death induced by oncogenes, DNA damage, or the Bcl-2 homologue Bak. *J. Cell Biol.* **136,** 215–227.
Mishell, B., and Shiigi, S. (1980). Cell mediated cytolytic responses. *In* "Selected Methods in Cellular Immunology" (A. Bartlett, ed.), pp. 124–128. W. H. Freeman, San Francisco.
Mosmant, T. (1983). Rapid colorimetric assay for cellular growth and survival: Application of proliferation and cytotoxicity assays. *J. Immunol. Methods* **65,** 55–63.
Newmeyer, D. D., Farschon, D. M., and Reed, J. C. (1994). Cell-free apoptosis in *Xenopus* egg extracts: Inhibition by bcl-2 and requirement for an organelle fraction enriched in mitochondria. *Cell* **79,** 353–364.
Ormerod, M. (1990). "Flow Cytometry: A Practical Approach." IRL Press, Oxford.
Pennington, M. W., and Thornberry, N. A. (1994). Synthesis of a fluorogenic interleukin-1 beta converting enzyme substrate based on resonance energy transfer. *Pept. Res.* **7,** 72–76.
Sambrook, J., Fritsch, E., and Maniatis, T. (1991). "Molecular Cloning: A Laboratory Manual," 2nd ed. Cold Spring Harbor Laboratory Press, Cold Spring Harbor, NY.
Takahashi, A., and Earnshaw, W. (1996). Ice-related proteases in apoptosis. *Curr. Opin. Genet. Dev.* **6,** 50–55.
Telford, W. G., King, L. E., and Fraker, P. J. (1994). Rapid quantification of apoptosis in pure and heterogeneous cell populations using flow cytometry. *J. Immunol. Methods* **172,** 1–16.
Thornberry, N. A., Bull, H. G., Calaycay, J. R., Chapman, K. T., Howard, A. D., Kostura, M. J., Miller, D. K., Molineaux, S. M., Weidner, J. R., Aunins, J., et al. (1992). A novel heterodimeric cysteine protease is required for interleukin-1 beta processing in monocytes. *Nature* **356,** 768–774.
Wang, X., Pai, J. T., Wiedenfeld, E. A., Medina, J. C., Slaughter, C. A., Goldstein, J. L., and Brown, M. S. (1995). Purification of an interleukin-1 beta converting enzyme-related cysteine protease that cleaves sterol regulatory element-binding proteins between the leucine zipper and transmembrane domains. *J. Biol. Chem.* **270,** 18044–18050.
Wyllie, A. H. (1980). Glucocorticoid-induced thymocyte apoptosis is associated with endogenous endonuclease activation. *Nature* **284,** 555–556.
Wyllie, A., and Morris, R. (1982). Hormone induced cell death: Purification and properties of thymocytes undergoing apoptosis after glucocorticoid treatment. *Am. J. Pathol.* **109,** 78–87.
Xiang, J., Chao, D., and Korsmeyer, S. (1996). Bax-induced cell death may not require interleukin-1β-converting enzyme-like proteases. *Proc. Natl. Acad. Sci. U.S.A.* **93,** 14559–14563.
Yuan, J. Y., Shaham, S., Ledoux, S., Ellis, H. M., and Horvitz, H. R. (1993). The *C. elegans* cell-death gene *ced*-3 encodes a protein similar to mammalian interleukin-1 beta converting enzyme. *Cell* **75,** 641–652.

16

Methods in *Drosophila* Cell Cycle Biology

Fabian Feiguin, Salud Llamazares, and Cayetano González
European Molecular Biology Laboratory
Heidelberg 69012
Germany

I. Introduction

The fruit fly, *Drosophila melanogaster*, provides an excellent model system for the study of cell division in a higher eukaryote. Most of the techniques used in these studies are now fairly standard and have been thoroughly described in previous publications (Roberts, 1986; Ashburner, 1989; González and Glover, 1993; Goldstein and Fyrberg, 1994). Therefore, we have decided to include in this chapter only those protocols that either have been significantly optimized compared with previous versions, or are altogether new. Needless to say, optimization is an entirely subjective concept that the reader should treat with caution. We consider our protocols optimized when, in our hands, they are either easier to follow, or more reliable (we hardly ever manage to improve both parameters at once) than previously published versions. Thus, optimized does not always equal better for everyone. For this reason, references to previous protocols have been included, when available.

Current Topics in Developmental Biology, Vol. 36

0070-2153/98 $25.00

The chapter is organized around three major issues: fluorescent *in situ* hybridization, immunofluorescence, and primary cultures. As far as fluorescent *in situ* hybridization is concerned, we include protocols for the application of this technique to polytene and diploid cells both in squashed and whole-mounted tissues. The section dealing with immunofluorescence techniques includes protocols for the staining of embryos, third-instar larval brains, and mature oocytes. Finally, a newly developed version of a protocol for primary culture of cells from third-instar larval brains is also included.

II. Fluorescent *in Situ* Hybridization to Polytene Chromosomes: Protocol 1

The following protocol is a modified version of the classic protocol for *in situ* hybridization of polytene chromosomes (see references quoted in Ashburner, 1989). The modification affects exclusively the detection system, which is based on a fluorescence signal instead of the colored precipitate produced by an enzymatic reaction. Thus, the advantages and drawbacks of this new version are related to the very nature of the fluorescent signal itself.

The main drawback of this protocol is the short-lived nature of the fluorescent signal. Unlike the products of enzymatic reactions, fluorescence fades away over time, and bleaches out quickly when observed under the microscope. Therefore, fluorescence-based preparations cannot be made permanent. Nevertheless, the fluorescent system has two major advantages that largely compensate for this relatively minor inconvenience. The first advantage is a noticeable increase in sensitivity. Under optimal hybridization conditions it is possible to produce preparations in which the background fluorescence signal is virtually null, thus achieving extremely high signal-to-noise ratios. The combination of such high-quality preparations with any of the highly sophisticated fluorescence microscope systems now available, such as confocal microscopes or cool charge-coupled device (CCD) cameras, can push the limits of detection to levels unreachable with enzymatic detection methods. The second advantage is the increased resolution. This is particularly useful when more than one target sequence is being studied at the same time. The fluorescent protocol can be used for double or triple hybridization with multiple probes that can easily be identified by using the appropriate nonoverlapping fluorochromes. In that case, the location of the different target sequences can be resolved without the constraints inherent to transmitted light microscopy.

A. Preparation of the Probe for *in Situ* Hybridization

Make up the reaction mixture containing:

5 μl of 5× oligolabeling buffer minus dT

1 μl biotin-16-uridine triphosphate
approximately 100 ng of DNA (boiled)
1 μl Klenow fragment of DNA polymerase (1 unit)

Make up to a final volume of 25 μl with water.

Efficient labeling is achieved in about 2 hours at 37°C or overnight at room temperature. Because of the very high incorporation rates there is no need to remove unincorporated nucleotides. This solution can be kept at −20°C for many months. Before hybridization, the probe must be diluted in 1 volume of water and 2 volumes of 2× hybridization buffer. It should then be boiled for 3 minutes and quenched on ice before it is applied to the chromosomes.

B. Preparation of Polytene Chromosomes for *in Situ* Hybridization

1. Salivary glands are dissected out from third-instar larvae in 0.7% NaCl.
2. Transfer the glands to a drop of 45% acetic acid for 30 seconds and then transfer to a 1:2:3 mixture of lactic acid:glacial acetic acid:water and incubate for 5 minutes.
3. Cover with a siliconized coverslip, sandwich between two sheets of blotting paper, and tap with the blunt end of a pair of tweezers to squash the cells.
4. Monitor the squashing procedure by examining the chromosomes by phase-contrast microscopy (a 40× dry objective is ideal for this purpose). When the chromosome spreading is optimum, leave to fix for 1 to 2 hours at room temperature or overnight at 4°C.
5. To remove the coverslip, immerse the preparation in liquid N_2 until boiling stops, level off the coverslip with the flick of a scalpel, and immerse the slide immediately in a jar containing 70% ethanol for at least 3 minutes. Slides can be kept at this stage for a long time. This is an appropriate step at which to accumulate all the slides so that they all can go into the next steps at the same time.
6. When all the slides are ready, they are transferred into absolute ethanol for 3 minutes and air dried. Dry slides can be kept in a dry place, covered from dust, for many months.
7. Before denaturation, the slides are treated by immersion in a jar containing 2× SSC at 65°C for 30 minutes, followed by two immersions in 70% ethanol for 5 minutes and absolute ethanol for 5 minutes and air dricd as before.
8. To denature the DNA, place the slides in freshly made 70 m*M* NaOH for 3 minutes. It is extremely important that this solution is made fresh every time.
9. Transfer the slides to 70% ethanol for 3 minutes, absolute ethanol for 3 minutes, and air dry as before.

10. Place 20 μl of the denatured probe on top of the chromosomes and cover with a coverslip.
11. Hybridization is carried out in a humid chamber overnight at 58°C.
12. After hybridization the slides are washed in the following series:
    2× SSC at 53°C for 2 minutes. Check that the coverslip falls off.
    4× SSC at room temperature for 5 minutes.
    4× SSC containing 0.1% triton X-100 at room temperature for 5 minutes.
    4× SSC at room temperature for 5 minutes. Make sure that the last wash does not contain any triton X-100. These washing solutions can be kept and reused in step 16.
13. Incubate the slides in 2% (v/v) fluorescein isothiocyanate (FITC)-labeled avidin in phosphate-buffered saline (PBS) for 30 minutes.
14. Wash in the following series:
    4× SSC at room temperature for 5 minutes.
    4× SSC containing 0.1% triton X-100 at room temperature for 5 minutes.
    4× SSC at room temperature for 5 minutes. Make sure that the last wash does not contain any triton X-100.
    If necessary, the signal can be enhanced 5- to 10-fold by incubation for 30 minutes with 1:100 FITC anti-avidin D followed by the same series of washes.
15. The preparation is now ready to be mounted for microscope examination. Remove the excess liquid with a paper tissue and add a drop of mounting medium containing 1 μg/ml propidium iodide to counterstain DNA.

There are several mounting media that can be used. The simplest one is a solution of 2.5% propyl gallate in 85% glycerol. This solution is cheap and easy to make, but it does not set, thus requiring the coverslip to be sealed. This can be achieved by applying nail varnish on the edges. Other mounting mediums such as Permount or Gelvatol can also be used.

III. Fluorescent *in Situ* Hybridization to Diploid Cells: Protocol 2

In situ hybridization to diploid cells provides the only route toward mapping and characterizing the behavior of heterochromatic sequences of *Drosophila*, which are both underrepresented and aggregated in polytene cells.

We include here two protocols for *in situ* hybridization to diploid cells from squashed and whole-mounted third-instar larval brains. For the purposes of genome mapping, squashed preparations are by far the best choice because they are easy to make and provide the highest possible resolution. The procedure can be carried out either in untreated brains or in brains in which the number of mitotic chromosomes has been artificially increased by incubation with colchicine.

When more functional studies are to be carried out, the *in toto* protocol should be used. Whole-mounted tissues preserve the native three-dimensional arrangement of subcellular organelles and are free of the artifacts produced by squashing. These are essential requirements that have to be met to study problems like chromosome pairing, chromosomal domains in interphase nuclei, attachment sites, and many others.

A. Fluorescent *in Situ* Hybridization to Squashed Diploid Cells

Brains are obtained from third-instar larvae by pulling from the mouth parts with a pair of tweezers while holding the larvae by their middle. The brain (ventral ganglion plus optic lobes) usually comes out together with the salivary glands and other internal tissues. Carefully remove these because brain preparations that are free from other tissues are essential to achieve good squashes. For most purposes, dissection is carried out in saline (0.7% NaCl), which is also appropriate for short-term culture. Short-term culture permits the study of the effects of drugs on the cell cycle, as well as the incorporation of tracers like BrdU, which is used to follow DNA synthesis.

1. Dissect out larval brains in 0.7% NaCl and incubate in 0.5 μg/ml colchicine in 0.7% NaCl in a dark, humid chamber for 2 hours.
2. Apply hypotonic shock by washing in 0.5% trisodium citrate for 10 minutes.
3. Place the dissected brains on a microscope slide, add a drop of 45% acetic acid, and leave for 30 seconds. Remove the liquid using tissue paper, add a drop of 60% acetic acid, and cover with a 18 × 18 mm^2 coverslip. Do not squash yet, but wait for 3 minutes.
4. Squash between two sheets of blotting paper by pressing with two fingers on opposite corners. Keep pressing for at least 10 seconds and release pressure gently. Repeat, pressing on the other two corners.
5. Immerse the end of the slide carrying the coverslip in liquid N_2. When the nitrogen ceases to boil, remove the slide and level off the coverslip with the flick of a scalpel.
6. Dehydrate by successive immersion of the slide in 70% ethanol for 3 minutes, 100% ethanol for 3 minutes, and air dry before use.
7. Bake the slide at 58°C for 1 hour in a dry oven.
8. Immerse slides in H_2O gently to remove the coverslips. Denature the chromosomal DNA by boiling for 3 minutes. The slides should be kept in hot (>80°C) water until the probe is applied.
9. Apply 10 μl of denatured probe and carry out the hybridization at 58°C in a humid chamber overnight.

10. After hybridization, carry out the washing and FITC staining as described before.

B. Fluorescent *in Situ* Hybridization to Whole-Mounted Diploid Tissues

1. Dissect brains in saline.
2. Fix for 10 minutes in formaldehyde 3.7% in saline.
3. Transfer to an Eppendorf tube and add about 500 μl of 37% formaldehyde and 200 μl n-heptane.
4. Incubate for 20 minutes in spinning wheel at room temperature.
5. Remove all the heptane and as much formaldehyde as possible, add 500 μl of 1% triton X-100 in PBS, and incubate for 20 minutes.
6. Remove supernatant and wash once in PBS.
7. Add 50 μl of probe and cover with 50 μl parafilm oil.
8. Denature in boiling water bath for 10 minutes and incubate overnight at 58°C.
9. Add 4× SSC under the oil and remove as much oil as possible.
10. Wash for 10 minutes in 4× SSC.
11. Wash for 10 minutes 4× SSC + 0.1% triton X-100.
12. Wash for 10 minutes in 4× SSC.
13. Incubate in 1:50 FITC-avidin in 4× SSC for 30 minutes at room temperature and repeat washes as in steps 10 to 12.
14. Incubate in 1 μg/μl propidium iodide in 4× SSC for 10 minutes.
15. Wash for 10 minutes in 4× SSC.
16. Mount in glycerol-propylgallate.

IV. Immunostaining Embryos: Protocol 3

Most of the activity that takes place during the first few hours of *Drosophila* development is accounted for by a series of very rapid nuclear cycles, some of which take as little as 10 minutes. This high mitotic activity renders the early *Drosophila* embryo a very useful system to study many aspects of cell cycle and chromosome behavior. Often, to study these processes, it is necessary to apply immunofluorescence-based techniques to visualize the subcellular organelles or the molecules under study.

Besides its own cell membrane, the *Drosophila* embryo is surrounded by the vitelline membrane and the chorion. These are impermeable to most chemicals and, therefore, must be removed for the fixatives to penetrate the embryo. The removal of the chorion is technically very simple and is harmless to the embryo. The removal of the vitelline membrane, on the other hand, can be carried out only after fixation of the embryo. The need to sort out this compromise between vitelline membrane removal and embryo fixation is the feature specific to the protocols for immunostaining *Drosophila* embryos.

Although the basic procedure remains unchanged, many of the details of the protocols for immunostaining embryos have been subjected to extensive modifications, and others have been eliminated altogether. Thus, from the very early protocols that involved many steps, used cumbersome solutions, and took a long time, modern ones have evolved that are much simpler and just as efficient. Here we describe three versions of a highly simplified protocol that, nevertheless, works just as efficiently as the most sophisticated alternatives.

The three versions included allow for three different kinds of fixation to be carried out. The reason for this is simple; different antigens require different fixation conditions and the only criterion to decide which fixation suits best the needs of a particular antigen is to try them all. The first recommended fixative is 4% formaldehyde. The fixative is best made up fresh from a stable 10% solution (high grade, Polysciences, Inc.). The second alternative is 37% formaldehyde. This is probably the best choice to preserve microtubule integrity. The third one is methanol fixation, which provides a good alternative for preserving microtubules, although it gives variable results.

1. Allow flies to lay eggs on 3% agar plates containing 4% fruit juice.
2. Brush the embryos from the collecting trays in 0.7% saline and place them on a nylon gauze in a Millipore filtration funnel.
3. Dechorionate by passing commercial bleach over the embryos for 3 minutes.
4. Wash thoroughly with water.

They are two alternatives at this point. The embryos can be processed immediately after any of the fixation protocols described later. Alternatively, they can be poured onto a dry plastic Petri dish, and covered with water so that they can be kept alive and at the same time be easily observed under the dissection microscope. This facilitates the selection of those embryos that have reached a particular developmental stage.

If formaldehyde fixation (4% or 37%) is chosen, proceed to step 5. If methanol fixative is to be used, go directly to step 6.

5. Transfer the embryos into a glass vial containing 1 ml of either 4% or 37% formaldehyde and 4 ml n-heptane. Incubate in spinning wheel for 20 minutes.
6. Place the embryos into a microcentrifuge tube containing 500 μl methanol and 500 μl heptane. Invert the tube several times. Most embryos will lose their vitelline membrane and sink to the bottom of the vial.
7. Remove all heptane and as much methanol as possible.
8. Add fresh methanol. If the embryos were fixed with formaldehyde, they do not need to be kept in methanol any longer. If they were not fixed

with formaldehyde, keep them in methanol for a further 2 hours at room temperature or overnight at 4°C.

9. Rehydrate in PBS.
10. Block any residual fixative by incubating the embryos in 10% fetal calf serum (FCS), 0.3% Tween in PBS for 1 hour. If RNase treatment is to be carried out, an aliquot of a boiled solution of the enzyme should be added at this time at a concentration of 2 mg/ml.
11. Incubate the preparation with the primary antibody in 10% FCS, 0.1% Tween in PBS either at 4°C overnight, or for 4 hours at room temperature.
12. Wash several times with 0.1% tween in PBS over a 1-hour period.
13. Incubate with the secondary antibody for 4 hours at room temperature or overnight at 4°C.
14. Wash as before, but using PBS alone for the last two or three washes.
15. To mount, place a drop of approximately 50 μl of mounting medium on a microscope slide and transfer the embryos into the drop. Move them around with a pair of twizers until they get embedded, cover with a coverslip, and seal the preparation with nail varnish.

V. Immunostaining Larval Neuroblasts: Protocol 4

The larval brain is the tissue of choice for the study of cell division during late development. Most mitotic mutants of *Drosophila* have been identified by screening squashed preparations of third-instar larval brains.

We include here two protocols for immunostaining cells from the larval central nervous system. The first uses a squashed preparation and the second uses whole brains. Squashing is intrinsically disrupting. Moreover, it is not compatible with microtubule preservation. Therefore, the usefulness of squashed preparations for immunostaining is actually very limited, and their use should be restricted to antigens that localize to well defined and stable structures (e.g., kinetochore). The only advantages of squashed preparations are that they are much easier to get and that they do not require the sophisticated image acquisition set-ups needed to study whole-mounted tissues.

The procedure for immunostaining intact preparations of larval brain was developed with the prospect of viewing the resulting preparation by scanning confocal microscopy or any other system that facilitates optical sectioning of thick specimens, thus permitting a study of the three-dimensional relationships between components of the mitotic cell. A new version of the old protocol that allows for microtubule staining in the absence of taxol is described here. Because the specimen to be stained is very thick, it is important to take special care in permeabilizing the tissue to facilitate antibody penetration, and to wash thoroughly after incubation with antibodies to reduce background staining. For the same reason, special care has to be taken in mounting the specimen for microscopy. It is essential to mount brains in a position that facilitates observation of the region to be studied. For general purposes, neuroblasts located on the thoracic region of the ventral side of the ganglion provide the best material. To observe these cells, the brains should be mounted with their ventral sides upward, bringing the neuroblasts as close as possible to the microscope objective. Intermediate techniques that make use of semisquashed preparations have most of the drawbacks of both squashed and *in toto* preparations and none of the advantages. Therefore, their use is not recommended.

A. Immunostaining Squashed Preparations of Larval Brains

1. Dissect out larval brains in 0.7% NaCl.
2. Place the brain in a microcentrifuge tube containing 3.7% formaldehyde in PBS, and incubate at room temperature for 1 hour.
3. Remove the formaldehyde, add 10% FCS, 0.3% triton X-100 in PBS, and incubate for 1 hour at room temperature.
4. Place the brains on a microscope slide, add a drop of 45% acetic acid, and leave for 30 seconds. Remove the liquid using tissue paper, add a drop of 60% acetic acid, and cover with an 18×18 mm^2 coverslip. Do not squash yet, but wait for 3 minutes.
5. Squash between two sheets of blotting paper by pressing with two fingers on opposite corners. Keep pressing for at least 10 seconds and release pressure gently. Repeat, pressing on the other two corners.
6. Immerse the end of the slide carrying the coverslip in liquid N_2. When the nitrogen ceases to boil, remove the slide and lever off the coverslip with the flick of a scalpel.
7. Dehydrate by successive immersion of the slides in 70% ethanol for 3 minutes, 100% ethanol for 3 minutes, and air dry before use.
8. Carry out immunostaining as described for *in toto* preparations (see V.B).

B. Immunostaining Whole-Mount Preparations of Larval Brains

1. Dissect out larval brains in 0.7% NaCl.
2. Incubate for 30 minutes in 3.7% formaldehyde followed by 60 minutes in 37% formaldehyde.
3. Transfer to 0.3% triton X-100, 10% FCS, in 0.7% NaCl, and keep there for about 20 minutes.
4. Incubate with the first antibody in 2.5 mg/ml RNAase, 0.3% triton X-100, 10% FCS, in 0.7% NaCl overnight.

5. Wash in 0.3% triton X-100, 10% FCS, in 0.7% NaCl at least four times for 10 minutes.
6. Incubate with second antibody in 0.3% triton X-100, 10% FCS, in 0.7% NaCl at least for 2 hours.
7. Wash in 0.3% triton X-100, 10% FCS, in 0.7% NaCl at least four times for 10 minutes, and then in 10% FCS in 0.7% NaCl for a few minutes.
8. If required, incubate in propidium iodide 1 μg/μl in 0.7% NaCl for 5 minutes.
9. Transfer the brains into a drop of approximately 50 μl of mounting medium on a microscope slide, move them around with a pair of twizers until they are embedded, cover, and seal with nail varnish.

VI. Immunostaining of Female Meiotic Spindles: Protocol 5

Female meiosis remained elusive to immunocytochemical analysis for many years. The main reasons for this are the difficulty in removing the vitelline membrane of stage 14 oocytes and the need to avoid unwanted activation of the oocyte. These difficulties were overcome by Theurkauf and Hawley (1992), who published the first characterization of the wild-type *Drosophila* female meiotic spindle. The technique has been widely used ever since to study different aspects of the wild-type process as well as to characterize the alterations brought about by mutation in different genes required for female meiosis (Fig. 1).

The published protocol works very well, its only drawback being the technical difficulty of vitelline membrane removal. We have circumvented this limitation by developing the alternative protocol presented here, which is very simple. Moreover, whereas the older protocol makes use of formaldehyde fixation, ours uses methanol as the only fixative. Therefore, at the very least, the new protocol provides a useful alternative with which to check for the optimal conditions for antigen preservation. It also offers a simple, working solution for those who cannot get the other protocol to work.

1. Dissect ovaries from adult females in absolute methanol and transferred to a 10-ml plastic tube containing about 2 ml of fresh methanol. We usually accumulate about 10 to 20 ovaries.
2. Sonicate the contents of the tube. We use a Sonifier B-12 from Branson Sonic Power Company fitted with a cone-shaped probe of about 3 to 4 mm in diameter at the bottom. Sonication is applied in five cycles of 1 second each.
3. Transfer oocytes without chorion and vitelline membrane to fresh methanol and keep at room temperature for a further 2 hours.
4. Block any residual fixative by incubating in 10% FCS, 0.1% tween in PBS for 1 hour. If RNase treatment is to be carried out, an aliquot of a

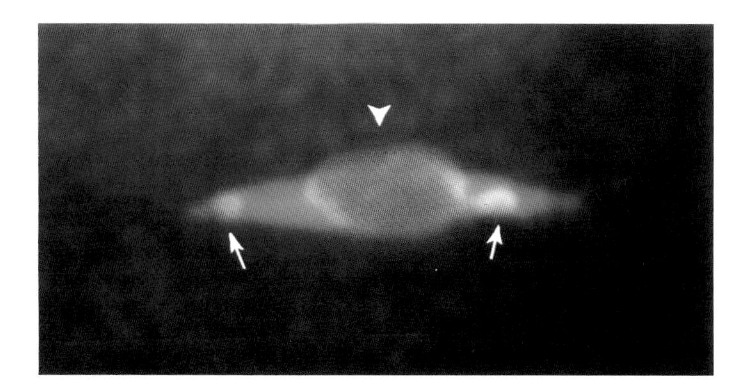

Fig. 1. *Drosophila* female meiosis. Chromosomes (red) and microtubules (green) in an oocyte arrested in metaphase I. Chromosomes were stained with propidium iodide. Microtubules were visualized by incubation with a mouse antibody against α-tubulin, followed by a fluorescently labeled second antibody that specifically recognizes mouse IgG. The arrows indicate the nonrecombinant fourth chromosomes, which are prematurely ejected toward the poles. The arrowhead points at the central mass of recombinant chromosomes. Fixation and immunofluorescence were carried out following protocol 5.

boiled solution of the enzyme should be added at this time at a concentration of 2 mg/ml.

5. Incubate the preparation with the primary antibody in 10% FCS, 0.1% Tween in PBS, either at 4°C overnight or for 4 hours at room temperature.
6. Wash several times with 0.1% Tween in PBS over a 1-hour period. Best results may be achieved by more extensive washing.
7. Incubate with the secondary antibody for 2 hours at room temperature or overnight at 4°C.
8. Wash as in step 4, but using PBS alone for the last two or three washes.
9. To mount, place a drop of approximately 50 μl of mounting medium on a microscope slide and transfer the oocytes into the drop. Move them around with a pair of tweezers until they are embedded, cover, and seal.

VII. Preparing *Drosophila* Neuroblast Primary Cultures: Protocol 6

The following protocol describes the preparation of primary cultures from third-instar larval brains. It has been modified from a previous protocol described by Wu *et al.* (1983). Although the starting material contains many different cell types, the only cells that survive and proliferate to any reasonable extent are neuroblasts. These can be seen as single cells that settle down during the first 24 hours of culture, after which they start to divide. Neuroblast proliferation results in nests of cells that resemble the ganglion mother cells and ganglion cells produced in the living animal. The entire process can be observed in real time under the microscope. We believe that there is a large scope for practical applications of these protocols to be developed in the future.

A. Preparation of Coverslips: Cleaning and Coating

Coverslips (11-mm round from Fisher Scientific) must be thoroughly cleaned and coated before use. To clean them, they are placed in porcelain racks and rinsed twice, 10 minutes each, in distilled water. They are then immersed in concentrated HNO_3 for 18 to 36 hours, rinsed in tissue-culture-grade water (four changes over 2 hours), and sterilized by dry heat (225°C for 6 hours). They can then be stored in 35-mm plastic Petri dishes arranged so that they do not overlap one another. Coating is achieved by applying a solution of 1 mg/ml poly-L-lysine hydrobromide in tissue-culture-grade water. They are incubated for about 18 hours, rinsed twice for 2 hours each with sterile water, and air dried.

B. Dissecting the Larval Brains, Cell Dispersion, and Plating

1. Surface-sterilize third-instar larvae by brief immersion in 70% ethyl alcohol and rinse three times with sterile distilled water.
2. Dissect the brains and ventral ganglions from larvae by pulling from the mouthparts with a pair of tweezers while holding the larvae from the middle. Dissection is carried out in a drop of calcium- and magnesium-free buffer (800 mg of NaCl, 20 mg of KCl, 5 mg of NaH_2PO_4, 100 mg of $NaHCO_3$, and 100 mg of glucose in 100 ml of distilled water).
3. Transfer the dissected brains to an Eppendorf tube containing 400 μl of dissection buffer plus 0.2 mg/ml of collagenase (type I, Sigma Chemical Co., St. Louis, MO, U.S.A.) and incubate for 30 minutes.
4. Remove the buffer–collagenase and rinse the brains three times (5 minutes each) in culture medium.
5. Dissociate into single cells by repeated pipetting up and down with a siliconized plastic micropipette (yellow tips). Cell density is determined in a hemocytometer and the cells plated onto polylysine-treated coverslips. After 2 to 3 hours, to allow for attachment, the coverslips are transferred to dishes filled with culture medium by flipping them over in such a way that the cells face the bottom of the dish.

Our standard plating density is 170,000 cells per milliliter in 35-mm plastic Petri dish, but lower and higher plating densities (40,000 to 400,000) can be used.

C. Feeding and Maintenance of Cultures

Cultures are kept at 25°C in a closed incubator with atmospheric concentration of gases. We routinely feed the cultures twice a week by replacing about one third of the medium at each feeding. Plating medium consist of Schneider's medium containing extra glucose (1000 mg/l), 5% FCS, 5% fly extract, and insulin 5 μg/ml (Sigma No. I5500).

D. Video Microscopy Analysis of Neuroblast Division

For long-term observation of living cells under the microscope, a special chamber is necessary to give the cells a constant environment in terms of hygroscopy, pH, and temperature. We use a special Petri dish with a hole in the bottom and a glass coverslip attached (Glass Bottom Microwells Plastek cultureware from the MatTek corporation, Ashland, MA, U.S.A.; part No. P35G-1.5-14-C, uncoated No. 1.5). For observations at 100× magnifications, the cells are grown directly on the glass coverslip attached at the bottom of the dish. The surface of this coverslip was previously treated in the same way as the normal glass coverslips. Observations at lower magnification can be easily performed in this chamber, introducing the coverslips with the attached cells facing down. In both cases a constant 25°C temperature is controlled by an objective heating ring (Bioptechs, Inc., Butler, PA, U.S.A.).

Neuroblast division is followed with a Zeiss Axiovert 135 inverted microscope equipped with 40× (numeric aperture [NA] 1.0), 63× (NA 1.4), and 100× (NA 1.4) Plan Apochromatic objectives. The microscope has a 100-W, 12-V lightbulb. A green filter is installed in the light path to reduce photo damage of the cells. To record phase-contrast and Nomarski images, a long-distance condenser is used. To increase sensitivity and reduce photodamage, the light is reduced and the camera gain increased with the CCD camera control (for VECDIC we work with polarizer, analyzer, 1.6× optovar, appropriate condenser, and Wollaston prisms specific to the NAs of the objectives). The recording set-up consists of a Sony Hyper HAD camera connected to a Hammatsu CCD camera C 2400 control panel, a black-and-white video monitor (live image), a Power Macintosh 8500 equipped with an image grabber (LG3 image grabber, Scion Co.), a time-lapse video recorder (Panasonic model AG6720), and a second black-and-white monitor (processed image). Time-lapse video recording is essential to record slowly occurring events. Images are collected at a speed slower than video rate and played back at video rate (34 frames per second). Normally, VHS tapes are used, but for more critical recordings it is advisable to use higher-quality tapes (SVHS). It is also possible to record images directly into the computer through an image grabber. Pictures can be stored on an optical drive. We use a 230-MB storage drive (Fujitsu M2512 A).

Acknowledgments

The protocol to establish primary cultures of neuroblasts was developed in collaboration with Carlos Dotti. We are most grateful to Carlos Dotti, Cristiana Mollinary, and Gaia Tavosanis for helpful comments on the manuscript.

References

Ashburner, M. (1989). "*Drosophila*: A Laboratory Manual." Cold Spring Harbor Laboratory Press, Cold Spring Harbor, NY.

Goldstein, L. S. B., and Fyrberg, E. A. (1994). "Methods in Cell Biology." Academic Press, New York.

González, C., and Glover, D. M. (1993). Techniques for studying mitosis in *Drosophila*. *In* "The Cell Cycle: A Practical Approach" (P. Fanty and R. Brooky, eds.), pp. 163–168. Oxford IRL Press, Oxford.

Roberts, D. B. (1986). "*Drosophila*: A Practical Approach." Oxford IRL Press, Oxford.

Theurkauf, W. E., and Hawley, R. S. (1992). Meiotic spindle assembly in *Drosophila* females: Behavior of nonexchange chromosomes and the effects of mutations in the *nod* kinesin-like protein. *J. Cell Biol.* **116,** 1167–1180.

Wu, C. F., Suzuky, N., and Poo, M. M. (1983). Dissociated neurons from normal and mutant *Drosophila* larval central nervous system in cell culture. *J. Neurosci.* **3,** 1888–1899.

17

Single Central Nervous System Neurons in Culture

Juan Lerma, Miguel Morales, and María de los Angeles Vicente
Instituto Cajal, C.S.I.C.
28002 Madrid, Spain

I. Introduction

The possibility of growing neurons in culture has been crucial in understanding many of the properties of nervous system activity at the cellular and molecular levels. Indeed, our knowledge of neuronal development, plasticity, gene expression, neurotoxicity, pharmacology, and the like has greatly improved with the use of neuronal cultures. In particular, the understanding of synaptic transmission between central neurons has progressed spectacularly thanks to the application of the patch-clamp technique to cultured neurons.

In the mid-1970s, to study interaction between cardiac myocytes and sympathetic neurons, Furshpan *et al.* (1976) developed a culture system they referred to as *microculture*. By using cocultures of a single neuron and a small number of myocytes in an area only a fraction of millimeter in diameter, they found that confinement of the growing neurites to a small number of target cells increased the probability of synapse formation and, therefore, the intensity of synaptic action. In addition, they observed that single neurons growing in microislands developed autapses, that is, synapses made by a neuron on itself (Van der Loos and Glasser, 1972). Therefore, with a minimal equipment requirement, cell microculture offers the possibility to dissect out synaptic interactions that are difficult to study *in situ*. Among other advantages, this culture system allows for rapid changes in the concentration of added drugs, and it lacks polysynaptic activity.

Microcultures, however, were not directly applicable to central nervous system (CNS) neurons because CNS glia can migrate on nonadhesive substrates, forming bridges between microislands. Subsequent work by Segal and Furshpan

Current Topics in Developmental Biology, Vol. 36

0070-2153/98 $25.00

(1990), Bekkers and Stevens (1991), and Segal (1991) succeeded in microculturing CNS cells. Segal and Furshpan (1990) first adapted the method to CNS neurons by using agarose as a substrate, which limited the proliferation of glial processes satisfactorily. The use of agarose as a nonadhesive substrate was crucial in forming isolated microcultures. However, the reason why agarose repels cell process growth is unknown.

The purpose of this chapter is to describe how to prepare microcultures of hippocampal cells essentially free of glial cells. This culture system consists in seeding immature neurons on islands of permissive substrate, in such a way that they grow in physical isolation from other neurons. It results in microislands containing just one or a few neurons that facilitate the study of synaptic transmission under controlled perfusion conditions. We include some examples of the activity of neurons grown under these microculture conditions.

II. Microculture Preparation: Steps to Follow

Hippocampal microcultures are established from dissociated hippocampi from day 17 or 18 rat embryos. Neurons from other parts of the brain may also be grown in microculture conditions. Cells are dissociated using conventional procedures described in detail elsewhere (see Banker and Goslin, 1991).

A. Step 1

In the flow hood, cover Petri dishes or glass coverslips with a thin film of agarose (0.2%) the day before starting the culture procedure. Typically, a small volume of liquid agarose is deposited and spread on a 35-mm Petri dish. Once the whole surface has been covered, the excess of agarose should be removed. This results in a very thin and transparent layer of agarose. Because the agarose solution is heated to liquefy, further sterilization is not required, although it is advisable to let the agarose dry (5 minutes) under the hood's UV light. Leave the Petri dishes in sterile conditions in a humidified incubator until use.

B. Step 2

Two hours before plating, the agarose should be sprayed with a sterile solution containing permissive substrate. We have used a mixture of poly(D-lysine) (100 mg/ml) and laminin (16 μg/ml), but other substrates can also be used. This procedure results in a randomly distributed population of spots (0.1–1 mm in diameter), embedding the agarose with permissive substrate. Before drying, the spots can be visualized under phase-contrast optics. Select those plates containing 20–30 islands/mm^2.

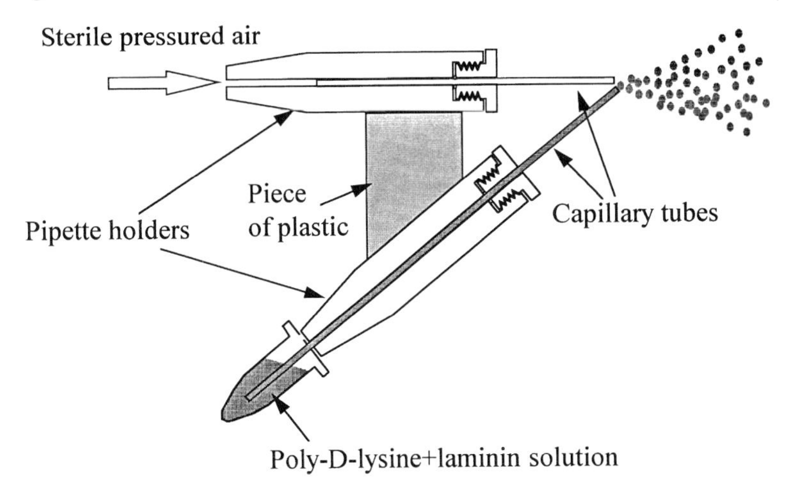

Fig. 1. A diagram of the microatomizer for spraying permissive substrate onto the agarose-covered dishes. It is made of two pipette holders (for glass capilaries of 1.2–1.5 mm in diameter) glued to a piece of plastic in such a way that the two pipette tips could be easily positioned to produce a fine mist of substrate.

To generate dots of substrate onto the agarose, several procedures can be used, such as a glass microatomizer (which can be purchased from Fisher) or the lift-off method, which consists of generating a 16-island pattern onto a glass cover-slip by using a patterned Silastic film that defines island boundaries. After coating the coverslip with polylysine and laminin, the film is peeled off in one piece, leaving the substrate on the islands (for details see Chien and Pine, 1991). However, spraying a fine mist of substrate onto the agarose is the most widely used method. The sprayer may be home-made by using two sterile glass capillaries like those used to construct patch pipettes. They are held in place (forming an angle of approximately 30°) with the aid of two pipette holders glued to a piece of plastic (Fig. 1). This has the advantage that glass capillaries can be easily sterilized and mounted in place just before use. In addition, their small dead volume helps to save substrate solution.

C. Step 3

Seed dissociated cells on these plates at a final concentration of $1-1.2 \times 10^5$ cells per plate. Usually this can be achieved by diluting the cells at $5-6 \times 10^4$ cells/ml of culture medium and adding 2 ml per Petri dish.

Cultures are then maintained in standard conditions (i.e., in a humidified incubator at 37°C and 5% CO_2) until use. The culture medium may be refreshed once a week by replacing just 0.6 ml with fresh medium. In our hands, larger replacements have catastrophic effects on cell survival. The first "refreshment" should not be done before 4 days of culture.

Although standard supplemented Dulbecco's modified Eagle's medium (DMEM) can be used, we have used the serum-free Neurobasal medium (Gibco, No. 320-1103) supplemented with B27 supplement (Gibco No. 680-7504), 2-mercaptoethanol (25 μM), glutamine (0.5 mM), and antibiotics. The composition of this medium and the supplement can be found in Brewer *et al.* (1993). Neurobasal–B27 produces a nearly pure neuronal population because glial growth is reduced. Some modifications may be introduced to allow long-term cultures; neuronal survival is a common problem in this culture system because the density is necessarily low. However, although Neurobasal medium sustains neuronal cultures for longer periods, we have observed some paucity of functional synaptic contacts in these cultures at days 7–14. Electrophysiologically measured synaptic transmission appears later (18–20 days in culture) in this medium than in neurons incubated in DMEM plus supplements. We do not know the reason for this, but it might be overcome by including a third part of medium conditioned for 48 hours by astrocytes in the refreshment medium. Therefore, the low density of glial cells in our cultures and the consequent lack of unknown glial factors may account for the delay in the appearance of measurable synaptic activity.

III. Examples of the Functionality of Microcultured Neurons

Figure 2 shows an example of hippocampal cells growing in drops of permissive substrate. Occasionally, neurites are seen to jump from island to island. Immunocytochemical analysis, however, revealed that these processes were exclusively axons, whereas dendrites remained mostly confined to the extension of the microisland. In the first example shown in Figure 2, cells were immunolabeled using an antibody raised against MAP2 protein, revealing the extension and spreading of the dendritic tree in 7-day microcultured cells. The second example shows an island containing four neurons after immunolabeling with an anti-TAU protein monoclonal antibody. This protein is specific for axons, and dendrites are essentially devoid of label. The area of neuritic growth is limited to the polylysine–laminin island, which is not visible. Note how processes escaping the island are immunoreactive for TAU but not for MAP2.

In hippocampal microcultures, both glutamatergic and gamma-aminobutyric acidergic central neurons can be found, and intercellular as well as autaptic transmission may be easily evaluated by recording the cell in voltage- or current-clamp conditions. The excitatory synaptic potential appears as a hump after the action potential. For assessing intercellular synaptic transmission in islands containing more than one neuron, the presynaptic neuron was studied in current-clamp conditions, whereas the postsynaptic cell was voltage-clamped at a hold-

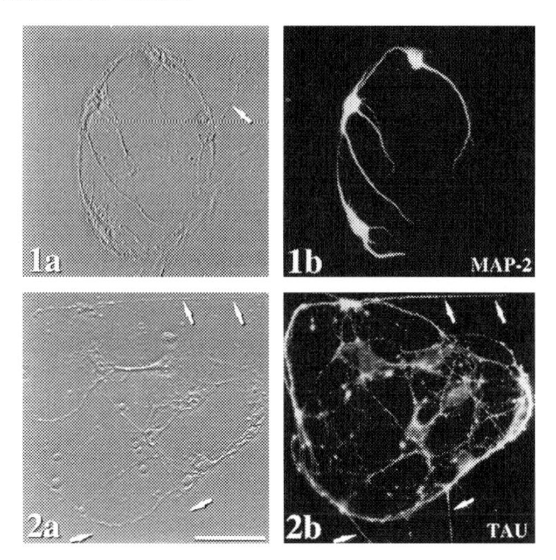

Fig. 2. In microcultured cells, dendrites are confined to the extension of the microisland of permissive substrate. On the left, examples of microislands containing three (1a) and five (2a) neurons viewed under differential interference contrast optics. On the right, the same islands labeled with an antibody against the microtubule-associated protein (MAP2) (1b) or TAU protein (2b). MAP2 specifically labels dendrites, whereas TAU exclusively labels axons. Note how dendrites are limited to the island whereas axons (arrows) may escape and form bridges between islands. The calibration bar is 50 μm.

ing potential of −70 mV. Action potentials were generated in the presynaptic cell by injecting pulses of depolarizing current. Synaptic responses were evoked in the postsynaptic neuron after presynaptic firing, exhibiting trial-to-trial fluctuations in their amplitudes. An example of excitatory transmission is illustrated in Figure 3. Excitatory synaptic responses were composed of two components, the fast non-NMDA and slower NMDA (N-methyl-D-aspartate) receptor-mediated (Fig. 4).

In the example shown in Figure 5, intercellular synaptic as well as autaptic inhibitory potentials were recorded in microcultures stablished from hippocampus. Inhibitory currents reversed at hyperpolarized potentials, as expected for chloride-permeable channels, and were inhibited by bicuculline. These examples demonstrate that, in functional terms, operation of autapses is indistinguishable from that of intercellular synapses. This microculture system has been similarly established for neurons obtained from different parts of the CNS and peripheral nervous system, and they develop functional synapses with a variety of neurotransmitter phenotypes. Sympathetic (Furshpan *et al.*, 1976; Chien and Pine, 1991), cerebellar (Liesi *et al.*, 1992), raphe nucleus (Gu and Azmitia, 1993; Johnson, 1994), and neocortical (Huettner and Baughman, 1988) neurons have also been successfully grown in microculture conditions.

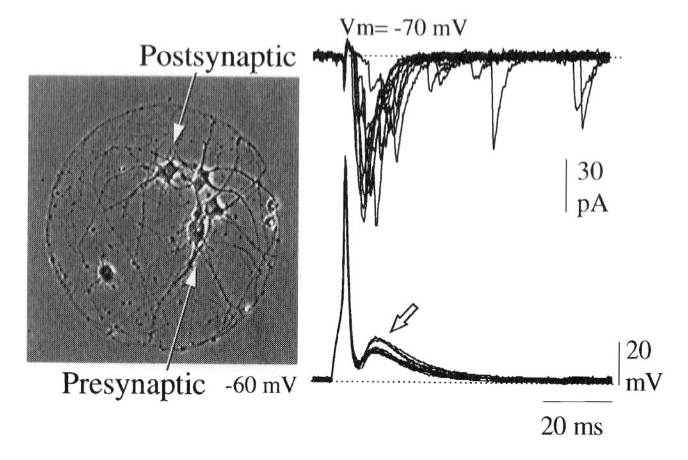

Fig. 3. Autaptic and intercellular excitatory synaptic transmission in microcultured hippocampal neurons. The island containing four neurons, as seen under phase-contrast microscopy, is depicted on the left. The presynaptic neuron was recorded under current-clamp conditions, showing a resting membrane potential of −60 mV approximately, whereas the postsynaptic neuron was voltage-clamped at a holding potential of −70 mV. The presynaptic cell was brought to fire by injecting brief depolarizing current pulses. Records are shown superimposed. Note the humps after action potentials (arrow), which correspond to autaptic excitatory postsynaptic potentials. Synaptic inward currents were observed in the postsynaptic neuron on presynaptic cell firing, revealing the existence of intercellular synaptic transmission. Note the fluctuation in amplitude of both autaptic and intercellular responses.

It is known that neurons in culture do not look like their counterparts *in situ.* This is particularly notorious in microcultured cells. However, it has been shown that the microisland environment retains many of the morphologic properties described in intact preparations (Landis, 1976; Johnson and Yee, 1995; Mennerick *et al.*, 1995). The microculture permits the structural inspection of an electrophysiologically characterized neuron. The ultrastructure of synapses in microcultured serotonergic neurons is similar to that of synapses formed by serotonergic neurons *in situ* (Johnson and Yee, 1995), and the morphology of autapses and synapses of microcultured hippocampal neurons shows membrane specializations reminiscent of the synaptic contacts formed in the adult hippocampus (Mennerick *et al.*, 1995). It might be considered that autapses are a sort of aberrant contact that occurs only in abnormal situations. It is unlikely, however, that the number of autaptic connections readily found in microcultures is larger than that *in vivo*. This issue has been approached in neocortical cells by Lübke *et al.* (1996). These authors found that 80% of the developing rat neocortical neurons establish autaptic contacts (2.3 autapses per neuron on average), indicating that autapsis is a rather normal way of transmission in the intact brain. Interestingly, similarly to synapses, autapses were mainly located on basal dendrites, close to the soma, either on dendritic spines or shafts. As illustrated in Figures 3 and 5, autaptic transmission in microcultures may be not only excita-

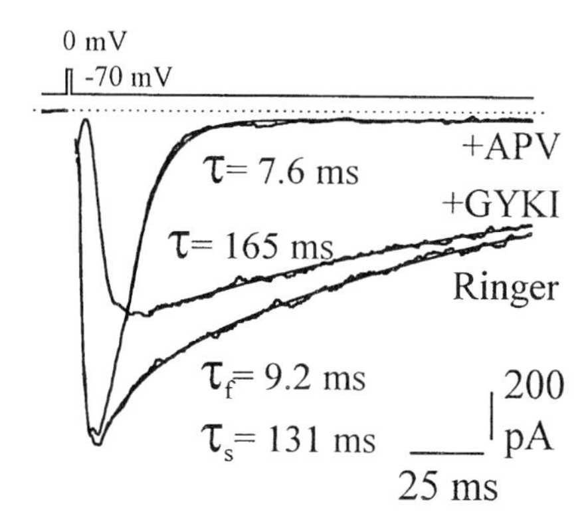

Fig. 4. Autaptic excitatory responses show dual components. Records correspond to a single cell microisland and are averages of five responses. The neuron was voltage-clamped at a holding potential of −70 mV. In a Ringer's solution lacking Mg^{2+} (to remove voltage-dependent block of NMDA channels) and containing glycine (10 μM, to allow activation of NMDA receptors), unclamped spikes were generated by changing the holding potential from the resting potential to 0 mV for 1 msec. For clarity, the induced inward Na^{+} spike has been removed. The response in Ringer's solution presented a decay with two time courses, which were fitted by exponentials (superimposed on the current) with the indicated time constants. It may be pharmacologically dissected out in two components, one rapid, detected by blocking NMDA receptors (D-2-amino-5-phosphonovaleric acid [APV]), and one slower, seen after blocking α-amino-3-hydroxy-5-methylisoxazole-4-propionate (AMPA) receptors (GYKI 53655).

tory but inhibitory (see also Bekkers and Stevens, 1991; Segal, 1991). In addition, the presence of spontaneous miniature autaptic potentials, exhibiting quantal fluctuations, has also been established for solitary neurons (Bekkers and Stevens, 1991; Segal, 1991).

From this overview, it is apparent for what microcultures are useful and which kind of questions may be asked in this preparation. Microcultures have been successfully used to approach specific questions on the mechanism of synaptic transmission (e.g., Rosenmund *et al.*, 1993; Tong and Jahr, 1994; Tong *et al.*, 1995). Voltage-sensitive dye recording has also been applied to neurons grown in microculture (Chien and Pine, 1991), taking advantage of the few number of cells in an island, which allows a complete record of the activity in every neuron in the network. Indeed, dye recording may be combined with extracellular stimulation, facilitating the study of synaptic interactions for long periods of time. Microcultures containing a single excitatory hippocampal neuron have served to demonstrate that sodium-dependent endogenous bursting underlies ictus-like epileptiform activity (Segal, 1994). These experiments were made possible by overcoming some experimental difficulties inherent to more complex systems. Although microcultured single neurons are a highly simplified system for studying epilepsy, it is obvious that they have a number of advantages for determining the activity of neurons underlying epileptic phenomena. Nevertheless, as recog-

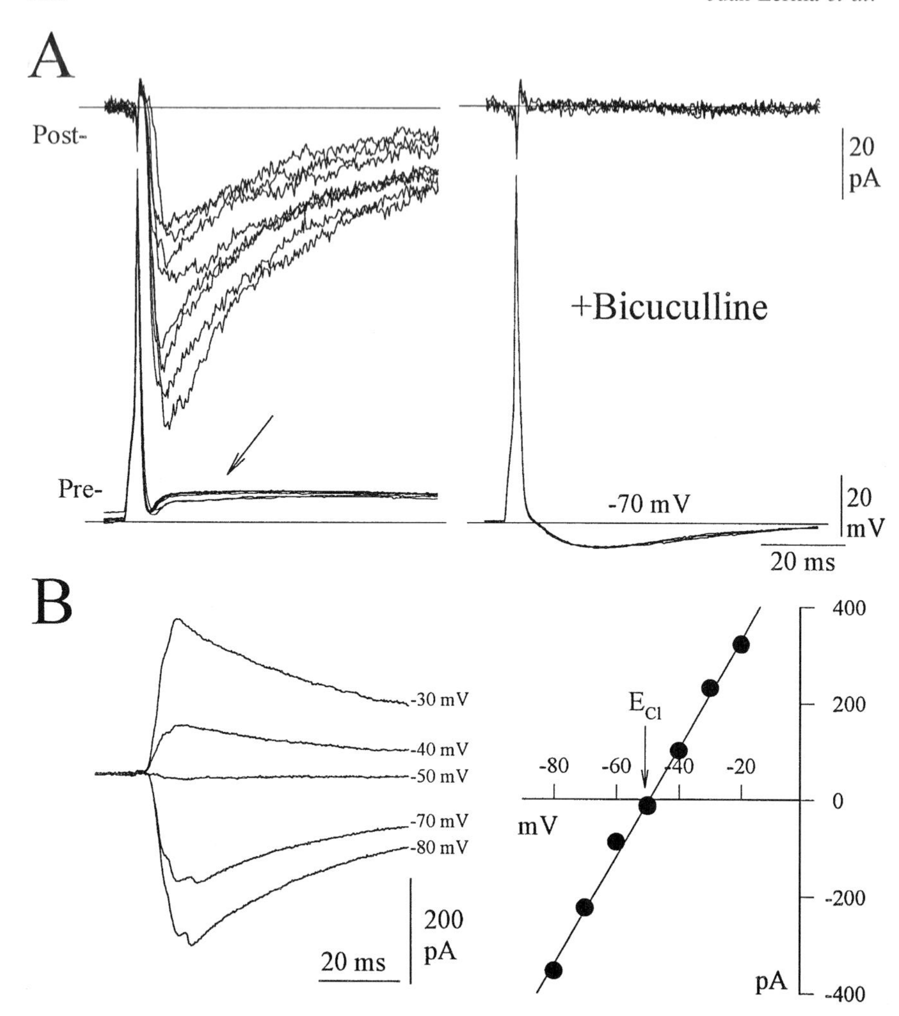

Fig. 5. Inhibitory synaptic transmission in microcultured hippocampal cells. (A) Experimental set-up as in Figure 4. The presynaptic cell was inhibitory on the postsynaptic neuron as well as on itself. A depolarizing autaptic inhibitory postsynaptic potential (arrow) did appear after each action potential in the presynaptic cell, as well as an inward current at the postsynaptic level. In both cases, synaptic responses were abolished by inclusion of bicuculline, the gamma-aminobutyric acid receptor antagonist. (B) Another cell showing inhibitory synaptic currents reversed at the chloride equilibrium potential (E_{Cl}). Postsynaptic currents were generated at different holding potentials (left). The current–voltage relationship was constructed and the reversal potential calculated (right). The arrow indicates the Nerst equilibrium potential for chloride, calculated by taking into account its extracellular (175 m*M*) and intracellular (23 m*M*) concentration.

nized by Segal (1994), there might be important differences between this culture system and the *in vivo* situation, including selective survival of certain neuronal phenotypes, altered glial ensheathment, and the like.

Glia–neuron microcultures have been used by Mennerick and Zorumski (1994) to measure the current induced by the uptake system in a voltage-clamped glial cell after the synaptic release of glutamate from a single neuron. These authors, after measuring both postsynaptic current and glial transporter responses, demonstrated that glial uptake helps to remove synaptically released glutamate, thus contributing to the termination of excitatory synaptic currents. Such a glial cell–neuron microculture appears to be an ideal system to study the release–uptake dynamics of fast neurotransmitters under excellent experimentally controlled conditions. It provides a framework for understanding the role of glia in both normal and pathologic processes.

In summary, microcultures offer several advantages over conventional mass cultures. The restricted nature of microculture synaptic networks allows studies of CNS synaptic transmission under acceptable experimental control, reducing the experimental hardware. The limited area the cells spread over facilitates perfusion with rapid systems, making it possible directly to demonstrate the presence on the membrane of functional receptors for neurotransmitters and to evaluate their possible role in mediating or modulating synaptic transmission (see Lerma *et al.*, 1997). Individual neurons can be continuously observed during maturation in culture, and the appearance of different receptor systems may be easily evaluated.

Acknowledgments

The authors thank Dr. M. A. Nieto for critical reading of manuscript, Dr. D. Leander (Elli Lilly & Co., Indianapolis) for the generous gift of GYKI 53655, and Dr. J. Avila from the Centro de Biología Molecular Severo Ochoa, CSIC-UAM, Madrid, for kindly providing the anti-TAU antibody. Work has been supported by grants to J. L. from the DGICYT (PB93/0150), FIS (95/0869), and the Biotech Program of the European Community (BIO2-CT93-0243).

References

Banker, G., and Goslin, K. (1991). "Culturing Nerve Cells," Cellular and Molecular Neuroscience Series (C. F. Stevens, ed.). The MIT Press, Cambridge, MA.

Bekkers, J. M., and Stevens, C. F. (1991). Excitatory and inhibitory autaptic currents in isolated neurons maintained in cell culture. *Proc. Natl. Acad. Sci. U.S.A.* **88,** 7834–7838.

Brewer, G. J., Torricelli, J. R., Evege, E. K., and Price, P. J. (1993). Optimized survival of hippocampal neurons in B27-supplemented Neurobasal™, a new serum-free medium combination. *J. Neurosci. Res.* **35,** 567–576.

Chien, C-B., and Pine, J. (1991). Voltage-sensitive dye recording of action potentials and synaptic potentials from sympathetic microcultures. *Biophys. J.* 697–711.

Furshpan, E. J., MacLeish, P. R., OLague, P. H., and Potter, D. D. (1976). Chemical transmission between rat sympathetic neurons and cardiac myocytes developing in microcultures: Evidence for cholinergic, adrenergic, and dual function neurons. *Proc. Natl. Acad. Sci. U.S.A.* **73,** 4225–4229.

Gu, X. I., and Azmitia, E. C. (1993). Integrative transport-mediated release from cytoplasmic and vesicular 5-hydroxtryptamine stores in cultured neurons. *Eur. J. Pharmacol.* **235,** 51–57.

Huettner, J. E., and Baughman, R. W. (1988). The pharmacology of synapses formed by identified corticocollicular neurons in primary cultures of rat visual cortex. *J. Neurosci.* **8,** 160–175.

Johnson, M. D. (1994). Synaptic glutamate release by postnatal rat serotonergic neurons in microculture. *Neuron* **12,** 433–442.

Johnson, M. D., and Yee, A. G. (1995). Ultrastructure of electrophysiologically-characterized synapses formed by serotonergic raphe neurons in culture. *Neuroscience* **67,** 609–623.

Landis, S. C. (1976). Rat sympathetic neurons and cardiac myocytes developing in microcultures: Correlation of the fine structure of endings with neurotransmitter function in single neurons. *Proc. Natl. Acad. Sci. U.S.A.* **73,** 4220–4224.

Lerma, J., Morales, M., Vicente, M. A., and Herreras, O. (1997). Glutamate receptors of the kainate type and synaptic transmission. *Trends Neurosci.* **20,** 9–12.

Liesi, P., Sappälä, I., and Trenkner. (1992). Neuronal migration in cerebellar microcultures is inhibited by antibodies against a neurite outgrowth domain of laminin. *J. Neurosci. Res.* **33,** 170–176.

Lübke, J., Markram, H., Frotscher, M., and Sakmann, B. (1996). Frequency and dendritic distribution of autapses established by layer 5 pyramidal neurons in the developing rat neocortex: Comparison with synaptic innervation of adjacent neurons of the same class. *J. Neurosci.* **16,** 3209–3218.

Mennerick, S., Que, J., Benz, A., and Zorumski, C. (1995). Passive and synaptic properties of hippocampal neurons grown in microcultures and in mass cultures. *J. Neurophysiol.* **73,** 320–332.

Mennerick, S., and Zorumski, C. F. (1994). Glial contributions to excitatory neurotransmission in cultured hippocampal cells. *Nature* **368,** 59–62.

Rosenmund, C., Clements, J. D., and Westbrook, G. L. (1993). Nonuniform probability of glutamate release at a hippocampal synapse. *Science* **263,** 754–756.

Segal, M. M. (1991). Epileptiform activity in microcultures containing one excitatory hippocampal neuron. *J. Neurophysiol.* **65,** 761–770.

Segal, M. M. (1994). Endogenous bursting underlies seizurelike activity in solitary excitatory hippocampal neurons in microcultures. *J. Neurophysiol.* **72,** 1874–1884.

Segal, M. M., and Furshpan, E. (1990). Epileptiform activity in microcultures containing small number of hippocampal neurons. *J. Neurophysiol.* **64,** 1390–1399.

Tong, G., and Jahr, C. E. (1994). Multivesicular release from excitatory synapses of cultured hippocampal neurons. *Neuron* **12,** 51–59.

Tong, G., Shepherd, D., and Jahr, C. E. (1995). Synaptic desensitizations of NMDA receptors by calcineurin. *Science* **267,** 1510–1512.

Van der Loos, H., and Glasser, E. M. (1972). Autapses in neocortical cerebri: Synapses between a pyramidal cell's axon and its own dendrites. *Brain Res.* **48,** 355–360.

18

Patch-Clamp Recordings from *Drosophila* Presynaptic Terminals

Manuel Martínez-Padrón and Alberto Ferrús
Instituto Cajal, C.S.I.C.
28002 Madrid, Spain

I. Introduction

Our current knowledge of synaptic transmission in the nervous system is the result of the convergence of countless studies performed in many different experimental preparations. No single preparation offers all the experimental advantages that would be desirable for the study of this highly complex process. One important problem is that nerve terminals, where synaptic transmission takes place, are usually too small for and not amenable to direct electrophysiologic recordings. In general, the events that take place in synaptic terminals have had to be inferred from recordings performed in the soma. However, it is generally admitted that the synaptic membrane is endowed with a wide and specialized repertoire of ion channels whose activity determines the modulation of neural signaling. Also, mounting evidence from the *in situ* expression of synaptic proteins and, most relevant, localization of specific protein isoforms (Ullrich *et al.*, 1995; Gardner and Kindler, 1996) indicate that synapses are very diverse. Nevertheless, detailed information on the general process of neurotransmitter release and its relation to Ca^{2+} entry has been obtained on some specialized synapses such as the squid giant axon, the chick cilliary ganglion, or the rat calyx of Held (Llinás *et al.*, 1981; Stanley, 1993; Borst and Sakmann, 1996).

The necessity of approaching synaptic studies from a multidisciplinary perspective justifies the effort to develop suitable methods in organisms where genetic and behavioral studies are also feasible. To that end, we have tried to develop a new preparation for synaptic transmission in *Drosophila*. The *Drosophila* larval body-wall muscle fibers are innervated by several axons that

Current Topics in Developmental Biology, Vol. 36

0070-2153/98 $25.00

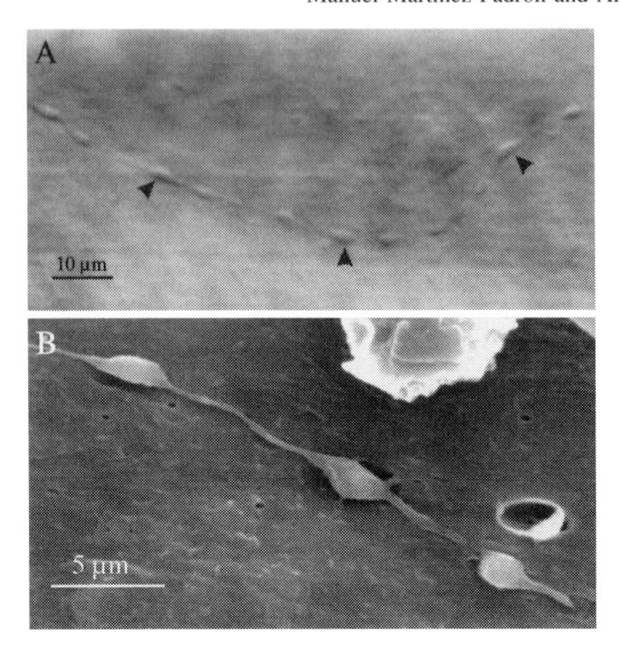

Fig. 1. Type III synaptic boutons. (A) Collagenase-digested boutons from an *ecd¹* mutant seen under Nomarski optics with a 40× water immersion objective, used to perform the electrophysiologic recordings. (B) Scanning electron micrograph of boutons innervating body-wall muscle fiber 12 of an *ecd¹* mutant, after a mild digestion treatment with collagenase. Notice the superficial location and accessible membrane of type III boutons, which allow patch recordings to be made.

produce morphologically distinct synaptic contacts on specific fibers. Several types of synaptic boutons, classed as types I to III, have been described (Atwood *et al.*, 1993; Jia *et al.*, 1993), which differ in their size, morphology, and vesicle phenotype. Type I boutons (up to 5 μm) mediate glutamatergic synaptic transmission (Jan and Jan, 1976; Johansen *et al.*, 1989), contain small, clear vesicles, and are surrounded by an elaborated subsynaptic reticulum provided by the muscle fiber. Type II boutons are small (<2 μm), express octopamin immunoreactivity (Monastirioti *et al.*, 1995), and contain a mixture of small, clear and large, dense core vesicles. Finally, Type III boutons (up to 1.7 × 6 μm; Fig. 1) contain mostly large, dense-core vesicles with a few small, clear ones and exhibit insulin-like and proctolin immunoreactivity (Gorczyca *et al.*, 1993; Anderson *et al.*, 1988) suggesting a peptidergic nature.

So far, we have succeeded in obtaining electrophysiologic recordings from type III boutons. For that purpose, we have taken advantage of a *Drosophila* mutation, *ecd*, which produces slightly enlarged type III nerve terminals. We have used the *ecd¹* allele, which is a temperature-sensitive recessive lethal. Mutant *ecd* larvae, when transferred to 30°C midway through the third instar, fail to pupate, remaining as larvae for an extended period of time and developing enlarged type III synaptic boutons. Type I boutons are deeply buried in the muscle and covered by several layers of subsynaptic reticulum, which precludes intraterminal recordings, whereas type II boutons are far too small for direct recordings.

II. Larval Preparation and Dissection

We use the larval preparation originally described by Jan and Jan (1976). Mature *Drosophila* third-instar larvae are pinned down dorsal side up onto a clear Sylgard-coated chamber using electrolithically sharpened tungsten pins. To improve the chances of finding large boutons, we tend to use larvae of the largest possible size. The dissection procedure is as follows:

1. Immobilize the larva through its head and tail using two tungsten pins. This is best done without any solution in the chamber to reduce the mobility of the larva.
2. Add *Drosophila* Ringer's (in m*M*, 100 NaCl, 5 KCl, 20 $MgCl_2$, 5 HEPES, and 115 sucrose, pH 7.3) to the chamber, and make an incision along the dorsal midline using the two main longitudinal tracheas as a reference. Then stretch open and fix the body wall by means of four additional pins inserted directly into the edges of the body wall.
3. Using fine forceps, carefully remove all internal organs, leaving only the central nervous system connected to the body wall muscle layer with the segmental nerves intact. Take special care to remove the tracheal system as completely as possible, because the very fine tracheas over the muscle fibers may later hamper visualization of synaptic boutons.
4. Perform a mild digestion of the preparation for 8–10 minutes in a solution containing 100 U/ml collagenase (Sigma type IA). Wash thoroughly with *Drosophila* Ringer's and transfer to the microscope.

III. Synaptic Boutons

Type III boutons have a localization mainly restricted to ventrolateral muscle fiber F12, and occasionally F13. They are unambiguously identified under Nomarski optics, with a 40× water immersion objective (see Fig. 1A), because of their characteristic elliptical morphology, medium size, and superficial location, and they are almost devoid of subsynaptic reticulum. Type III boutons can be observed in undigested preparations, but they bulge out and are easier to detect after treatment with collagenase, which also allows free access to the bouton's membrane (Fig. 1B). When looking for synaptic boutons, we normally search for the entrance point of the nerve to F12, which enters laterally through the middle point of the muscle and then branches to form a "T" shape that runs parallel to the inner edge of the muscle fiber.

Some considerations are important when attempting to record from synaptic boutons. After dissection, the muscle fibers contain many vacuoles that tend to obscure visualization of the synaptic boutons themselves. The state of the muscle fibers depends both on the genotype and the quality of the dissection, as well as on the extracellular solution used. In hemolymph-like physiologic solutions (Steward *et al.*, 1994), containing relatively high sucrose and magnesium concentrations, the vacuoles disappear after 10–15 minutes, leaving a smooth muscle surface where type III boutons are easier to find.

We have succeeded in recording from a number of different genotypes, including the CS wild strain. However, recordings are quite difficult and depend on the probability of finding boutons large enough for patching, which are rare. In general, for reasons unknown to us, not all genotypes respond in the same way to the collagenase treatment. In some strains, type III boutons come out and contrast nicely against the muscle surface after a short collagenase digestion (8 minutes). Other strains require longer digestion times that usually damage the muscle surface and make the muscle fibers break free from their insertion points before boutons are readily visible. We advise assaying a range of collagenase treatment times to find the most suitable conditions for each genotype. Recordings are far easier in the *Drosophila* mutant *ecd*, which presents slightly enlarged type III nerve terminals and, perhaps even more important, healthy and smooth muscle fibers that respond well to collagenase treatment.

IV. Electrophysiologic Recordings

Synaptic boutons are quite fragile, and we usually damaged them when attempting to establish the whole-cell configuration by breaking the membrane patch under the pipette. We therefore gain electrical access to the intracellular space using the nystatin perforated patch-clamp technique (Horn and Marty, 1988), which also produces more stable recordings, although it does not provide pharmacologic control of the intracellular space. The technical procedure is as follows:

1. High-resistance patch pipettes (10–20 MΩ) are pulled from thick-wall borosilicate glass (O.D 1.2 mm, I.D. 0.8 mm, World Precision Instruments), coated with a layer of Sylgard, and fire polished under ocular control with a microforge.

2. The tip of the pipette is filled by capillarity with a solution containing (in m*M*) 130 KCl, 4 $MgCl_2$, 10 HEPES, and 10 EGTA (pH 7.3 balanced with KOH) by introducing it into a drop of solution. While keeping the tip in the drop, the pipettes are back-filled with the same solution, to which we add a saturating concentration of Nystatin (200 μg/ml) from a stock solution of 50 mg/ml dis-

solved in dimethyl sulfoxide. At this point it is important to proceed quickly with the experiment because nystatin accumulation at the tip of the pipette will prevent seal formation later.

3. The pipette is mounted in a micromanipulator and moved close to the bouton while applying a constant mild positive pressure. We typically wait in close proximity to the bouton until the very first few crystals of nystatin reach the tip before trying to establish a seal, because the recording time is too short to allow time for the antibiotic to reach the membrane after the seal is made. On the other hand, if too much nystatin has already reached the tip, it will not be possible to make a seal.

4. Gigaseals of up to 30 GΩ form by applying gentle suction after the tip of the pipette is brought into contact with the terminal's membrane. The nystatin begins to perforate the membrane very soon after seal formation, thereby allowing recording of either the membrane potential (Fig. 2A) or ionic currents in whole-cell configuration (Figs. 2B and 3A).

5. To record single-channel currents (Fig. 3B), perforated vesicles (Levitan and Kramer, 1990) can be obtained by briefly switching to current-clamp mode and gently pulling the electrode away from the synaptic bouton. This procedure sometimes causes the bouton to break free from the arborization and move away with the pipette. If the membrane reseals at both ends of the bouton, whole-bouton currents can be recorded from single isolated terminals. It is possible to increase the probability of pulling a single bouton by making a larger-tipped pipette and polishing it down to the same high resistance, thus increasing the surface of the membrane–glass contact.

V. Technical Considerations

We have begun to use the *Drosophila* preparation to analyze the ionic currents present in the presynaptic membrane. Because synaptic boutons are connected to each other through fine, relatively long processes, the charging of the membrane capacitance displays a rapid, single exponential component that represents the charging of the bouton under the recording electrode, followed by a number of smaller, slower exponential components that correspond to the charging of the process (Jackson, 1992).

To calculate the electrode series resistance and bouton capacitance, we systematically compensate the pipette capacitance, and then acquire current records in response to a 10 mV depolarizing pulse, from −80 to −70 mV, that activates only passive currents. We obtain estimates of these parameters by fitting several exponential functions to the current transient, and then integrating the area under the fastest exponential component after subtraction of the leak current. The fastest component of the charging transient has a capacitance of around 1 pF, and a decaying time constant of 58 μsec. The associated series resistance is in the

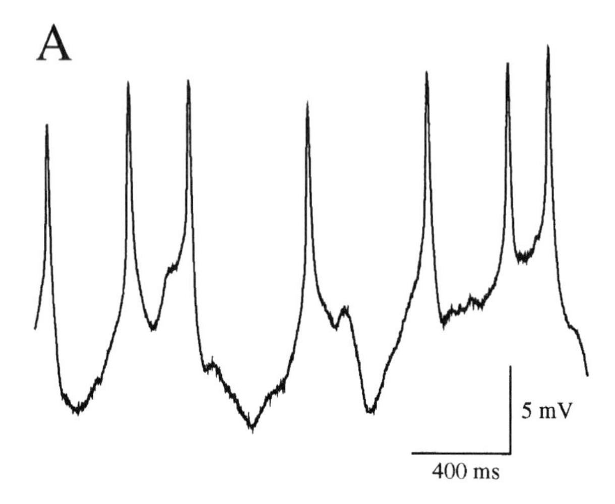

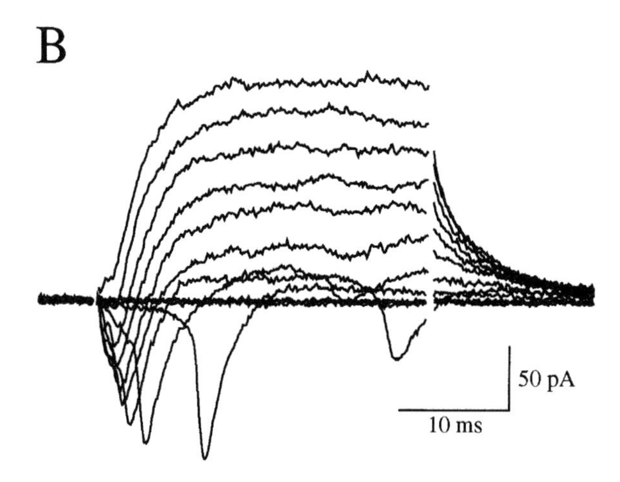

Fig. 2. Direct electrophysiologic recording from type III boutons. (A) Action potentials recorded under current-clamp conditions in type III boutons in response to a steady depolarizing current injection. (B) Activation of whole-terminal ionic currents on membrane depolarization to a variety of membrane potentials, from a holding potential of −40 mV. The macroscopic current consistently displayed a tetrodotoxin (TTX)-sensitive inward current carried by sodium, followed by a delayed outward current due to potassium ions.

range of 50 to 60 MΩ, or about three to four times larger than the pipette tip resistance before obtaining a seal. Without series resistance compensation, this average series resistance represents a voltage error of about 5 mV for a current magnitude of 100 pA, and about 1 mV with 80% compensation. The values of the time constant and series resistance thus calculated determine the speed and

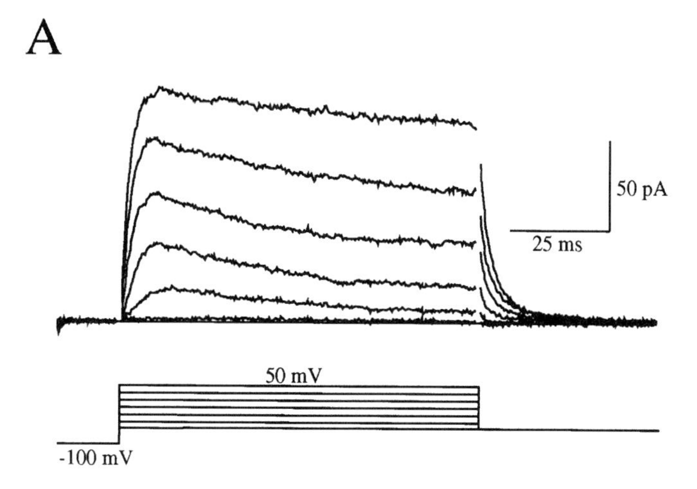

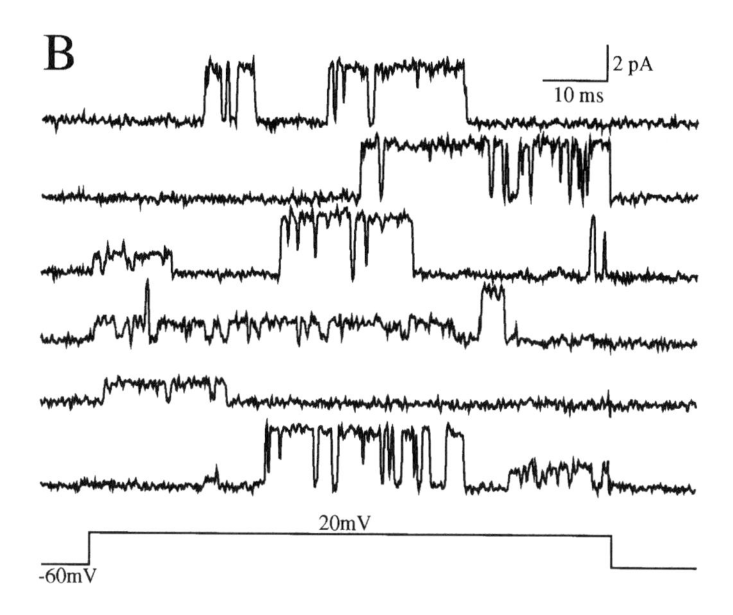

Fig. 3. Potassium currents at the synaptic terminals. Several K^+ currents and channels are present in the membrane of type III synaptic terminals. (A) Current traces in response to depolarizing voltage steps from a hyperpolarized membrane potential (−100 mV) showing that the macroscopic K^+ current contains two different kinetic components: a rapidly inactivating and a noninactivating component. (B) Single voltage-activated K^+ channels recorded from a perforated vesicle, excised from the membrane of a CS type III bouton, containing at least two K^+ channels with different conductances.

quality of the voltage-clamp system for the synaptic bouton immediately under the pipette. The nerve terminals are charged rapidly enough to study the kinetics of ionic currents, provided that the ionic channels under study are present in the bouton itself.

We have evidence that this is in fact the case for several potassium currents (Fig. 3A), because a large fraction of the currents is still present in isolated terminals that have been pulled out from the muscle fiber. Conversely, sodium currents do not appear to be located within the terminal itself. In the absence of tetrodotoxin (TTX), we typically observe repetitive inward currents in response to membrane depolarization, suggesting that action potentials are being produced in poorly space-clamped areas of the membrane, outside the synaptic terminal proper (see Fig. 2B). Also, these inward currents are not present in isolated boutons.

Type III synaptic boutons have a mixed population of dense-core synaptic vesicles, and they have been reported to stain positively with antibodies against a number of peptides, including insulin and proctolin. Direct stimulation of type III boutons with an intracellular electrode under current-clamp conditions does not produce any apparent membrane current in the muscle fiber. Also, the amplitude and time course of the evoked potential (due to type I boutons) in response to nerve stimulation is not affected when the intensity of the stimulus is increased enough to recruit the axon of type III boutons (assessed by direct recording of the bouton). These experiments appear to indicate that type III boutons are not involved in fast synaptic transmission. Considering their vesicle content and immunoreactivity profile, they may release neuromodulators that affect the electrical properties of muscle fibers or transmitter release from other terminals.

VI. Perspectives

The described method allows studies of normal and mutant conductance as well as single ion channels *in vivo*. We have applied this preparation to study voltage-dependent K^+ currents. Using the proper genotypes, one can manipulate the expression of specific proteins in this terminal. In particular, the new generation of position-specific enhancers and inducible systems (see Sun *et al.*, 1995; Brand *et al.*, 1994, for references) would open the field of synaptic physiology to direct molecular approaches.

Type III boutons are not fast signaling synapses to the muscle. It is obvious that gaining access to type I boutons would represent a greater step toward the desired convergence of molecular, behavioral, and physiologic methodologies. The major problem to be solved with these boutons is their deep location in the muscle infoldings and the surrounding subsynaptic reticulum. However, the strategy followed with type III might be successful with type I as well, namely, the use of a mutation to render the bouton accessible to the electrode. Alternatively, a piezoelectrically driven micromanipulator could be used to break through the subsynaptic reticulum with a beveled-tip electrode. This system, if successful, would allow voltage-clamp studies of the terminal, although not membrane seals for patch-clamp. These, among many other possibilities, should be tested if the investigator is convinced that the effort is worthwhile. With previous studies in *Drosophila*, this has been the major handicap to overcome.

References

Anderson, M. S., Halpern, M. E., and Keshishian, H. (1988). Identification of the neuropeptide transmitter proctolin in *Drosophila* larvae: Characterization of fiber-specific neuromuscular endings. *J. Neurosci.* **8,** 242–255.

Atwood, H. L., Govind, C. K., and Wu, C. F. (1993). Differential ultrastructure of synaptic terminals on ventral longitudinal abdominal muscles in *Drosophila* larvae. *J. Neurobiol.* **24,** 1008–1024.

Borst, J. G. G., and Sakmann, B. (1996). Calcium influx and transmitter release in a fast CNS synapse. *Nature* **383,** 431–434.

Brand, A. H., Manoukian, A. S., and Perrimon, N. (1994). Ectopic expression in *Drosophila. Methods Cell Biol.* **44,** 635–654.

Gardner, C. C., and Kindler, S. (1996). Synaptic proteins and the assembly of synaptic junctions. *Trends Cell Biol.* **6,** 429–433.

Gorczyca, M. G., Augart, C., and Budnik, V. (1993). Insulin-like receptor and insulin-like peptide are localized at neuromuscular junctions in *Drosophila. J. Neurosci.* **13,** 3692–3704.

Horn, R., and Marty, A. (1988). Muscarinic activation of ionic currents measured by a new whole-cell recording method. *J. Gen. Physiol.* **92,** 145–159.

Jackson, M. B. (1992). Cable analysis with the whole-cell patch clamp: Theory and experiments. *Biophys. J.* **61,** 756–766.

Jan, L. Y., and Jan, Y. N. (1976). Properties of the larval neuromuscular junction in *Drosophila melanogaster. J. Physiol.* **262,** 189–214.

Jia, X. X., Gorczyca, M., and Budnik, V. (1993). Ultrastructure of neuromuscular junctions in *Drosophila*: Comparison of wild type and mutants with increased excitability. *J. Neurobiol.* **24,** 1025–1044.

Johansen, J., Halpern, M. E., Johansen, K. M., and Keshishian, H. (1989). Stereotypic morphology of glutamatergic synapses on identified muscle cells of *Drosophila* larvae. *J. Neurosci.* **9,** 710–725.

Levitan, E. S., and Kramer, R. H. (1990). Neuropeptide modulation of single calcium and potassium channels detected with a new patch clamp configuration. *Nature* **348,** 545–547.

Llinás, R., Steinberg, I. Z., and Walton, K. (1981). Presynaptic calcium currents in squid giant synapse. *Biophys. J.* **33,** 289–322.

Monastirioti, M., Gorczyca, M., Rapus, J., Eckert, M., White, K., and Budnik, V. (1995). Octopamin immunoreactivity in the fruit fly *Drosophila melanogaster. J. Comp. Neurol.* **356,** 275–287.

Stanley, E. F. (1993). Single calcium channels and acetylcholine release at a presynaptic nerve terminal. *Neuron* **11,** 1007–1011.

Steward, B. A., Atwood, H. L., Renger, J. J., Wang, J., and Wu, C. F. (1994). Improved stability of *Drosophila* larval neuromuscular preparations in haemolymph-like physiological solutions. *J. Comp. Physiol. A* **175,** 179–191.

Sun, Y. H., Tsai, C. J., Green, M. M., Chao, J. L., Yu, C. T., Jaw, T. J., Yeh, J. Y., and Bolshakov, V. N. (1995). *white* as a reporter gene to detect transcriptional silencers specifying position-

specific gene expression during *Drosophila melanogaster* eye development. *Genetics* **141,** 1075–1086.
Ullrich, B., Ushkaryov, Y. A., and Sudhof, T. C. (1995). Cartography of neurexins: More than a 1000 isoforms generated by alternative splicing and expressed in distinct subsets of neurons. *Neuron* **14,** 497–507.

Index

F

G

H

Contents of Previous Volumes

Volume 31

Cytoskeletal Mechanisms during Animal Development

Guest edited by David G. Capco

Section I
Cytoskeletal Mechanisms in Nonchordate Development

Section II
Cytoskeletal Mechanisms in Chordate Development

Volume 32

Volume 33

Volume 34

Volume 35